TL45.

Happy Christmas, Jimmy

With love

Sandy

1990

THE LETTERS OF ERASMUS DARWIN

Other books by Desmond King-Hele:

Doctor of Revolution: the life and genius of Erasmus Darwin (Faber, 1977)
(Editor) *The Essential Writings of Erasmus Darwin* (MacGibbon & Kee, 1968)
Shelley: his Thought and Work (Macmillan, 1960)
Satellites and Scientific Research (Routledge & Kegan Paul, 1960)
Theory of Satellite Orbits in an Atmosphere (Butterworths, 1964)
Observing Earth Satellites (Macmillan, 1966)
The End of the Twentieth Century? (Macmillan, 1970)
Poems and Trixies (Mitre, 1972)

Erasmus Darwin (1731–1802)

The Letters of

ERASMUS DARWIN

Edited by Desmond King-Hele

CAMBRIDGE UNIVERSITY PRESS

Cambridge

London New York New Rochelle

Melbourne Sydney

Published by the Press Syndicate of the University of Cambridge
The Pitt Building, Trumpington Street, Cambridge CB2 1RP
32 East 57th Street, New York, NY 10022, USA
296 Beaconsfield Parade, Middle Park, Melbourne 3206, Australia

First published 1981

Printed in Great Britain at the
University Press, Cambridge

British Library Cataloguing in Publication Data

Darwin, Erasmus
The letters of Erasmus Darwin.

1. Darwin, Erasmus
2. Naturalists – England – Biography
3. Physicians – England – Biography
I. Title II. King-Hele, Desmond
509'.2'4 QH31.D3 80–41796

ISBN 0 521 23706 8

CONTENTS

LIST OF ILLUSTRATIONS

Frontispiece
Erasmus Darwin (1731–1802)

The illustrations appear between pages 192 and 193

INTRODUCTION

Darwin's life and letters

Erasmus Darwin, who lived from 1731 until 1802, was by profession a physician, widely acclaimed as the finest doctor of his time in England. By nature Darwin was a large and powerful-looking man, cheerful and healthy, and, although he stammered, he was a witty and persuasive talker. He was very sociable and had a passion for science and technology. He founded three Midland scientific societies, the most important being the Lunar Society of Birmingham, which arose from his friendship with Matthew Boulton and William Small. The Lunar Society was a major intellectual driving-force of the Industrial Revolution in Britain, which has led to the modern world of technology. Darwin was a compulsive inventor and with the encouragement of his Lunar friends, particularly Josiah Wedgwood, James Watt and James Keir, he produced working inventions ranging from a speaking-machine to a horizontal windmill, and dozens of designs on paper. Darwin's most potent talent, however, was his deep scientific insight, which enabled him to make several important discoveries in biology and geophysics and to propose many scientific ideas which time has shown to be correct. He won most fame late in life, in a quite different sphere, as a poet: his long poem *The Botanic Garden* captivated the literary world in the 1790s. Coleridge called him 'the first *literary* character in Europe', the Napoleon of literature, as it were.

Today Darwin earns most credit among scientists for recognizing and describing biological evolution, specifying the essentials of plant nutrition and of photosynthesis, and explaining how clouds usually form. And he has attracted the attention of literary critics not so much for his poems as for his immense influence on the English Romantic poets, Wordsworth, Coleridge, Shelley and Keats.

Darwin was one of the liveliest of letter writers: his exuberant personality and bubbling ideas overflow into his letters, and he usually wrote as he felt, in a rough and vigorous style, free of the formality that deadens many eighteenth-century letters. *The Botanic Garden* shows us the literary artificer; in his letters we see the real Darwin.

As well as being lively, Darwin's letters are of considerable historical interest. The Lunar group included the men who were leading Britain and the world into the industrial age, and his letters to them illuminate some of the sources of these enterprises. The 272 letters in this volume include 43 to Josiah Wedgwood, 36 to Matthew Boulton, 24 to James Watt, six to Sir Joseph Banks, and four to Benjamin Franklin. Other letters have biographical interest, such as the 17 letters to his son Robert (Charles Darwin's father); the four long letters to R. L. Edgeworth; and the eleven letters to John Barker, Darwin's partner in an enterprise for a slitting-mill. Darwin's inventions figure in about 30 letters, while his literary leanings declare themselves in four long letters written in verse. Among many letters of medical interest, there are two to the Duke of Devonshire, ten to Tom Wedgwood, and a dozen to James Watt, as well as those to individual doctors such as Beddoes, Jenner and Withering.

The subjects arising in a volume of letters are inevitably somewhat random, and important topics may remain unmentioned. As a framework for the reading of the letters, it may be useful to have a brief narrative account of Darwin's life and work. This occupies the next four pages and is a reprint, with minor revisions, of the first chapter of my biography of Darwin, *Doctor of Revolution* (Faber, 1977).

Erasmus Darwin was born in 1731 at Elston near Newark, the fourth son of an early-retired lawyer. He went to Chesterfield School from 1741 to 1750 and then spent three years at Cambridge University, where he studied medicine. This was followed by two years at the Edinburgh Medical School, and in 1756 he qualified as a doctor. He failed to attract any patients at Nottingham, so he migrated to Lichfield, where he had the good fortune – or perhaps skill? – to cure a prominent local man whom all other doctors had condemned to death. His future career assured by this *coup de théâtre*, he settled at Lichfield: he lived there for twenty-five years, giving medical advice for a fee to the rich and freely to the poor.

At the end of 1757 Darwin married Mary Howard, and soon after took a pleasant house at the edge of the Cathedral Close. The marriage proved very happy, marred only by Mary's ill-health. They had three sons who survived infancy. The eldest, Charles, was a brilliant medical student, but he died tragically at the age of nineteen from an infection, after cutting his hand while dissecting. The second son, Erasmus, became a lawyer. The third, Robert, followed his father and was a successful doctor at Shrewsbury.

Before 1760 Erasmus became friendly with Matthew Boulton and Benjamin Franklin. Boulton was then a manufacturer of buckles and other 'toys', but by his energy and enterprise he became the leading manufacturer of England and 'the father of Birmingham'. Franklin was already famous as a man of science for his electrical discoveries. In 1765 Franklin introduced Dr William Small to Boulton and Darwin. Affable and learned, Small had recently returned from teaching in America, where Thomas

Jefferson, one of his pupils, declared that Small's example 'fixed the destines of my life'. The meetings of Darwin, Boulton and Small in the 1760s led to the monthly gatherings of friends at the time of the full moon, later characterized as the Lunar Society of Birmingham.

In the early 1760s Darwin was wildly enthusiastic about making a steam carriage with twin cylinders. Though Boulton was too cautious to take up this idea, Darwin's head of steam was not wasted: James Watt visited him in 1767 and they at once became friends. Watt was gloomy by nature and needed cheering: who better to do so than the energetic and optimistic steam-enthusiast Darwin? No wonder Watt, though usually reticent, praised Darwin so highly: 'It will be my pride, while I live, that I have enjoyed the friendship of such a man'.

Another man who enjoyed that friendship was Josiah Wedgwood. In 1765 Wedgwood and James Brindley began promoting what was virtually a new transport system, the Trent-and-Mersey canal, and Darwin was one of their keenest helpers. Later Darwin designed a horizontal windmill, which was used to grind colours at Wedgwood's pottery for thirteen years, until displaced by steam. The Darwin and Wedgwood families grew up together, and Robert Darwin married Susannah Wedgwood in 1796.

One of Erasmus's student friends at Edinburgh, James Keir, settled near Lichfield in 1767, and later set up an alkali factory. Keir was a pioneer of the chemical industry, and he devised an ingenious process for obtaining soda from salt in dilute solution.

On his daily rounds as a doctor, Darwin was bumped and joggled over the vile roads in his carriage, so it is not surprising that he was keen to construct comfortable carriages. One of his designs, specially resistant to overturning, greatly impressed Richard Edgeworth, and inspired him to become an inventor. He visited Darwin in 1766 and became a lifelong friend.

This group of friends – Boulton, Darwin, Small, Wedgwood, Watt, Keir and Edgeworth – formed the nucleus of the early Lunar Society; Joseph Priestley was a later addition to their number. 'Intellectually the most effective provincial group that has ever come together in England', C. D. Darlington has called them. Individually they were among the most energetic and intellectually able men of the era, and their meetings, by some strange alchemy, channelled their talents fruitfully, producing a revolutionary forward thrust in science and technology.

Darwin also moved in a non-Lunar orbit, within the literary circle of the Cathedral Close at Lichfield, presided over by Canon Seward. Darwin was an accomplished versifier and he helped Seward's daughter Anna to success as a poet. She became known as the 'Swan of Lichfield'. There were also visits from Lichfield's 'Great Cham of Literature', Samuel Johnson, but he and Darwin did not get on well: there was not room for two heavyweight sages in one Close.

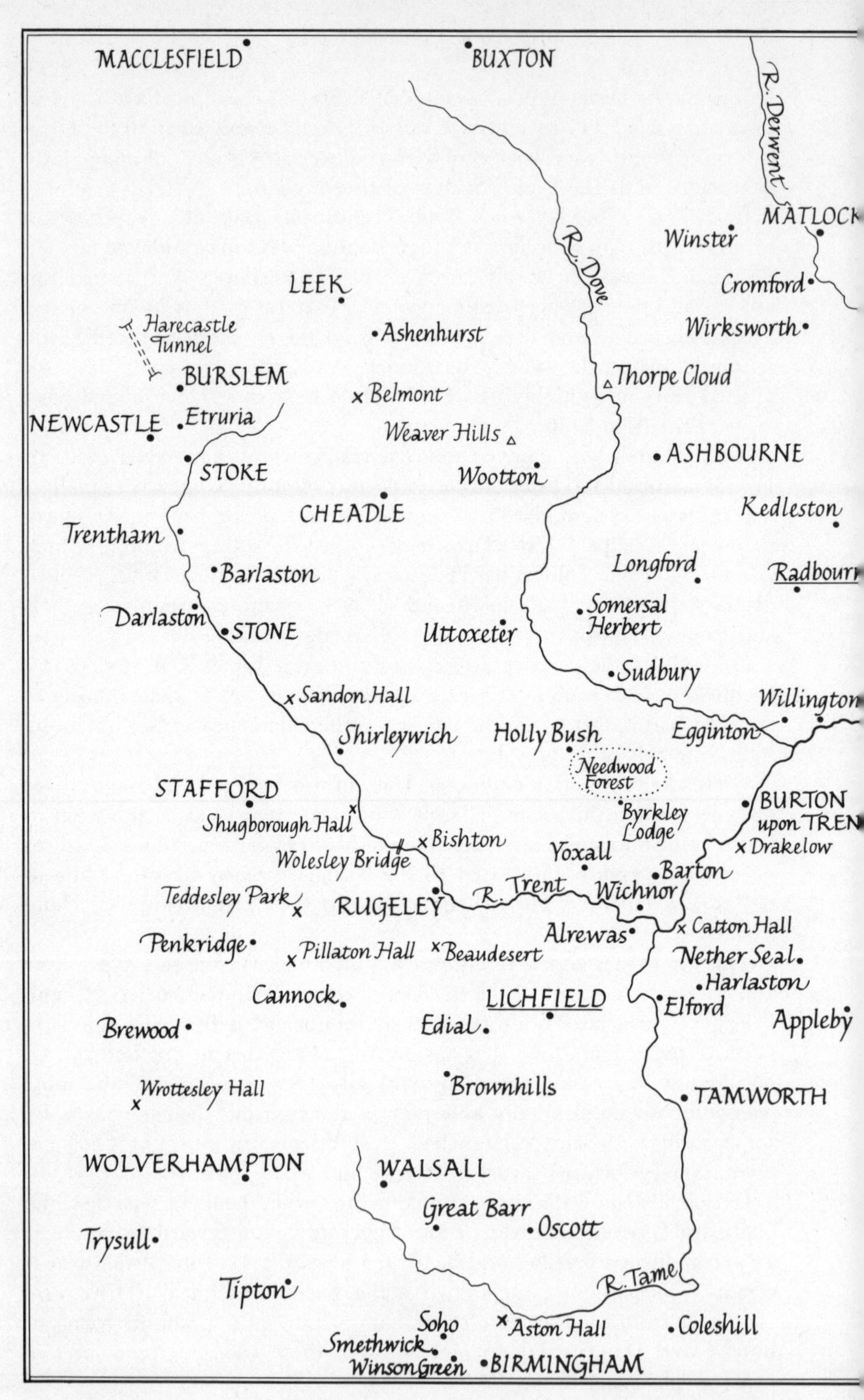

Sketch map of the Midlands, showing places mentioned in the letters.

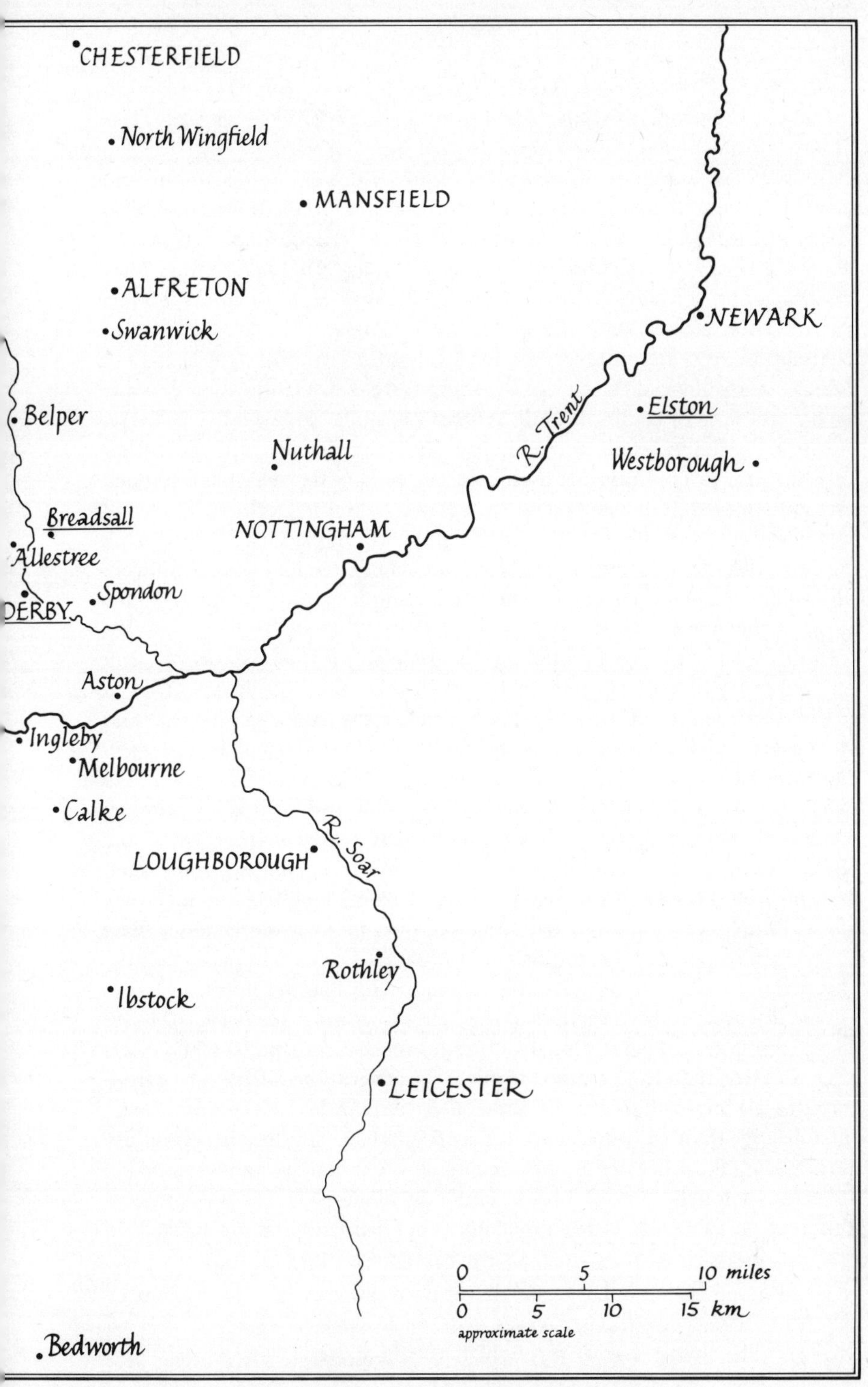

Places where Erasmus Darwin lived are underlined.

During the 1760s Darwin's reputation as a doctor continued to grow, but he had no cure to offer his wife Mary, and she died in 1770, aged thirty. Her tributes to her husband were recorded by Anna Seward.

In the 1770s Darwin renewed his friendship with Benjamin Franklin and gave his approval to the American Revolution. He continued living at Lichfield, unmarried; but he had two illegitimate daughters, Susan and Mary Parker, who were brought up at his home. In 1777 he was busy creating a fine botanic garden on the western outskirts of Lichfield, and he began writing verses about the plants.

About 1778 he fell in love with the lively and beautiful Mrs Elizabeth Pole, 31 years old and the wife of a leading Derbyshire landowner, Colonel Sacheverel Pole of Radburn Hall. At first he could only woo her in verse; but in 1780 Colonel Pole conveniently died, and Darwin, though rather ugly, fat and nearly fifty, brushed aside more eligible rivals and married Mrs Pole in 1781. They moved to Radburn and later to Derby. The second Mrs Darwin's rare talent for creating happiness was exercised fully during the 21 outwardly quiet years of their marriage. They had seven children, who acknowledged their good fortune in having been brought up in such a happy atmosphere.

In the years of his second marriage, Darwin was less active socially, and devoted much of his energy to writing books. In the 1780s he spent seven years on two lengthy translations of Linnaeus, the *System of Vegetables* and the *Families of Plants*. After that he wrote his five original books, each epoch-making in its own way. The first was his encyclopedic poem *The Botanic Garden*, which entranced the reading public and the literary pundits: 'Dr Darwin has destroyed my admiration for any poetry but his own', wrote Horace Walpole. Part II of the poem, *The Loves of the Plants*, was published first, in 1789, because it was lighter – 968 rhyming couplets about the sex life of plants, nicely personified and backed by copious scientific notes. Part I, *The Economy of Vegetation*, published in 1792, is a compendium of scientific knowledge in 1224 rhyming couplets, again with lengthy notes.

Was it wrong of those readers of the 1790s to admire *The Botanic Garden*? The poem is rambling and unstructured: but Darwin's polished couplets have a lustre that time cannot tarnish, as generations of literary critics confirm. To this skill in verse he added three other talents: 'A greater range of knowledge than any other man in Europe', as Coleridge put it; originality in proposing new theories and foreseeing new technology; and an amazing lightness of touch – he is forever laughing at himself. All three talents illuminate his forecast of mechanically propelled cars and aircraft:

> Soon shall thy arm, UNCONQUER'D STEAM! afar
> Drag the slow barge, or drive the rapid car;
> Or on wide-waving wings expanded bear
> The flying chariot through the fields of air.

Fair crews triumphant, leaning from above,
Shall wave their fluttering kerchiefs as they move;
Or warrior bands alarm the gaping crowd,
And armies shrink beneath the shadowy cloud.

The pictures of the plants in Part II are continuously playful. For example in *Lychnis*, where the males and females are on different plants, each female saucily peeps out when mature,

In gay undress displays her rival charms,
And calls her wondering lovers to her arms.

Science so pleasantly presented was much to the taste of the leisured reading public. As L. T. C. Rolt has remarked, Darwin performed 'the astonishing feat of rendering in imaginative terms which any literate man could then appreciate the sum of human knowledge at that time. It was an achievement which was never repeated or even attempted'.

His next book was quite different – a massive tome on animal life, *Zoonomia*, in two volumes totalling 1400 pages, published in 1794–6. By this time his fame as a doctor had become almost legendary, a status scarcely credible today when the mystique of the doctor has largely evaporated. His book was a great success, and was widely translated and republished. We cannot value the book so highly today, because its basic theory is wrong. But we can still admire some of Darwin's bold experimental cures, his linking of body and mind, and his pioneer work in treating mental illness humanely instead of brutally.

We can also admire the riveting chapter entitled 'of generation', where Darwin picks out the main mechanisms of what we call biological evolution, noting the changes produced in animals over many generations by selective breeding, the inheritance of mutations, the role of 'lust, hunger and security' in determining the favoured variations, and the operation of sexual selection. He suggests that 'all warm-blooded animals have arisen from one living filament', during a time span of 'millions of ages'.

In the England of the 1790s, reeling from the shock of the French Revolution, evolution – even without the initial *r* – was unacceptable: denying that God created species was near-treason. But Erasmus had one influential disciple, his son Robert, who generated a pro-evolutionary atmosphere in his own family and helped to prepare the mind of his son Charles for evolution.

After the massive *Zoonomia* came a slim volume of 128 pages, entitled *A Plan for the Conduct of Female Education in Boarding Schools* (1797), written for his daughters Susan and Mary Parker, who had started a school at Ashbourne in Derbyshire. He totally opposed the conventional idea that girls should become feather-brained simpering misses: he wanted to see them 'sound in body and mind', and took a significant step towards modern ideas of sexual equality.

Darwin's next book, *Phytologia* (1800), is an impressive 600-page survey of vegetable life and all that sustains it, from carbon dioxide down to manures, full of good ideas ranging from sewage farms and biological control of insects to artesian wells and designs for new drill-ploughs. Darwin does best of all on plant nutrition. He describes photosynthesis more fully than anyone had done before, specifying the essential ingredients (carbon dioxide, water and sunlight) and the end-products (oxygen and sugar). He was also the first to recognize that nitrogen and phosphorus were essential plant nutrients.

Darwin died in 1802, at the age of 70, and his second long poem, *The Temple of Nature*, was published posthumously in 1803. It is, in my view, his finest achievement. He traces the evolution of life from microscopic specks in primeval oceans through plant life, fishes, amphibia, reptiles, land animals and birds. The couplets are just as bright as in *The Botanic Garden*; but what is really astonishing is to see him calmly and correctly announcing the processes of evolution more than fifty years before *The Origin of Species* and a hundred years before the details were generally agreed. He also points to the pressures shaping evolution, painting a brilliant picture of the struggle for existence among animals, insects, fish and plants.

Darwin's books run to one and a half million words: they show how hard he worked and how varied were his accomplishments, but they do not reveal his contributions to technology and physics. Among his Lunar Society friends he was famous for his mechanical inventions, which included new kinds of carriages, the speaking-machine, and 'electrical doubler', an office copying-machine, a wire-drawn river ferry and a canal lift, of a design proved in operation a hundred years later.

Darwin did not devote much time to what we would call scientific research, but he still managed to achieve more than most scientists today can hope for. He had five papers published in the *Philosophical Transactions* of the Royal Society, of which he became a Fellow when he was 29 years old. Two of the papers are historic: one, in 1785, gives the first clear exposition of the principle of the artesian well; the second, in 1788, propounds the principle of adiabatic expansion, fundamental to all heat engines, and then applies it to explain the formation of clouds: 'When large districts of air are raised . . . they become so much expanded by the great diminution of pressure over them . . . that . . . the air by its expansion produces cold and devaporates'.

Perhaps the most surprising of Darwin's achievements is his profound influence on the English Romantic poets. Wordsworth was nineteen and Coleridge seventeen when *The Loves of the Plants* appeared. Three years later the completed *Botanic Garden* captured Wordsworth's sympathy by its exact description of nature and fervour for the French Revolution, and it shot straight into Coleridge's subconscious: *The Ancient Mariner* and 'Kubla Khan' are riddled with verbal echoes of Darwin.

Twenty years later, when Darwin's reputation had declined, he began to fascinate Shelley and Keats. With Keats he shared a relish for 'the realm of Flora and old Pan'. Shelley was Darwin's keenest disciple and had from him the idea of infusing nature poetry with science, as in 'The Cloud', as well as admiring him for being a sceptic in religion and a radical in politics. Darwin has also been called the chief progenitor of science fiction, through his influence on Mary Shelley's *Frankenstein*, which arose from a discussion between Byron and Shelley about Darwin's ideas on the generation of life. His literary reputation remained high for many years: in the 1860s Craik's popular *History of English Literature*, though critical of Darwin, still gave him twice as much space as Shakespeare, and six times as much as Byron.

In brief, Erasmus Darwin was a giant, who achieved more in a wider range of intellectual disciplines than anyone since.

The letters

My aim has been to record in this book all known letters written *by* Darwin; relevant excerpts from letters *to* Darwin are quoted in the Notes.

Darwin's medical prescriptions are outside the scope of the collection, but a list of prescriptions (probably very incomplete) is given as an Appendix.

There has been no previous collection of Darwin's letters, but excerpts have appeared in numerous books. The most important of these are, in chronological order: J. E. Stock, *Life of Thomas Beddoes* (1811); R. L. Edgeworth, *Memoirs* (1821); J. Muirhead, *James Watt* (1854); E. Meteyard, *Life of Josiah Wedgwood* (1865); J. K. Moilliet, *Life of James Keir* (1868); and, above all, Charles Darwin, *Life of Erasmus Darwin* (1879), which has extracts from thirty letters. More recently, the fullest selections have been those in R. E. Schofield, *Lunar Society of Birmingham* (1963); in *The Essential Writings of Erasmus Darwin* (1968); and in D. King-Hele, *Doctor of Revolution* (1977).

Of the 272 letters published here, 187 are transcribed from manuscripts, of which 162 are in Darwin's own writing. Apart from 23 autograph letters privately owned, the present locations of the manuscripts are as follows, in alphabetical order.

Number of letters		
Autograph	*copies*	*Location*
6		American Philosophical Society Library, Philadelphia, Pa.
39		Birmingham Reference Library
2		Birmingham University Library
3		Bodleian Library, Oxford
3	3	Botany Library, British Museum (Natural History), London
	1	Bristol Central Reference Library
5		British Library, London

Number of letters		
Autograph	*copies*	*Location*
4	4	Cambridge University Library
4		Chatsworth, Derbyshire
6		Darwin Museum, Down House, Downe, Kent
1		Derby Central Library
9	1	Fitzwilliam Museum Library, Cambridge
3		Harvard University Library, Cambridge, Mass.
4		Historical Society of Pennsylvania, Philadelphia, Pa.
15	2	Keele University Library, Keele, Staffs.
2		Linnean Society, London
1		Liverpool City Libraries
2		New York Public Library
1		Nottinghamshire Record Office
2		Royal Society Library, London
3		Royal Society of Arts Library, London
1		Royal Society of Medicine Library, London
	1	John Rylands University Library of Manchester
11		William Salt Library, Stafford
4		Sandon Hall, Stafford
2		St John's College Library, Cambridge
2	13	University College Library, London
2		Wellcome Institute Library, London
2		Yale University Library, New Haven, Conn.

In addition, there is one autograph MS letter in the possession of Dr Ralph Colp Jr of New York, which he will be publishing separately.

The other letters are taken from printed sources, as specified at the end of each letter. Unfortunately I have not been able to locate the originals of the letters quoted by Charles Darwin in his *Life of Erasmus Darwin*, or those quoted by Eliza Meteyard in her *Life of Josiah Wedgwood*; but it is possible that these manuscripts will be found in the future.

I have searched widely for manuscripts of Darwin's letters in Britain, but I apologize in advance for failing to find any that I have missed. For the USA, such apologies are sure to be needed: I found only six sources and there may be many more. I believe, however, that I have located the great majority of the letters of Erasmus Darwin that have escaped the ravages of time, and that the present volume provides a balanced and reasonably complete picture of Darwin the letter-writer in all his moods.

The letters are printed chronologically. They are listed chronologically, with dates and recipients, on pages xxv–xxx. The recipients are listed alphabetically on pages xxxi–xxxii, and asterisks mark the 48 recipients whose portraits may be found among the illustrations. A map of the Midland region, with all the Midland place-names mentioned in the letters, appears on pages x–xi.

The letters are designated alphabetically within each year. Thus letter

66B is the second in 1766; 86C is the third in 1786; 00A is the first in 1800 (but note that P follows N, with O omitted). This method seems preferable to serial numbers, because the date of any letter cited in the notes is immediately apparent.

Editorial principles

In transcribing the manuscripts, my aim has been to print the letters as readably as possible without losing the flavour of the original. In general I have retained Darwin's spelling and capitalization, and most of his punctuation. But I have restored or modernized most of his abbreviations, as follows:

M^{r} M^{rs} D^{r} S^{r} L^{d} are transcribed as Mr Mrs Dr Sir Lord

y^{t} ye y^{r} wh p^{r} are transcribed as that the your which per

& &c comp^{li} Bro^{r} are transcribed as and etc compliments Brother

I also restore words in which many letters are abbreviated into a flourish, with the exception of obe^{d} ser^{v}, which is transcribed as 'obed. serv.'. Darwin's normal signature (see photograph, between p. 192 and p. 193) is transcribed as 'E Darwin'.

I retain Darwin's adventurous spelling. Though he is often inconsistent, there are some words which (judged by modern usage) he consistently mis-spells. The most important are:

agreable; chuse; knowlege; least (for 'lest'); medecine; recieve; sais.

Editorial interventions are signalled by square brackets, the most frequent being [?], to indicate that the preceding word is scarcely legible and may be in error. A missing but apparently intended word is inserted as [word]. A word apparently unintended is marked as [~~word~~]. Usually, Darwin's mis-spellings are retained without comment; but if the error is very misleading, the correct spelling is inserted in square brackets: e.g. 'ancient Room [Rome]'.

Curly brackets indicate a torn or mutilated manuscript, e.g. {conjectural words} or {*torn, gap of two lines*}. (Fortunately, Darwin himself never uses square or curly brackets.) A row of asterisks indicates an omission of unknown length.

I follow Darwin's erratic punctuation to the limits of commonsense and perhaps beyond. I have made some changes, however. At the end of a sentence Darwin often writes a dash followed by a full stop, which overrules the dash: on these occasions I omit the dash. Sometimes he writes a full stop (or comma), then puts in a dash and adds another phrase: on these occasions I omit the full stop (or comma). When he inserts a dash after a full stop before starting a new sentence, I have made my own judgement: most of the dashes seem useless and are omitted; some signal a change of topic, a semi-new-paragraph, as it were, and these are retained; sometimes the dash is followed by a long blank, and the next sentence is then printed as a new paragraph. Darwin's full stops are sometimes elongated into what looks like a comma; while the tails of some of his commas have faded with the passage of time, so that they look like full stops. Wherever necessary, the sense and grammar are my guides in choosing between a comma and a full stop. Punctuation added by me (rather than reinterpreted) is enclosed in square brackets.

Darwin usually starts sentences with capital letters but occasionally forgets, or compromises by writing an abnormally large lower-case letter, not reproducible in print. I have allowed readability to prevail by giving an initial capital to every new sentence (after a full stop). Darwin bestows initial capitals on many words (usually nouns): I have preserved this idiosyncrasy, which sometimes denotes emphasis and sometimes seems just random. I have not recorded deletions, which are rare in Darwin's letters, and usually of no significance.

In the letters transcribed from manuscript copies, not in Darwin's hand, my editing has been more severe. I have altered words that appear to be wrong, marking them with [?] if there is doubt; and I have altered punctuation etc to conform with the conventions already described.

Letters taken from printed sources have been treated in the same way as the manuscript copies. In general, the unknown copying conventions of the original transcriber have inevitably been preserved. The dating of the letter given in the printed source is retained without comment unless there is reason for doubt.

Darwin's diagrams are incorporated into the texts of the letters, most of them by photocopying from the originals. When photocopying of the manuscript was not permitted, the diagrams were either traced or redrawn.

After the text of the letter, the name and address of the recipient are given, if they appear on the manuscript. Words written on by the recipient are recorded (preceded by the word *Endorsed*) only if they are relevant or significant.

After each letter further information is given under four headings.

Original The location of the original manuscript is given, if known; otherwise the words 'Not traced' appear.

Printed If the letter, or a substantial excerpt, has previously been printed, a reference is given. No attempt is made to record all printed versions, the books mentioned on p. xv being given preferentially as references.

Text The source of my text is stated.

Notes I wished to avoid disfiguring and falsifying the texts of the letters with superscript note numbers. I hope that notes on specific points can still quite easily be found, however, since most of the letters are short and the notes follow a logical order, namely: (1) information on the recipient of the letter (where necessary); (2) a note on the dating (if it is doubtful); (3) attempted explanations of obscurities, identification of places, people, etc., in the order in which they arise in the text. All Midland place names mentioned in the letters are marked on the map on pp. x–xi; if a name is on the map, it is not mentioned in the notes. Dates of birth and death of individuals are given whenever possible. Doubtful dates of birth may appear in one of three forms: (a) 1733/4, if the date seems fairly certain to be either 1733 or 1734; (b) 1733?, if there is some (unconfirmed) documentary basis for 1733; (c) *c.*1730, implying 'probably between 1725 and 1735'.

The heading for each letter gives the name of the recipient and the date. If the recipient, the date (or any part of it) is supplied editorially, this is indicated by square brackets.

Abbreviations

The following abbreviations are used for books often cited (the place of publication is London unless otherwise stated).

Alum. Cantab.
J. A. Venn, *Alumni Cantabrigienses.* Part I, to 1751: 4 vols. Part II, 1752–1900: 4 vols. Cambridge University Press, 1922 and 1944.

Alum. Oxon.
J. Foster, *Alumni Oxonienses, 1715–1896.* 4 vols. Parker, Oxford, 1888.

C. Darwin, *Autobiography*
The Autobiography of Charles Darwin, edited by Nora Barlow. Collins, 1958.

C. Darwin, *Life of E. Darwin*
Charles Darwin, *The Life of Erasmus Darwin*, with an essay on his scientific works by Ernst Krause. Murray, 1879. (2nd ed. 1887.) (First edition reprinted by Gregg International, Farnborough, Hants., 1971.) In the first edition, and many catalogues, Krause is given as the author of this book.

DNB
Dictionary of National Biography (to 1900). 22 vols. Oxford University Press.

Doctor of Revolution
D. King-Hele, *Doctor of Revolution: the Life and Genius of Erasmus Darwin.* Faber, 1977.

Edgeworth, *Memoirs*
R. L. and Maria Edgeworth, *Memoirs of Richard Lovell Edgeworth.* 2 vols. Hunter, 1820. (Reprinted by Irish University Press, Shannon, 1969.)

Essential Writings
The Essential Writings of Erasmus Darwin, edited by D. King-Hele. MacGibbon and Kee, 1968.

Gent. Mag.
Gentleman's Magazine. Vols. 1–100, 1731–1830.

Glover, *Derby*
S. Glover, *History of the County of Derby.* 2 vols. Mozley, Derby, 1829.

Lunar Society
R. E. Schofield, *The Lunar Society of Birmingham*. Oxford University Press, 1963.
Meteyard, *Wedgwood*
E. Meteyard, *The Life of Josiah Wedgwood.* 2 vols. Hurst and Blackett, 1865–6. (Reprinted by Cornmarket Press, 1970.)
Moilliet, *Keir*
J. K. Moilliet, *Life and Correspondence of James Keir*. Privately printed, 1868.
Muirhead, *James Watt*
J. P. Muirhead, *The Origin and Progress of the Mechanical Inventions of James Watt*. 3 vols. Murray, 1854.
Namier & Brooke, *Commons*
L. Namier and J. Brooke, *The House of Commons, 1754–1790*. 3 vols. HMSO, 1964.
Nichols, *Leicester*
J. Nichols, *History and Antiquities of the County of Leicester*. 4 vols. Nichols, 1815.
OED
Oxford English Dictionary. 13 vols. Oxford University Press.
Pearson, *Galton*
K. Pearson, *The Life, Letters and Labours of Francis Galton.* 4 vols. Cambridge University Press, 1914–30.
Phil. Trans.
Philosophical Transactions of the Royal Society of London. Vols. 1–90, 1666–1800.
Scott, *Admissions*
R. F. Scott, *Admissions to the College of St John the Evangelist in the University of Cambridge*, Parts III and IV, Deighton Bell, Cambridge, 1903.
Seward, *Darwin*
A. Seward, *Memoirs of the Life of Dr Darwin.* Johnson, 1804.
Seward, *Letters*
A. Seward, *Letters*. 6 vols. Constable, Edinburgh, 1811.
Stock, *Beddoes*
J. E. Stock, *Memoirs of the Life of Thomas Beddoes*. Murray, 1811.
Wedgwood Circle
B. and H. Wedgwood, *The Wedgwood Circle*, 1730–1897, Studio Vista, 1980.
Wedgwood, *Letters*
Josiah Wedgwood, *Letters*. Edited by K. E. Farrer. 3 vols. Privately printed, 1903–6. Reprinted by Morten, Manchester, 1974.

Style of references

References to books having several volumes give the volume number in small Roman numerals, and the page number, e.g. Wedgwood, *Letters* iii 273. (The publisher, place of publication if not London, and date, are also given, except for the books listed above.)

References to periodicals normally give the volume number, in bold type, and page numbers, with the date in brackets: e.g. *Phil. Trans.*, **78**, 43–52 (1788).

References to Darwin's poems give the Canto number in capital Roman numerals and line numbers, e.g. *Economy of Vegetation* III 329–36.

Acknowledgments

I am deeply indebted to Dr Hugh Torrens of the University of Keele for his willing help throughout the project. His enthusiasm spurred me to attempt the work of collecting and editing the letters; he located a number of letters I might never have found; and his deep knowledge of eighteenth-century scientific biography often rescued me from an impasse in identification. He provided hints and information for so many of the notes that a listing would be impracticable: instead I gladly and sincerely thank him for his continual aid and encouragement.

Through the kindness of Lord Gibson-Watt, I was granted access to his magnificent private collection of Watt family papers at Doldowlod, where I found twenty previously unrecorded letters from Darwin to James Watt, providing documentary evidence of their close friendship and cooperation over 30 years. I am most grateful to Lord Gibson-Watt for giving me the opportunity to publish these letters.

I also wish to thank the Earl of Harrowby for his kind assistance over the letters from the Harrowby MSS at Sandon Hall, and Professor Richard Keynes for making available letters in his possession.

The Royal Society made a grant from its History of Science Fund which enabled me to assemble the texts of the previously-known letters. I thank the Royal Society for this help, and particularly Mr N. H. Robinson, the Society's Librarian, and his staff, for their interest and assistance.

The transcribing and typing of a collection of manuscripts is a heavy labour, and I should like to record my gratitude to Mrs Valerie McMillan, whose skilful and accurate typing of my transcripts, including Darwin's many mis-spellings, has greatly eased my task.

The manuscripts of the letters came from 30 different sources (listed on p. xv) and it is a pleasure to acknowledge the willing cooperation of the many Librarians, Archivists and Keepers of the manuscripts. Without their help this volume would not exist. Although it is invidious to mention names, there are some whose help has gone far beyond what could reasonably be expected. To them, in alphabetical order, I offer special thanks: Mr D. G. C. Allan, Librarian of the Royal Society of Arts; Mr A. Andrews, Archivist at the Birmingham Reference Library; Dr Whitfield J. Bell, Librarian of the American Philosophical Society; Dr Ian Fraser, Archivist at the Keele University Library; Mr Peter Gautrey, Assistant Librarian at the Cambridge University Library; Mr F. B. Stitt, Librarian at the William Salt Library, Stafford; and Mr Philip Titheradge, Custodian of Down House.

For vital help on special subjects, I particularly thank Dr Christopher Lawrence, of the Wellcome Institute for the History of Medicine, who provided the substance, and often the wording, of many of the medical notes; Dr Sydney Smith, the kindest and most knowledgeable of Darwinians; Mr J. Gould, whose study of the Wichnor ironworks was most valuable; and many other correspondents, named in the notes, who have helped to identify people mentioned in the letters. However, the responsibility for any errors in the specialist notes is mine alone.

Most of my research has been done in the Cambridge University Library and I am most grateful to the staff of that unique institution: I requested hundreds of eighteenth and nineteenth century books and periodicals, and every single one materialized within a few minutes.

Finally, I should like to express my grateful thanks to Dr Alan Winter of the Cambridge University Press for his consistent help and encouragement during the preparation of this book.

I have been most fortunate in obtaining permission for publication from the owners and keepers of nearly all the known manuscripts. For their courtesy in allowing me to transcribe manuscripts and publish printed versions, I should like to offer my grateful thanks to

American Philosophical Society, Philadelphia
Avon County Library, Bristol
Birmingham Public Libraries
Birmingham University Library
Bodleian Library, Oxford
Matthew Boulton Trust, Birmingham
Trustees of the British Library, London
Trustees of the British Museum (Natural History)
Syndics of the Cambridge University Library
Derbyshire County Library
Trustees of the Devonshire Collections, Chatsworth
Down House and the Royal College of Surgeons of England
Syndics of the Fitzwilliam Museum, Cambridge
The Rt Hon. the Lord Gibson-Watt PC
The Rt Hon. the Earl of Harrowby and the Harrowby MSS Trust
Harvard University Library
Historical Society of Pennsylvania, Philadelphia
Keele University Library
Sir Geoffrey Keynes and Professor R. D. Keynes, FRS
Mrs V. M. Kindersley and the Nottinghamshire Record Office

Linnean Society, London
Liverpool City Libraries: Hornby Library
The New York Public Library; Astor, Lenox and Tilden Foundations
Mr D. W. H. Neilson and Derby Central Library
The late Professor E. S. Pearson FRS and University College London
Royal Society of Arts, London
Royal Society of London
Royal Society of Medicine, London
John Rylands University Library of Manchester
Master and Fellows of St John's College, Cambridge
Trustees of the William Salt Library, Stafford
Messrs Josiah Wedgwood and Sons Ltd, of Barlaston, Stoke-on-Trent
Trustees of the Wellcome Institute
Yale University Library

For permission to reproduce photographs of the specified portraits in their possession, I am most grateful to the following:

Trustees of the British Museum (Charles Blagden; Thomas Cadell; Jonas Dryander; Charles Greville; Joseph Johnson)
Trustees of the Chatsworth Settlement (Georgiana, Duchess of Devonshire *by* J. Reynolds; 5th Duke of Devonshire *by* F. Batoni)
Mr George Darwin and the Master and Fellows of Darwin College, Cambridge (Erasmus Darwin *by* J. Wright)
Down House and the Royal College of Surgeons of England (Elizabeth Pole *by* J. Wright)
The Rt Hon. the Earl of Harrowby (Lady Harrowby *by* J. Reynolds; 1st Lord Harrowby; Rt Hon. Dudley Ryder)
Keele University Library (Robert W. Darwin, son)
National Portrait Gallery (Thomas Beddoes *by* S. T. Roche; Thomas Day *by* J. Wright; R. L. Edgeworth *by* H. Hone; William Hayley *by* H. Howard; Edward Jenner *by* J. Northcote; Anna Seward *by* T. Kettle; Sir John Sinclair *by* H. Raeburn; James Watt *by* C. F. von Breda)
Mr D. W. H. Neilson (Phoebe Horton *by* T. Lawrence)
Royal Society of London (Sir Joseph Banks *by* T. Phillips; Benjamin Franklin *by* J. Wright; James Keir; Joseph Priestley; William Watson *by* L. F. Abbot)
Science Museum, London (Matthew Boulton *by* W. Beechey)
Dr H. S. Torrens (J. C. Lettsom, Richard Pulteney, John Walcott, William Withering)
Josiah Wedgwood and Sons Ltd., Barlaston (Thomas Bentley *by* J. Wright; Thomas Byerley; Josiah Wedgwood *by* J. Reynolds)
(When the photograph is of a painting, the artist's name is given, if known.)

The sources of the other portraits are:

Thomas Brown: frontispiece of D. Welsh, *Life of Thomas Brown* (Edinburgh, 1825)

Joseph Cradock: frontispiece of J. Cradock, *Memoirs* (1828)

James Currie: frontispiece of W. W. Currie, *Memoir of James Currie* (1831)

Robert W. Darwin (brother); Lucy Galton; S. T. Galton: K. Pearson, *Life . . . of Francis Galton* (1914)

Thomas Garnett: *Illustrated London News*, **114**, 846 (1899)

James Hutton (*by* H. Raeburn): J. L. Caw, *Scottish Portraits* (Edinburgh, 1902) Portfolio IV, page 54.

Edward Jerningham: *European Magazine*, **25**, 411 (1794)

Samuel More: *European Magazine*, **36**, 363 (1799)

Richard Polwhele: frontispiece of R. Polwhele, *Traditions and Recollections* (1826)

Peter Templeman: frontispiece of *Trans. Society of Arts*, Vol. 17 (1799)

Gilbert Wakefield: frontispiece of G. Wakefield, *Memoirs* (1804)

Thomas Wedgwood: frontispiece of R. B. Litchfield, *Tom Wedgwood* (1903)

For permission to reproduce facsimiles of the manuscripts of Darwin's letters, I thank the Syndics of the Cambridge University Library.

CHRONOLOGICAL LIST OF LETTERS

The list is arranged by letter number, date and recipient

49A [1749 March] Susannah Darwin
49B 1749 December 28 Samuel Pegge

50A 1750 December 11 William Burrow

54A 1754 November [23?] Thomas Okes

56A 1756 [early September] Albert Reimarus
56B 1756 September 9 Albert Reimarus
56C 1756 [October?] Albert Reimarus

57A 1757 March 23 William Watson
57B 1757 December 24 Mary Howard

62A 1762 [October 30?] Matthew Boulton

63A 1763 July 1 Matthew Boulton
63B 1763 December John Barker
63C 1763 December 16 John Barker
63D 1763 December 17 John Barker
63E 1763 December 22 John Barker
63F 1763 December 23 John Barker
63G 1763 December 29 John Barker

64A [1764 January 4?] John Barker
64B [1764 January 4?] John Barker
64C 1764 March 19 John Barker
64D 1764 March 24 John Barker
64E [1764?] Matthew Boulton
64F 1764 November 3 Matthew Boulton

65A 1765 October [16–21] Thomas Bentley
65B 1765 December 12 Matthew Boulton

66A 1766 March 8 Peter Templeman
66B 1766 March 11 Matthew Boulton
66C 1766 April 27 Josiah Wedgwood
66D [1766 Summer] Matthew Boulton

67A 1767 February 7 Unknown man
67B 1767 July 2 Josiah Wedgwood
67C 1767 July 29 Matthew Boulton
67D 1767 August 18 James Watt
67E 1767 November 8 Josiah Wedgwood
67F 1767 November [12–18] Josiah Wedgwood
67G [1767 late] Josiah Wedgwood

68A 1768 [February] Josiah Wedgwood
68B 1768 June 14 Josiah Wedgwood
68C 1768 July 31 Matthew Boulton
68D 1768 August 26 Thomas Feilde

69A 1769 February 4 Peter Templeman

* Not available for publication.

ALPHABETICAL INDEX OF RECIPIENTS

* indicates that a portrait may be found among the illustrations, between pages 192 and 193

THE LETTERS

49A. To Susannah Darwin, [March 1749]

Dear Sister,

I reciev'd yours about a fortnight after the date that I must begg to be excused for not answering it sooner: besides I have some substantial Reasons, as having in mind to see Lent almost expired, before I would vouch for my Abstinence throughout the whole: and not having had a convenient oppertunity to consult a Synod of my learned friends about your ingenious Conscience, and I must inform you we unanimously agree in the Opinion of the Learned Divine you mention, that Swine may indeed be fish but then they are a devillish sort of fish; and we can prove from the same Authority that all fish is flesh whence we affirm Porck not only to be flesh but a devillish Sort of flesh; and I would advise you for Conscience sake altogether to abstain from tasting it; as I can assure You I have done, tho' roast Pork has come to Table several Times; and for my own part have lived upon Puding, milk, and vegetables all this Lent; but don't mistake me, I don't mean I have not touch'd roast beef, mutton, veal, goose, fowl, etc., for what are all these? All flesh is grass! Was I to give you a journal of a Week, it would be stuft so full of Greek and Latin as translation, Verses, themes, annotation, Exercise and the like, it would not only be very tedious and insipid but perfectly unintelligible to any but Scholboys.

I fancy you forgot in Yours to inform me that your Cheek was quite settled by your Temperance, but however I can easily suppose it. For the temperate enjoy an ever-blooming Health free from all the Infections and disorders luxurious mortals are subject to, the whimsical Tribe of Phisitians cheated of their fees may sit down in penury and Want, they may curse mankind and imprecate the Gods and call down that parent of all Deseases, luxury, to infest Mankind, luxury more distructive than the Sharpest Famine; tho' all the Distempers that ever Satan inflicted upon Job hover over the intemperate, they would play harmless round our Heads, nor dare to touch a single Hair. We should not meet those pale thin and haggard countenances which every day present themselves to us. No doubt men would still live their Hunderd, and Methusalem would lose his Character; fever banished from our Streets, limping Gout would fly the land, and Sedentary Stone would vanish into oblivion and death himself be slain.

I could for ever rail against Luxury, and for ever panegyrize upon abstinence, had I not already encroach'd too far upon your Patience, but it being Lent the exercise of that Christian virtue may not be amiss, so I shall proceed a little furder – [The remainder of the letter is hardly legible or intelligible, with no signature.]

P.S. – Excuse Hast, supper being called, very Hungry.

Original Not traced. In the possession of Charles Darwin, 1879.
Printed C. Darwin, *Life of E. Darwin*, pp. 9–11.

Text From above source. (The comment in square brackets is Charles Darwin's.)
Notes
(1) Susannah Darwin (1729–1789) was Erasmus's favourite sister, and one of his earliest poems is in praise of her:

My dearest Sue
Of lovely hue
No sugar can be sweeter;
You do as far
Excel Su-gar
As sugar does saltpetre.

(From Pearson Papers, 578, page 34, University College London.) Susannah was housekeeper for Erasmus at Lichfield in 1756–7; and when he was a widower, from 1770 to 1781, she acted both as housekeeper and as foster-mother to his son Robert.
(2) Erasmus was writing from Chesterfield School in response to a skittish letter from Susannah about her abstinence at Lent: the letter, which includes 'A Diary in Lent', is printed in C. Darwin, *Life of E. Darwin*, pp. 7–9. Erasmus was seventeen when he wrote this letter, and he already has at his command the bantering tone that animates so many of his later letters. He also reveals a rather cynical attitude towards his future profession and a complete disregard for correct spelling (Methuselah borrows the end of Jerusalem). Darwin's praise of temperance, though part of the banter, was not altogether insincere: the preaching of temperance became part of his medical stock-in-trade – not surprisingly when so many of his richer patients grossly overindulged.
(3) Although Susannah's 'Diary in Lent' is dated 'Feb. 20, 1748', she is observing the pre-1752 convention of keeping the old year-number until 25 March of the New Year. She says that the first day of Lent was 8 February, as it was in 1749. (It was 24 February in 1748.) Easter day was 26 March in 1749, and Erasmus probably wrote his letter about ten days before that.

49B. To Samuel Pegge, 28 December 1749

Wrote from Elstone to Chesterfield in Christmas Hollydays, 1749

Eras: D'Arwin of Nottinghamshire to Sam[ll]: Pegge of Kent sendeth greeting.

Here Muse! – assume your Wings – to Rother these –
Send Pegge all Health, Prosperity and Ease:
Or on a Broomstick cleave the Clouds, and bear
That Source of Health and Ease, some Xmas Cheer.
Carry him this Pudding rich with India's Plumb,
With these Mince-Pyes attract his willing Thumb:
Thus spoke the dying Pigg, "Let all abroad
The bright Black-pudding smoak upon his Board:

While snaky Sausages their volumes roll,
And hiss and spit before the burning Coal."
Then let the Ham's delicious Red be seen,
Spread on the greasy Quintessence of Cream:
This Moon-like Cheshire Cheese, then Clio, bear;
Encreasing Moons their waining Horns repair,
But Cheese, unlike Prometh's immortal Breast,
Will grow no more, when hungry Vultures feast.
O, and by all means, let friend Samuel share
These Pass[?]-shot Patriges, this well-run Hare.
Then take this promiss'd Book, at once to glut
His Head extensive, and extensive Gut.
'Twas Homer's Rule (you know) at every Feast,
To treat the [deathless gods?] with like Repast.
Is this sufficient Load, good Muse, dost think?
If not, here take this Cask of Xmas drink. –
What say'st? – "can't carry such substantial things"!
What Muse for Samuel can refuse her Wings? –
What, what, "Your airy shadows can't sustain
Earth's firm and solid Dross." – excuses vain! –
Well, since you won't, on you alone depend
The sad Miscarriage – I lament my Friend.
– But go you shall – and mind – asks He how I
These Winter Days the lazy Hours employ,
Tell him – stay, let me see – I don't know what,
Tell him I rise – and Betty where's my Hat? –
Is Mr Shootall come? – bless me, how late! –
Unloose the Doggs – my Breakfast – almost eight! –
Equipt with Pouch and Powder, e'er 't be Light,
Doggs, Guns, and Horses rush into my Sight:
Supper and Cards conclude the idle Day,
And thus his Brother sleeps his Time away.

Then, Muse, enquire if still around his Head,
Rose-blooming Health her golden Pinions spread:
How pass the solitary Days, and then
If bright Apollo smile upon his Pen. –
Yes this is all – and tell him in ye end
Tell him sincerely Darwin is his Friend.

Now Muse, assend, I'll have no more delay,
Expand your Wings, and stretch, and soar away.

Dear Brother Samuel

I'm sorry the Muse could not bring you the present design'd, but to You I know the Will is as good as the Deed; and I will bring the book with me when we return:

caetera desunt
December 28, 1749.

Original Not traced. There is a typescript copy at University College London, Pearson Papers, 577, pages 6–7.
Printed Unpublished.
Text From typescript copy at University College London.
Notes
(1) Samuel Pegge (1733–1800) was one of Darwin's friends at Chesterfield School, where Erasmus, now 18, was in his last year. Pegge went on with Darwin to St John's College, Cambridge, in 1750. Pegge's father, Samuel Pegge the elder (1704–1796), was a well-known antiquary and probably a friend of Darwin's father, since both were members of the Gentlemen's Society at Spalding. Pegge the elder was educated at Chesterfield School and St John's College, Cambridge (where he was a Fellow from 1726 to 1731): Pegge's example may have induced Robert Darwin to direct three of his sons along the same educational path. Pegge the elder was vicar of Godmersham, near Ashford, Kent from 1731 to 1751: this explains Darwin's greeting to 'Samuel Pegge of Kent'; sending the Muse to 'Rother', a river 15 miles south-west of Godmersham, is presumably poetic licence – especially since Pegge was apparently at Chesterfield. In his later career Samuel Pegge the younger displayed a wide range of talents as a musical composer, poet, antiquary, barrister and historian of the Royal Household. See *DNB* for both Pegges.
(2) This letter in verse confirms other indications that the young Darwin was already adept at turning out rhyming couplets. The picture of him spending his days idly on shooting and hunting with dogs, though rather surprising, parallels the behaviour of his grandson Charles, who at a similar age was told: 'You care for nothing but shooting, dogs and rat-catching, and you will be a disgrace to yourself and all your family.' (C. Darwin, *Autobiography*, p. 28).
(3) My insertion in line 22 – [deathless gods?] – fills a blank in the typescript. Presumably Darwin wrote the words in Greek, and the typist omitted them. In Greek, 'deathless gods' would be θεοί ἀθάνατοι, which does not scan (either in the nominative or accusative); but Darwin may well have been subtle enough to write Greek words which only scan after translation.
(4) 'Caetera desunt' – 'the rest is missing' – was a customary apologia for an unfinished school writing exercise. (The typescript copy in the Pearson Papers has 'coetere desunt'.)

50A. To William Burrow, 11 December 1750

St John's, Cambridge
Mr Burroughs December 11th 1750
Master of the Chesterfield Grammar School

Honoured Sir,

Mr Adison, I think, in a paper upon Love lays down, that it is much more difficult for one really under the Influence of that Passion, to testify it to the Fair, than He who only feigns it out of Interest: whether this may proceed from Awe, Respect, Fear of Offending, or whatever else, I'm persuaded the same will hold good in the return of gratitude: when first, Sir, I took up my pen to write to you, the more I thought, the more I was confounded: and the difficulty of the Task appeared equal to the pleasure of it: in Letters of this kind I concluded, were I too warm, it might be looked upon as feigned and complemental: if reserved, as forgetful and ungrateful: I found it painful not to let my words glow like the Heart that dictated, yet could not but be under some Apprehensions how it might be recieved: that you see Sir (as I think is Tully's turn), I can not cease to be in Pain without being in some measure uneasy: at length relying on the favourable reception my rhime-compositions were ever honoured with from your Hands, I chose rather to throw the following into Verse, since Truth is still Truth, adorned or unadorned, in Prose or Poetry: I could have wished it had been equal to the subject on the one Hand, and to the Spirit of the Writer on the other, and I should not have feared excelling in the sublime and passionate. If it be not so correct as might have been expected from the importance of the subject, ascribe it to the want of a Sayle, Pegge and Fairfax to have revised it, e'er it was submitted to your perusal from

Yours
with Sincerity
Eras. Darwin

Original Not traced. There is a typescript copy at University College London, Pearson Papers, 577, page 2.
Printed Unpublished.
Text From typescript copy in Pearson Papers.
Notes
(1) William Burrow (1683–1758), whose surname was sometimes spelt Burrough, Burrows or Burroughs, entered St John's College, Cambridge in 1702 and became MA in 1709. He was headmaster of Chesterfield School from 1722 to 1752 and rector of North Wingfield, Derbyshire from 1739 to 1758. Under his guidance Chesterfield became the leading school in the north of England. One of his pupils said Burrow 'was very skilled in Greek and was a very good and understanding teacher'. Darwin's flattering letter was written at the end of his first term at Cambridge and, although it cannot be accepted at its face value, he apparently thought well of Burrow. For further details of Burrow, see *Alum. Cantab.*, Part I, i 264, and J. C. V. Kendall and M. P. Jackson, 'A History of the Free Grammar School, Chesterfield' (1965) (typescript book, available at Derbyshire County Library, Matlock).
(2) 'Mr Adison . . . in a paper on Love' refers to Addison's essay 'On Love and Marriage' (J. Addison, *Essays*, Ward Lock, 1882, pp. 318–21): 'It is easier for an artful man who is not in love to persuade his mistress he has a passion for her . . . than for one who loves with the greatest violence'.

(3) There are several Latin sayings about pleasure never being quite free of pain. Perhaps the best known is from Ovid, *Metamorphoses* VII 453–4:

nulla est sincera voluptas,
Sollicitumque aliquid laetis intervenit.

(No pleasure is unalloyed; some anxiety always interferes with our joys.) Darwin credits Cicero with the authorship, but only tentatively.

(4) The letter as printed here is intended as an introduction to an elaborate poem of 130 lines in rhyming couplets, 'In Imitation of the 5th Satire of Persius'. The six satires of the Roman poet Persius Flaccus (AD 34–62), written in the style of Horace, were popular in the Middle Ages but are little known now. Darwin's translation is too long to reproduce here, and mostly rather tame. But at the end he compares Burrow with a 'stately Pine' round which 'young Scions grow'. These 'pupil-plants', such as 'the Kayes and Fairfax[es]' will, he says, grow into 'the Burrows of the rising Age' (Pearson Papers, 577, page 5).

(5) Sayle, Pegge and Fairfax were boys at Chesterfield School, and were presumably better Latin scholars than Darwin. ('Kaye' in the poem may be a misprint for 'Sayle'.) William Sayle (1732?–1799) went up to St John's, Cambridge in 1751, the year after Darwin, who in 1753 sent him a poem subtitled 'An Ode to Mr Wm. Sayle, Retford, Nottinghamshire' (Pearson Papers, 577, pages 73–4). Sayle became a country clergyman, being vicar of Stowey in Somerset from 1774 to 1799. (See Scott, *Admissions*, Part III, p. 604.) There is a probable further reference to Sayle in 1781 (letter 81H). For Pegge, see letter 49B. Fairfax was probably Robert Fairfax (1732–1803), who went up to Christ Church, Oxford in 1751. Fairfax came from Newton Kyme, near York, and apparently (like many country gentlemen) lived there all his life. (See J. Foster, *Pedigrees of the County Families of Yorkshire*, privately printed, 1874, Volume 1; and *Alum. Oxon.* ii 444.)

54A. To Thomas Okes, [23?] November 1754

* * *

Yesterday's post brought me the disagreeable news of my father's departure out of this sinful world. He was a man, Okes, of more sense than learning; of very great industry in the law, even after he had no business, nor expectation of any. He was frugal, but not covetous; very tender to his children, but still kept them at an awful kind of distance. He passed through this life with honesty and industry, and brought up seven healthy children to follow his example. He was 72 years old, and died the 20th of this current. 'Blessed are they that die in the Lord!'

That there exists a superior ENS ENTIUM, which formed these wonderful creatures, is a mathematical demonstration. That HE influences things by a particular providence, is not so evident. The probability, according to my notion, is against it, since general laws seem sufficient for that end. Shall we say no particular providence is necessary to roll this Planet round the Sun, and yet affirm it necessary in turning up <u>cinque</u> and <u>quatorze</u>, while shaking a box of dies? or giving each his daily bread? – The light of Nature affords us not a single argument for a future state; this is the only one – that it is possible with God; since he who made us out of nothing can surely re-create us; and that he will do this is what we humbly hope. I like the Duke of Buckingham's epitaph –

"Pro Rege saepe, pro Republica semper, dubius, non improbus, vixi: incertus, sed intarbatus morior. Christum advenero, Deo confido benevolenti et omnipotenti, Ens Entium miserere mei!"

Erasmus Darwin

Original Not traced. In possession of Thomas Verney Okes, surgeon at Cambridge, in 1808.

Printed *Gentleman's Magazine*, **78**, 869 (Oct. 1808); and C. Darwin, *Life of E. Darwin*, pp. 14–15.

Text From *Gentleman's Magazine*, as specified above.

Notes

(1) After 3 years as an undergraduate, Darwin left Cambridge in the summer of 1754 to complete his medical training at the Edinburgh Medical School. His father's death, on 20 November 1754, occurred during his first term at Edinburgh, and it was from Edinburgh that he wrote this letter to Okes, who was an undergraduate friend at Cambridge. Dr Thomas Okes (1730–1797) was at King's College, Cambridge from 1750: see *Alum. Cantab.* He practised as a physician at Bedford, at Lynn in Norfolk, and at Cambridge, and was for many years a surgeon in the Army before finally settling at Exeter before 1779. See *Medical Register* (1779). Okes continued to practise in Exeter, and there is a memorial tablet on the south wall of the nave of Exeter Cathedral.

(2) Darwin's use of 'disagreeable' in the first line strikes us as odd, but it merely meant 'unpleasant'. Darwin's reflections on life show that he was already something of a sceptic.

(3) The letter was written in November, presumably within three or four days of his father's death: the 23rd seems a reasonable guess.

(4) The epitaph of the Duke of Buckingham (1628–1687) may be translated as follows: 'For the King often, for the Republic always, wavering, not wicked, I have lived: uncertain, but untroubled I die. I shall come to Christ, I trust in a benevolent and omnipotent God, Ens Entium take pity on me!' 'Ens Entium', literally 'Being of Beings' or 'Essence of Essences', was a phrase often used by those who did not care to commit themselves to the word 'God'.

56A. To Albert Reimarus, [early September] 1756

* * *

I am very sorry to hear that D. took six guineas from the poor young man. He has nothing but what hard labour gives him; is much distressed by this thing costing him near £30 in all, since the house where he lay cheated him much... When he returns I shall send him two guineas. I beg you would not mention to my brother that I send this to him.

* * *

Original Not traced. In the possession of the Reimarus family 1879. Photographed by Charles Darwin 1879.
Printed C. Darwin, *Life of E. Darwin*, p. 18.
Text From above source.
Notes
(1) The recipient of the letter, Dr Johann Albert Heinrich Reimarus (1729–1814), was the son of the well-known German philosopher Hermann Reimarus (1694–1768), whose deistic philosophy had some influence on Darwin (see letter 69C). Albert Reimarus took his degree at Leyden in 1754 and then went on to the Edinburgh Medical School, where he became a close friend of Darwin, who wrote a poem in honour of Reimarus's graduation (Pearson Papers, 577, page 30, University College London). For the life of Albert Reimarus, see *Deutsche Biographie* (Leipsig, 1888) Vol. 27, pp. 704–9.
(2) This excerpt is from a letter written when Darwin, newly graduated, was making his abortive attempt to start a practice at Nottingham. It seems that Darwin 'sent a working man to a London surgeon, Mr D., for a serious operation. Reimarus and Dr Darwin appear to have had some misunderstanding with the surgeon, expecting that he would perform the operation gratuitously'. (C. Darwin, *Life of E. Darwin*, p. 18). Apparently Darwin then wrote anonymously to the surgeon, to complain about the charge: see next letter. I have not identified 'D.'

56B. To Albert Reimarus, 9 September 1756

[Nottingham Sept. 9th 1756]

* * *

You say I am suspected to be the Author of it [the anonymous letter], and next to me some malicious person somewhere else, and that I am desired as I am a gentleman to declare concerning it. First, then, as I am upon Honour, I must not conceal that I am glad there are Persons who will revenge Faults the Law can not take hold off: and I hope Mr D. will not be affronted at this Declaration; since you say he did not know the distress of the Man. Secondly, as another Person is suspected, I will not say whether I am the Author or not, since I don't think the Author merits Punishment, for informing Mr D. of a Mistake. You call the Letter a threatening Letter, and afterwards say the Author pretends to be a Friend to Mr D. This, though you give me several particulars of it, is a Contradiction I don't understand.

* * *

Original Not traced. In the possession of the Reimarus family 1879. Photographed by Charles Darwin 1879.
Printed C. Darwin, *Life of E. Darwin*, pp. 18–19.
Text From above source.
Notes
(1) The words in square brackets are as given by Charles Darwin, who also states, 'In a P.S. he adds that Reimarus might show the letter to Mr D.'

(2) This letter shows the young Erasmus getting himself into a tangle. His motives were good but his methods were unwise – and uncharacteristic. Perhaps the episode taught him that it was better to be blunt than devious.

56C. To Albert Reimarus, [October?] 1756

* * *

I believe I forgot to tell how Dr Hill makes his 'Herbal'. He has got some wooden plates from some old herbal, and the man that cleans them cuts out one branch of every one of them, or adds one branch or leaf, to disguise them. This I have from my friend Mr G———y, watch-maker, to whom this print-mender told it, adding, 'I make plants now every day that God never dreamt of'.

* * *

Original Not traced. In the possession of the Reimarus family 1879. Photographed by Charles Darwin 1879.
Printed C. Darwin, *Life of E. Darwin*, p. 17.
Text From above source.
Notes
(1) Charles Darwin refers to 'several letters' between Erasmus and Albert Reimarus in 1756. 'Various subjects were discussed between them, including the wildest speculations by Erasmus on the resemblance between the action of the human soul and that of electricity, but the letters are not worth publishing' (*Life of E. Darwin*, p. 17).
(2) Charles Darwin says this letter was written while Erasmus was at Nottingham, i.e. between September and November 1756.
(3) 'Dr Hill' was John Hill (1716?–1775), often known as Sir John Hill, apothecary, botanist, playwright, prolific author and controversialist. He was very talented and versatile, but also unscrupulous and impudent: Garrick said, 'such a villain sure never existed'. Among Hill's 76 books was *The British Herbal; an History of Plants and Trees native of Britain* (London, 1756), the work Darwin comments on. For further details of Hill, see the five-page account in *DNB*.
(4) I have not identified Mr G———y.

57A. To William Watson, 23 March 1757

Sir,

The inclosed papers were designed for the perusal of the Royal Society; being an endeavour to confute the opinion of Mr Eeles about the ascent of vapours, published in the last volume of their Transactions. But the author, having no electrical friend, whose sagacity he could confide in, has at length prevailed upon himself to be so free to send them to Mr Watson; to whom the world is so much indebted for the advancement of their knowledge in electricity.

Whence, Sir, if you should think, that these papers have truth, the great Diana of real philosophers, to patronize them, you will confer a favour upon me, by laying them before that learned Body. If, on the contrary, you should deem this confutation trifling or futil, I hope you will be humane enough to suppress them, and give me your objections; and by that means lay a still greater obligation on one, who has not the pleasure to be personally acquainted with you. From,

Sir,

Your very humble servant,

March 23, 1757. Erasmus Darwin,

Physician at Litchfield, Staffordshire.

Addressed To Mr William Watson, FRS.

Original Not traced. The MSS of Darwin's other papers for the Royal Society are in the Society's archives, but this one is not.

Printed Philosophical Transactions of the Royal Society, Volume 50, pp. 240–1 (1757).

Text From above source.

Notes

(1) William Watson FRS (1715–1787), who was knighted shortly before his death, was well known for his researches in electricity, medicine and natural history. He noted that moist air conducted electricity and he studied electrical discharges in gases at low pressures. In 1757 he was the leading English 'electrician'.

(2) Darwin's letter is a covering note for his two papers entitled 'Remarks on the Opinion of Henry Eeles, Esq, concerning the Ascent of Vapour'. His phrasing is polished, and his picture of himself lorn and lonesome, lacking an 'electrical friend', and throwing himself on the mercy of Watson, was designed to please the recipient of the letter: Watson recommended the papers to the Royal Society for publication, and they appeared in *Phil. Trans.*, **50**, 241–54, just after the letter. Darwin refuted Eeles's theory that vapours rise if electrically charged.

(3) Henry Eeles (1700–1781) was a 'combative crank': see J. L. Heilbron, *Electricity in the 18th and 19th centuries* (1979), p. 445.

(4) Although ultra-polite, Darwin may be writing tongue-in-cheek when he asks Watson to be 'humane enough' to suppress the paper if he doesn't approve.

57B. To Mary Howard, 24 December 1757

Darlaston, Dec. 24, 1757

Dear Polly,

As I was turning over some old mouldy volumes, that were laid upon a Shelf in a Closet of my Bed-chamber; one I found, after blowing the Dust from it with a Pair of Bellows, to be a Receipt Book, formerly, no doubt, belonging to some good old Lady of the Family. The Title Page (so much of it as the Rats had left) told us it was "a Bouk off verry monny muckle vallyed Receipts bouth in Kookery and Physicks". Upon one Page was "To make Pye-Crust," – in another "To make Wall-Crust," – "To make Tarts," – and at length "To make Love." "This Receipt," says I, "must be curious, I'll send it to Miss Howard next Post, let the way of making it be what it will." – Thus it is. "To make Love. Take of Sweet-William and of Rose-Mary, of each as much as is sufficient. To the former of these add of Honesty and Herb-of-grace; and to the latter of Eye-bright and Motherwort of each a large handful: mix them separately, and then, chopping them altogether, add one Plumb, two sprigs of Heart's Ease and a little Tyme. And it makes a most excellent dish, probatum est. Some put in Rue, and Cuckold-Pint, and Heart-Chokes, and Coxcome, and Violents. But these spoil the flavour of it entirely, and I even disprove of Sallery which some good Cooks order to be mix'd with it. I have frequently seen it toss'd up with all these at the Tables of the Great, where no Body would eat of it, the very appearance was so disagreable."

Then follow'd "Another Receipt to make Love," which began "Take two Sheep's Hearts, pierce them many times through with a Scewer to make them Tender, lay them upon a quick Fire, and then taking one Handful ——" here Time with his long Teeth had gnattered away the remainder of this Leaf. At the Top of the next Page, begins "To make an honest Man." "This is no new dish to me," says I, "besides it is now quite old Fashioned; I won't read it." Then follow'd "To make a good Wife." "Pshaw," continued I, "an acquaintance of mine, a young Lady of Lichfield, knows how to make this Dish better than any other Person in the World, and she has promised to treat me with it sometime," and thus in a Pett threw down the Book, and would not read any more at that Time. If I should open it again tomorrow, whatever curious and useful receipts I shall meet with, my dear Polly may expect an account of them in another Letter.

I have the Pleasure of your last Letter, am glad to hear thy cold is gone, but do not see why it should keep you from the concert, because it was gone. We drink your Health every day here, by the Name of Dulcinea del Toboso, and I told Mrs Jervis and Miss Jervis that we were to have been married yesterday, about which they teased me all the Evening. I heard nothing of Miss Fletcher's Fever before. I will certainly be with Thee on Wednesday evening, the Writings are at my House, and may be dispatched that night, and if a License takes up any Time (for I know nothing at all about these Things) I should be glad if Mr Howard would order one, and by this means, dear Polly, we may have the Ceremony over next morning at eight o'clock, before any Body in Lichfield can know almost of my being come Home. If a License is to be had the Day before, I could wish it may be put off till late in the Evening, as the Voice of Fame makes such quick Dispatch with any News in so small a Place as Lichfield. I think this is much the best scheme, for to stay a few Days after my Return could serve no Purpose, it would only make us more watch'd and teazed by the Eye and Tongue of Impertinence.

I shall by this Post apprize my Sister to be ready, and have the House clean, and I wish you would give her Instructions about any trivial affairs, that I cannot recollect, such as a cake you mentioned, and tell her the Person of whom, and the Time when it must be made, etc. I'll desire her to wait upon you for this Purpose. Perhaps Miss Nelly White need not know the precise Time till the Night before, but this as you please, as I [*illegible*]. You could rely upon her Secrecy, and it's a Trifle, if any Body should know. Matrimony, my dear Girl, is undoubtedly a scrious affair (if any Thing be such), because it is an affair for Life. But, as we have deliberately determin'd, do not let us be frighted about this Change of Life; or however, not let any breathing Creature perceive that we have either Fears or Pleasures upon this Occasion: as I am certainly convinced, that the best of Confidants (tho' experienced on a thousand other Occasions) could as easily hold a burning cinder in their Mouth as anything the least ridiculous about a new married couple! I have ordered the Writings to be sent to Mr Howard that he may peruse and fill up the blanks at his Leizure, as it wilt (I foresee) be dark night before I get to Lichfield on Wednesday. Mrs Jervis and Miss desire their Compl. to you, and often say how glad she shall be to see you for a few Days at any Time. I shall be glad, Polly, if thou hast Time on Sunday night, if thou wilt favour me with a few Lines by the return of the Post, to tell me how Thou doest, etc. – My Compl. wait on Mr Howard if He be returned. – My Sister will wait upon you, and I hope, Polly, Thou wilt make no Scruple of giving her Orders about whatever you chuse, or think necessary. I told her Nelly White is to be Bride-Maid. Happiness attend Thee! adieu

from, my dear Girl,
thy sincere Friend,
E Darwin

P.S. – Nothing about death in this Letter, Polly.

Original Not traced. In the possession of Charles Darwin, 1879.
Printed C. Darwin, *Life of E. Darwin*, pp. 21–4.
Text From above source.
Notes

(1) Mary Howard (1740–1770) was married to Darwin on 30 December 1757. By all accounts they enjoyed a very happy married life, despite Mary's frequent illnesses, which took the form of seizures by paroxysms of pain and led to her death at the age of thirty. See letter 92A. Mr Howard is Mary's father, Charles Howard (1706–1771), a Lichfield solicitor, and a schoolfellow and friend of Samuel Johnson, who called him 'a cool and wise man'. Howard stood as surety at Johnson's marriage, as well as Darwin's. Charles Howard had married Penelope Foley in 1734: she came from a family noted for talent and early death – either by illness or execution (see Pearson, *Galton*, i 244–6) – and she seems to have passed

on both qualities. Penelope Howard herself died in 1748 at the age of 40, and only two of her six children survived infancy; but her grandson Charles Darwin showed extraordinary talent before his death at the age of 19 (see letter 78H).

(2) The Jervis family owned the manor house at Darlaston, near Stone, throughout the eighteenth century. Darwin's host, Mrs Jervis, may have been Grace Jervis (née Ward), widow of John Jervis (1693–1720), or her daughter-in-law Mary Jervis (née Wade), widow of John Jervis (1721–1755). Mrs Jervis the elder (if still alive) would have been aged 55–60; Mrs Jervis the younger (who lived till 1800) would have been about 25–30, and she had a two-year-old son, John Jervis (1755–1802). The famous admiral of the Napoleonic wars, John Jervis (1735–1823), created Earl of St Vincent in 1797, was a nephew of the elder Mrs Jervis, being the son of her husband's younger brother Swynfen. Darwin's 'Miss Jervis' may have been a daughter or niece of the elder Mrs Jervis. No daughter of the elder Mrs Jervis is recorded in *Burke's Landed Gentry* (see 18th edn (1972), Vol. 3, pp. 496–7), but the marriage in 1760 of 'Miss Jervis, sister to the late John Jervis of Darlaston' is mentioned in T. Harwood, *Erdeswick's Staffordshire* (1844), p. 34.

(3) Darwin had been living at Lichfield for just over a year when this letter was written, and his sister Susannah (see letter 49A) had been acting as housekeeper for him.

(4) The letter was written on Saturday 24 December. Darwin planned to return to Lichfield on the evening of Wednesday 28 December and be married on the 29th, though in fact the marriage was on Friday 30 December.

(5) The letter offers a good example of Darwin's confidence and bantering exuberance.

62A. To Matthew Boulton, [30? October] 1762

Dear Boulton

Dr Petit desires I would use my Interest with your Worship, to procure him a Thermometer or two – now why won't you sell these Thermometers, for I want one also myself. He is going to try thermometrical Experiments upon the cold Water at Scarborough, where he is gone to practis Medecine. Have you call'd down no Rain to Birmingham yet? What's become of Walker? – Dr Petit desires you'll write a Paper and become Member of the R.S.

I am told Dr Franklin has wrote a very nonsensical Paper on Fire, knowing nothing of Chemistry.

Adieu E Darwin.

Original Birmingham Reference Library, Matthew Boulton Papers, Darwin 7.
Printed Excerpts in *Lunar Society*, p. 28, and *Doctor of Revolution*, p. 49.
Text From original MS.
Notes

(1) Matthew Boulton (1728–1809), the future 'first manufacturer of England', had decided in 1760 to launch out from his existing business of buckle-making and to

establish a large new general manufactory on land he had bought at Soho on the outskirts of Birmingham. The Soho manufactory was not completed until 1766, but Boulton was already experimenting in the manufacture of scientific instruments, and his thermometers were much in demand. For biography of Boulton, see H. W. Dickinson, *Matthew Boulton* (CUP, 1937) and E. Delieb, *The Great Silver Manufactory* (Studio Vista, 1971), pp. 17–27.

(2) Darwin probably met Boulton soon after beginning his medical practice at Lichfield in November 1756, perhaps through Luke Robinson (1731–1764), a patient of his, who was Boulton's brother-in-law. By 1762 their friendship was of long standing, but this is the first letter between them that has survived.

(3) Dr J. L. Petit, FRS (1736–1780) had supported Darwin's candidature as a Fellow of the Royal Society. Darwin was elected in 1761 and Petit wanted Boulton to follow in his footsteps. In fact Boulton did not become FRS until 1785.

(4) Petit wrote to Boulton on 28 October 1762 to request some thermometers (see *Lunar Society*, p. 28) and asked Darwin to pass on his request to Boulton. This letter is Darwin's covering note to Boulton, presumably written about two days later.

(5) 'Walker' may be Adam Walker (1731–1821), the itinerant lecturer, who later fascinated Shelley (see *DNB*); or, more probably, Zaccheus Walker (1737–1808), who acted as manager for Boulton and married Boulton's sister Mary (1731–1768). Zaccheus was the eldest son of Robert Walker (1709–1802), the virtuous curate of Seathwaite in Dunnerdale, styled 'Wonderful' Walker and celebrated by Wordsworth in *The Excursion* (VII 316–60) and a Memoir (*Poetical Works*, OUP, 1904, pp. 910–15).

(6) Benjamin Franklin (1706–1790), the leading man of science of the day, had visited Birmingham in 1758 and 1760, and had become friendly with both Darwin and Boulton. See letter 72A.

63A. To Matthew Boulton, 1 July 1763

Lichfield
Jul.1.1763

Dear Boulton

As you[r] are now become a sober plodding Man of Business, I scarcely dare trouble you to do me a Favour in the nicknachatory, alias philosophical way: I have got a most exquisitly fine Balance, and a very neat Glass Box, and have all this day been employ'd in twisting the Necks of Florence-Flasks – in vain!

Now if you like Florence Wine, I begg leave to make you a present of one Bottle, or two, if the first does not answer, to drink success to Philosophy and Trade, upon condition that you will procure me one of their Necks to be twisted into a little Hook according to the copper Plate on the reverse of this Paper: It must be truly hermetrically seal'd, air-tight, otherwise it will not answer my End at all.

Harrison's Clock at last has not a progressive Ballance, but an interrupted one, like the common Time-pieces, except in Weight and Workmanship.

I am extreemly impatient for this new Play-Thing! as I intend to fortell every Shower by it, and make great medical discovery as far as relates to the specific Gravity of Air: and from the Quantity of Vapor. – Thus the Specific Gravity of the Air, should be as the Absolute Gravity (shew'd by the Barometer) and as the Heat (shew'd by Boulton's Themometer). Now if it is not always found as these two (that is as one and inversely as the other) then the deviations at different Times must be as the Quantity of dissolved Vapour in the Air.

The common Hygrometers only shew when the Air is in a Disposition to recieve or part with moisture, not at all the Quantity it contains. – I begg my Compl. to Mrs Boulton and Mr and Mrs Fothergale, and am, dear Boulton, your affectionate Friend

Erasmus Darwin

Addressed For Mr Boulton, on Snow-Hill, Birmingham.

Original Birmingham Reference Library, Matthew Boulton Papers, Darwin 8.
Printed Excerpts in *Lunar Society*, p. 29, and *Doctor of Revolution*, p. 53.
Text From original MS. (*Diagram* From original.)
Notes
(1) 'Harrison's Clock' refers to the famous chronometer No. 4 of John Harrison (1693–1776), made in 1762. This chronometer won Harrison the £20 000 reward offered by the Admiralty for the accurate determination of longitude at sea.
(2) In the fourth paragraph of this letter Darwin states what is now usually called the ideal gas law, that density is proportional to pressure divided by absolute temperature. In the last sentence of the paragraph he also shows a precognition of the law of partial pressures. The first of these fundamental gas laws is usually credited to J. A. C. Charles (1787), and the second to John Dalton (1801).
(3) Mrs Boulton was Boulton's second wife Anne (née Robinson) (1733–1783): Anne's elder sister Mary (1727–1759) had been his first wife. Through these

marriages and the death of Anne's brother Luke in 1764, Boulton came into possession of all the wealth of Anne's father, the prosperous Lichfield merchant Luke Robinson (1683–1749). Never was a legacy better spent, for the money enabled Boulton to build the Soho manufactory, 'the eighth wonder of the world' and a power-house of the Industrial Revolution.

(4) Mr 'Fothergale' is John Fothergill (*c.*1710–1782), Boulton's partner in the 'toy manufacturing' company, Boulton and Fothergill: 'toys' covered most small manufactured articles, buttons, buckles, chains, jewellery, plate, snuff-boxes etc. See E. Robinson, *Univ. of Birmingham Hist. Journ.*, **7**, 60–79 (1959).

63B. To John Barker, December 1763

Dear Sir

I think it would be improper to write to Mr Adams again till we are determined whether we shall use it as a Slit-Mill or not. I will write to Mr Offley to night if you have got any Information about the present Miller's Lease, or let it alone till next Post as you think fit. I like your Scheme of employing them both if Mr [Garbett?] approves of it, who will be here the 22d but we shall be rival'd by better Works at King's Mill and at Burton Mills, and at Wiln. We shall look foolish to risk being refused again by Offley, when perhaps Mr G. will not approve the Scheme.

But I will write to Night, or next Post, or when you please, only say what you would have wrote.

From sincerely
yours E D

Addressed For Mr Barker. (*Endorsed* Dec 1763.)

Original William Salt Library, Stafford, S.MS 478.
Printed Unpublished.
Text From original MS.
Notes

(1) John Barker (*c.*1720–1781) was a prosperous Lichfield business man, variously described in documents as a draper, a merchant, a banker and a gentleman. He was bailiff of Lichfield in 1753, and a commissioner of the Lichfield and Staffordshire Turnpike Trust in the 1750s and 1760s; and he purchased Harlaston Manor, near Elford, from Sir John Egerton in 1772. There are two letters from Barker, to Boulton (1779) and Garbett (1780), in Birmingham Reference Library (Matthew Boulton Papers B.I 107, and Boulton and Watt Papers). After Barker's death, his widow continued the bank, and *Bailey's British Directory* of 1784 records 'Catherine Barker, Banker' as a resident of Lichfield.

(2) This letter is the first of ten hurried and sometimes obscure notes from Darwin to Barker about a business enterprise involving the mills on the river Trent at Alrewas and Wichnor, not far from Lichfield. This enterprise, hitherto rather a mystery, has recently been elucidated in a valuable article by J. Gould, 'The

Lichfield Canal and the Wychnor Ironworks', *Trans. South Staffs. Arch. and Hist. Soc.* (to be published). In 1758 James Brindley (see letter 72B) was commissioned to survey a canal from Liverpool to Hull, and in 1760 his plan for the route from Burslem to Wilden Ferry south of Derby was published (there is a copy in Burton Public Library). This stretch of the planned canal passed near Alrewas and Wichnor, and also had a branch to Lichfield. Barker and Darwin knew of the plan, and they went into partnership with the Birmingham merchant Samuel Garbett (see note 4) and the Elford paper manufacturer Robert Bage (see note 5), with the idea of setting up a mill for slitting and rolling iron at Wichnor. The mill was to be supplied with water by a wide cutting from the nearby river Trent, a cutting which was on the line of the proposed canal and could later be sold to form part of the canal, thus also ensuring that the forge would be on the future canal and well placed for trade. There was a need for slitting-mills, i.e., mills for cutting iron into nails (see *Victoria County History of Staffordshire* ii 114–24); but the project was ambitious and risky. Despite the problems, the Wichnor forge was established, probably in 1765; the cutting was made and became part of the Grand Trunk canal; and the forge operated for nearly 20 years, apparently under Barker's management until his death in 1781. Soon after that, it was sold, probably at a loss.

(3) The plan for the canal cut is in Lichfield Record Office (Turnpike Trustees Minute Book D.15/2/2) and reads, 'A Plan of Mills, Lands, Pool and Canal Houses granted by John Levett Esq. to Mr John Barker and Company at Whichnor Bridges in the County of Stafford'. It is signed by Barker, Garbett, Darwin, Bage and Levett: there is no date, but 1765 is probable. The total area involved was about five acres and the length of canal about half a mile. The canal cut, now part of the Trent-and-Mersey canal, can be seen by taking the roads both left and right off the A38 at Wichnor Bridges. See letter 64D for Levett.

(4) Samuel Garbett (1717–1805), the 'Mr G.' of this letter, was one of the early pioneers of the Industrial Revolution, a leading ironmaster of the day and very much the most experienced of the partners. In the 1740s with Dr John Roebuck (1718–1794), he operated a precious metal refinery in Steelhouse Lane, Birmingham. In 1746 they started the first factory for making sulphuric acid by the lead chamber process: it was 'a pivotal event in 18th-century economic history', in the words of A. and N. L. Clow, *The Chemical Revolution* (Batchworth, 1952), p. 133. Roebuck later transferred the factory to Prestonpans, near Edinburgh. Then Roebuck set up the famous Carron Iron Works, again with Garbett as a partner, and later (without Garbett) tried in vain to develop Watt's steam engine. (For further details of the Garbett–Roebuck projects, see the article on Roebuck in *DNB*.)

(5) Robert Bage (1728–1801), the fourth man in John Barker's Company, owned a paper mill at Elford, near Lichfield, and made the best paper in the country, it was said. The historian William Hutton called him 'one of the most amiable of men and . . . one of the best', and Sir Walter Scott said: 'His integrity, his honour, his devotion to truth, were undeviating and incorruptible'. Hutton also mentions the Wichnor forge: 'In 1765 Bage entered into partnership with three persons (one of them the celebrated Dr Darwin) in an extensive manufacturing firm. The speculation proved unfortunate and Bage lost a considerable sum of money. He turned for consolation to literary studies'. (Quotations from L. Jewitt (ed.), *The*

Life of William Hutton, Warne, 1872, pp. 170–2.) According to Scott, Bage lost £1500. Bage's 'literary studies' took the form of novels, which were much admired, then and now, particularly *Hermsprong* (1796). See *DNB* for Bage.

(6) The ten letters from Darwin to Barker are mainly concerned with the mills at Wichnor and Alrewas, and with persuading the Wichnor miller to move to Alrewas. The first sentence of this letter refers to Wichnor Mill. John Offley (1717–1784) was MP for Orford and was the elder brother of Lawrence Offley, who, with David Garrick, had been one of Samuel Johnson's three pupils at Edial, near Lichfield, in 1736. John Offley owned extensive estates at Wichnor, and had probably indicated that he wished to sell (see letter 64D). For a biography of Offley, see Namier & Brooke, *Commons* iii 223. Mr Adams was probably Offley's agent.

(7) The 'better Works' were all further down the river Trent, north-east of Wichnor. Wiln. is probably Willington or Wilden.

(8) This letter was written some days before 22 December, and is probably, but not certainly, earlier than the next two, which are dated 16 and 17 December.

63C. To John Barker, 16 December 1763

Dear Barker

I intend seeing you to Night, I think the Miller of Whichnor, should be immediately talk'd with on the Subject, mentioning that Aldrewas will be used as a Corn-Mill, and {may?} affect him that way, if He does not agree with us to change.

E D

Addressed For Mr Barker. (*Endorsed* 16 Decr 1763.)

Original William Salt Library, Stafford, S. MS 478.
Printed Unpublished.
Text From original MS.
Notes

Darwin and Barker were asking the Wichnor miller to move to Alrewas, which would otherwise compete with him as a corn-mill.

63D. To John Barker, 17 December 1763

Your first Letter is wrote so ill that part of it is not legible – I see no Hast in acquainting Mr Hall. But approve much of your going over to the Rugely Agent: and would offer you my Chaise, but it would starve you to Death, that would have you take one: The following Note you will, or will not send to Mr Cobb as you think proper, by my Boy.

That Aldrewas Mill being used as a Corn-Mill will affect the Whichnor Millers Trade, must be used as a principle argument.

Addressed For Mr Barker. (*Endorsed* 17 Dcr 63.)

Original William Salt Library, Stafford, S. MS 478.
Printed Unpublished.
Text From original MS.
Notes
The first paragraph is very obscure. I have not identified Mr Hall or the Rugeley agent. Mr Cobb is probably Francis Cobb (1724–1807), the Lichfield banker and later Receiver-General for Staffordshire. He was a leading citizen of Lichfield and has a memorial in the Cathedral. See A. L. Reade, *Johnsonian Gleanings*, Part VII, p. 172 for further details.

63E. To John Barker, 22 December 1763

Dear Barker

Mr Bage will send to Hadley this afternoon, to meet him at Lichfield on Saturday twelve o'Clock, and then if you think proper to send to Mr Adcock to meet them about instructions for the Lease.

A Cut may be made from the Mill-Damm along the Side of the River to near Coney-Ford for about 50 Pounds and a Bridge over the Turnpike 10 Pounds more, or there need not be a Brid[g]e but a shallow Wash in its stead.

By this means another foot fall or perhaps two may be procured and hence there may be Water enough for both Mills. The same might be done by cutting the Mill Tail deeper, but it would annually fill up, and be a constant Expense.

The Fall now is only 2 ft 9 Inch exactly measured. I think removing the Miller is absolutely adviseable. Please to explain this to Mr Adcock as the Thought is indeed his, and a good one, I believe.

Mr Adcocks notion of taking the Water after the Mill and a Cut across from the main Water to the Mill-Tail Water can not possibly be of Service except you was to drown a thousand acres of Meadow Ground, which by this other means will be made something dryer.

Please to instruct Mr Adcock to stipulate that the Taxes (if any shall hereafter be paid by the Tenant, for Mr Adams pays them now, tho' according to his Lease, Mr Hadley ought to pay them) that they shall be paid by Mr Hadley. Whichnor Mill seems a much cheaper Take[?] than Aldrewas in many respects: almost 20£ a year cheaper.

Perhaps I shall see you tomorrow. You'l contrive some means to send to Mr Adcock.

Addressed For Mr Barker. (*Endorsed* This letter was wrote on 22 or 23 Decr.)

Original William Salt Library, Stafford, S. MS 478.

Printed Unpublished.
Text From original MS.
Notes
(1) Of the two dates in Barker's endorsement, only 22 December is acceptable, because Darwin refers to 'tomorrow' and 'Saturday' as different days, and 23 December was a Friday.
(2) For Mr Bage, see letter 63B. Mr Hadley and Mr Adcock are probably agents of the Barker–Darwin consortium.

63F. To John Barker, 23 December 1763

Dear Sir

Whichnor will not be worth having unless we can have it entire.

If we do not take Aldrewas, we have no Chance of getting Wichnor entire.

And we can not lose much by Aldrewas as we can employ it in the Corn-Trade.

Our whole Expence in procuring Wichnor will be

Rent 40	or 45£ per Ann.
Difference of Aldrewas paid to the Miller	20£ per Ann.
Deduction for Eel-fishery	15£ per Ann.
	50£ per Ann.

If Burton Mill has 7 ft fall it will be a better affair by ¼ or ⅙ and may come much cheaper. – But how do you know it is seven foot fall?

Would you take both? – I think a Messenger should stop Bage and Hadley as Adcock is to be at Newcastle: and defer the Agreement till Tuesday, in which Time G will have seen the Mill near Burton: or if He does not I will go over and see it myself on Sunday or Monday.

Original William Salt Library, Stafford, S.MS 478.
Printed Unpublished.
Text From original MS.
Notes
(1) The MS is endorsed '23 Decr. 1763' in Barker's writing.
(2) A third mill, at Burton, has now come into the picture.
(3) Newcastle is of course Newcastle-under-Lyme, here and in all subsequent letters, except 91E.
(4) As before, G is Garbett.
(5) The letter was written on Friday, and Sunday was Christmas day.

63G. To John Barker, 29 December 1763

Dear Barker,

I still fear Wichnor is a bad Mill for our purpose, and wish much that Downing may see it again before we close with the Wichnor Miller. For Aldrewas Mill lies so little above it, that if that Miller does not chuse to grind in Summer for half a Day, our Work can not go on. Mr Downing may see the Burton affair at the same Time. I go to Rugely tomorrow, but would rather not call as I can do no good, and you can write to him.

I wish you to send an Express to Downing if you think proper.

E D

Addressed For Mr Barker. (*Endorsed* 29 Decr. 1763.)

Original William Salt Library, Stafford, S.MS 478.
Printed Unpublished.
Text From original MS.
Notes

Mr Downing, who may be 'the Rugeley agent', is the only newcomer here. Hectic haste still seems to reign.

64A. To John Barker, [4? January 1764]

Mr Barker and Dr D engage to put Aldrewas Mill into such Order, as two Judges of the Business shall determine to be equall to Wichnor. One chose by Mr Woodhouse and one by Mr Barker.

And to let it to Mr Woodhouse for the Rent of 35£ a year (if that is the whole Sum He now pays to Mr Offley).

And to allow him a certain annual Charge for keeping it in repair: this to be determined by what Mr Offley has allow'd for 7 years past. But Mr Offley does not entirely repair Whichnor, for if He did how [h]as Mr Woodhouse laid out 70£! and has not Mr Offley raised his Rent 10£ per ann. for new repairs.

1. The 70 Pound Mr Woodhouse has laid out on Whichnor will thus be repaid him, by the proper repairs at Aldrewas.

2d. Mr Woodhouse will have a considerable advantage in the Eel-Fishery which at Aldrewas is valued at 15 25£ a year, and at Whichnor at only ¼ of the other: from the Eels at Whichnor getting over the Wier in Time of Floods, which is the Time of Eel-fishing and which are all taken at Aldrewas at the Floodgates; there being no Wier.

3d. Mr Woodhouse will have no Rival in the Corn-Trade, at Aldrewas: which Mills would otherwise be immediately work'd by Mr Barker and Mr Bage, with a thousand Pound capital: and must consequently affect the Markets of Burton, Derby and Lichfield.

4thly. Aldrewas Mill having a Wheel more than Whichnor will do more Business in the Corn-Trade.
5thly. Mr Woodhouse will have an advantage by grinding for the Parish, and neighbourhood of Aldrewas, without losing that of Whichnor.

Addressed For Mr Barker. (*Endorsed* Darwin propo. sent Adcock.)

Original William Salt Library, Stafford. S. MS 478.
Printed Unpublished.
Text From original MS.
Notes
(1) This draft agreement should be read with the next letter, which is very probably the continuation of it. The two items are printed separately here because the connection is not certain and the manuscripts are in different places.
(2) For the date, see next letter.
(3) The manuscript is very rough and largely crossed out, but still completely legible. Mr Woodhouse is evidently 'the Miller of Wichnor'.
(4) Darwin seems to use the symbols £ and ℔ (i.e. pounds weight) interchangeably. Both are transcribed as £.

64B. To John Barker, [4? January 1764]

Dear Barker

If you would have me write over again or add any thing to the above please to send me them very early in the morning, as I must go out by ten.

I would have you give Lord Suffolk a Copy of the Proposals, and of my Letters to Lord Vernon and to Mr Offley as that will instruct him what to say.

I shall not return till Friday Noon. I see no reason to offer the Miller more than what is mention'd in the Proposals.

You will please to add the Name of Sir Lister Holts Steward, and if you think proper, if Mr Offley should chuse Wyott, you may object to him, as concern'd in the Iron Way.

Original New York Public Library, Erasmus Darwin Misc. MSS.
Printed Unpublished.
Text From original MS.
Notes
(1) This is very probably a covering letter for the draft agreement in the previous letter, but they are separated in case it is not.
(2) There is no date on either the agreement or the letter, but the negotiations were proceeding with such haste that the letter seems likely to be within a week of the previous one (29 December 1763). Since Darwin was going to Rugeley on Friday 30 December, the Friday mentioned here is probably Friday 6 January 1764. On the journey he mentions, Darwin is more likely to have been away one night than two or three (see letter 96F) and, if so, the date is 4 January.

(3) Henry Howard, Earl of Suffolk (1739–1779), who was deputy Earl Marshal 1763–5, and Secretary of State for the North 1771–1779, was presumably involved as a landowner.

(4) George Venables-Vernon (1710–1780) was created Lord Vernon in 1762 after serving as MP for Lichfield (1731–1747) and Derby (1754–1762). He lived at Sudbury and owned a great deal of land in Staffordshire and Derbyshire.

(5) 'Wyott' is presumably John Wyatt (1700–1766), a fertile inventor who has never gained the credit he deserves as a pioneer of the Industrial Revolution. In the 1730s he became interested in machines for cotton spinning and he took out the patent on spinning cotton by rollers revolving at different rates. His attempts to exploit the invention were unsuccessful and he became bankrupt; it was left to Arkwright to put the invention into practice (see Letter 85B). Later Wyatt became an employee of Boulton and perfected a compound-lever weighing-machine which was widely used, as well as inventing the oval or double-headed lathe. Darwin seems to have regarded him as a rival inventor (see letter 64E). For details of Wyatt, see *DNB*.

(6) Sir Lister Holt, Bart. (1720–1770) lived at Aston Hall, near Birmingham. See *Complete Baronetage*, Vol. I, pp. 105–6.

(7) The following note is written on the manuscript: 'This Autograph of Erasmus Darwin MD is authentic and came into Capt. Fernyhough's hands from his Original Papers – from the Bank in Lichfield – John Barker Esqr – 1764. It seems that Dr Darwin speculated as a Partner with Mr Barker in the Whichnor Forges as an Iron Master – Co. Stafford.' Captain Thomas Fernyhough (1776–1844) was a collector, antiquary and genealogist, who compiled several collections of Staffordshire manuscripts and press cuttings now in the William Salt Library at Stafford. He also edited *Military Memoirs of Four Brothers* (Sams, London, 1829) and at his death was Governor of the Military Knights of Windsor. See R. Simms, *Bibliotheca Staffordiensis* (1894) and *Gent. Mag.* (N.S.) **21**, 217 (1844).

64C. To John Barker, 19 March 1764

Monday Night
Mar. 1[9] –64

Dear Barker

I was out of Town till this Evening. I have consulted Mr Howard and we are of opinion the Affair should be finish'd if possible while you are in Town, at least we wish you to stay one Post more. As I intend to send to Mr Adcock early tomorrow morning to desire him to meet me at Wichnor and in Company with him I intend to visit all the Tenents concern'd, and you will be certain to hear from me the next Post. I hope Mrs Barker is well, and am

Your affectionate Friend
E Darwin

Addressed For Mr Barker.

Original William Salt Library, Stafford, S. MS 478.
Printed Unpublished.

Text From original MS.

Notes

(1) This letter is endorsed by Barker, with the note 'Dr D– 18 March 64, to J.B– In London'. Darwin's dating is rather illegible, but is most probably '18'. However, 19 March was a Monday.

(2) 'Mr Howard' was Darwin's father-in-law (see letter 57B). Howard often acted as Chairman of the Commissioners of the Lichfield and Staffordshire Turnpike Trust.

(3) The letter shows that the affair of the Wichnor Mill was still active, and that, despite the problems, Darwin and Barker were now more friendly rather than less.

64D. To John Barker, 24 March 1764

Dear Barker

I met Mr Adcock with Mr Bage at Wichnor yesterday, the Miller I had wrote to but the Letter lay at the Inn, not having been sent up to him, so He was absent at Lichfield. Mr Shorthose occupys about ¾ of the Ground we go through, and is very obliging, and says He is willing to comply, Mr Adcock mention'd a Guinea an Acre. Mr Burton occupies the other part and tho' He was absent, by the Account of the Neighbours, Mr Adcock seem'd to make no difficulty of his complying, as indeed no Reasonable Objection can be made, since the Cut will coast along the bottom of his piece below the Road, after it leaves the Lane, and can be no kind of Detriment to him but the meer loss of Ground.

The Miller I saw at Lichfield at my Return. I offer'd him one hundred pounds advantage in the Exchange of the Mills, and that this should be determined by four Millers, of Character, two to be chosen by him and two by us. He says, his uncle at Eginton, Mr Harrison, will not change: in short He appears to me to want to make a better Market of us. I mention'd to the tenants that we should offer him 100£ Advantage, and they thought it quite a sufficient Sum.

It seems some of the Tenants do not grind with him, as He has taken seventeen Pound for Toll out of five Pecks, whereas seven Pound would have been more than is by Law allow'd, this is a Secret I find.

Mr Adcock is to see the Miller again on Tuesday, but think your staying in Town can now no longer be of consequence. For I see no Probability of this Man coming to Terms, and (to tell you my opinion) I think we shall stand at two [too] high a Rent already. As 2000£ will be expended before we can begin to work at all.

If you mention any part of this Letter {to} Mr Offley, please to make my most respectful Compliments to him.

Mr Levet and his Brother have been to the Inn at Wichnor and enquired much about Mr Offley's Estate there, what extent it was, if good Land, and if it was not let too cheap, and many other Questions, wonderfully inquisitive.

Mr Howard thinks you need not consult any Co[u]nsel about the Turnpike till you return hither.

from yours affect.
E Darwin

Addressed Mr Barker. (*Endorsed* Dr Darwin 24 March 1764.)

Original William Salt Library, Stafford, S. MS 478.
Printed Unpublished.
Text From original MS.
Notes
(1) This is the longest and most coherent of these ten letters to Barker, and also the last. Darwin's foray into business taught him much about the problems of letting mills, negotiating with landowners for a cut through their territory, etc, and the knowledge served him well when he became involved with Wedgwood in the promotion of the Grand Trunk Canal in 1765. He also learnt a lesson in finance – in the end he presumably lost several hundred pounds in the venture. He never again tried to act as entrepreneur.
(2) John Levett of Lichfield (1721–1799) purchased John Offley's estates at Wichnor in 1765 (see T. Harwood, *Erdeswick's Survey of Staffordshire*, Nichols, London, 1844, p. 318). So Darwin was right to take note of the 'wonderfully inquisitive' questions asked by Levett and his brother. It was from Levett, rather than Offley, that Barker and Company bought the land (see letter 63B). Levett went to live at Wichnor soon after. See Namier & Brooke, *Commons* iii 39.
(3) Mr Shorthose was probably the innkeeper at Wichnor. In 1793 the inn at Wichnor was described as 'convenient and spacious' and 'kept by Mr Robert Shorthose' (*Universal British Directory*, 1793).
(4) The last sentence, about the Turnpike, is a reminder that the planned canal cut near Wichnor (see letter 63B) would cross the Lichfield–Derby turnpike. There is a two-page agreement, dated 14 May 1764, in the Minute Books of the Lichfield and Staffordshire Turnpike Trust at Lichfield Record Office, which gives Darwin, Barker and Garbett permission to take 'a canal or ditch 14 feet wide' over or under the Turnpike road 'near Whichnor Bridge', by means of 'an arched bridge of stone'; and they agree to keep it in good condition.
(5) The Wichnor forge is not mentioned in any later letter of Darwin.

64E. To Matthew Boulton, [1764?]

Dear Boulton,

As I was riding Home yesterday, I concider'd the Scheme of the fiery Chariot, and the longer I contemplated this favourite Idea, the [more] practicable it appear'd to me. I shall lay my Thoughts before you, crude and indigested as they occur'd to me; telling you as well what I thought would not do, as what would do; as by those Hints you may be led into various Trains of thinking upon this Subject, and by that means (if any Hints can assist your Genius, which without Hints is above all others I am acquainted with) be more likely to improve, or disapprove. And as I am quite mad of this Scheme, I begg you will not mention it, or shew this Paper, to Wyot or any Body.

These things are required. First a Rotatory Motion. 2 easyly altering its Direction to any other Direction. 3 To be accelerated, retarded, destroy'd, revived instantly and easyly. 4 the Bulk, Weight, and Expence of the machine as small as possible in proportion to its Use.

Previous Observations

I am in doubt whether three or four wheels, will be better? whether the Power should act on one wheel only? On which wheel? Let there be three Wheels, and the Power move one of the hinder ones, as A, and the rest be loose. Then I say the wheel A (if there were but 2 wheels A and B) would describe some curve as AC: now tho' there is also a wheel at x yet if A only be moved it will endeavour to carry the Machine som little in that direction, unless the [the] wheel x be so turn'd by the Helm as to oppose, viz. by having its direction as the line ab, when the direction of the other wheels are cd. Now I believe this to [be] false Reasoning, since the Lever from A to x, being so much longer than that from B to A, I believe it may press a little upon that wheel but not at all so as to alter its direction. Hence if one wheel only is to move, perhaps it may be nearly the same thing which wheel it is whether A or x.

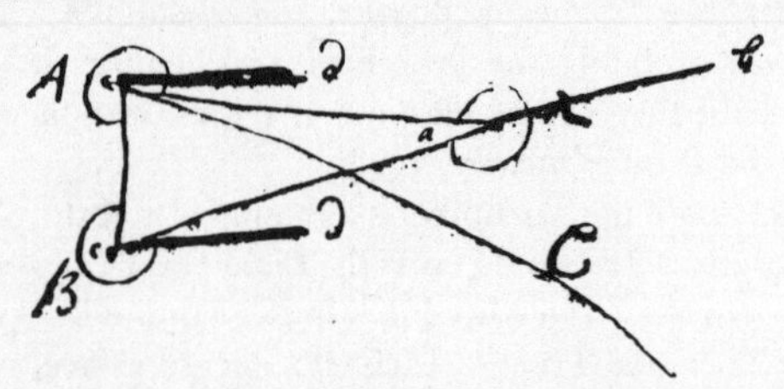

Second, suppose four wheel, could not one Hind one, and the opposite fore one be made to move, and so only two be loose? – No! the two wheels that move, a and c, must be of equal diameters, and they will move in equal Times (or the Velocities different, so as to conteract the inequallity of diameter) then let the Carriage move in a Circle, and it lies under just the same inconvenience as if both the hinder wheels were fix'd on one axis.

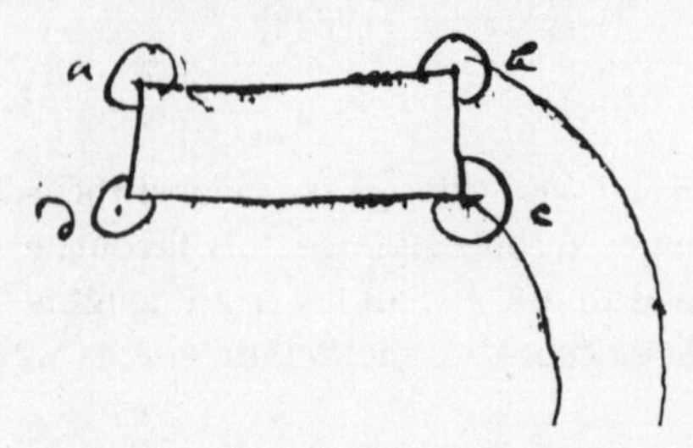

Hence I conclude that a three wheel'd Carriage is the best, that to move any one of those wheels, will answer as well as moving the first wheel or nearly so. And that two wheels can not be moved in any Carriage without great Friction in turning about.

Thus Then

Let there be two Cylinders, antagonists, at each End of the Beam, or Lever as it is call'd, which may be made of a Bar of Iron, the diameters of the Cylinders equall. And each equal . . . Inches. One Steam Cystern serves both, and is placed between them. NB. the agitation of the Water by the roughness of the roads will encrease the Quantity of Steam.

Remark. Suppose one Piston up, and the Vacuum made under it by the Jet d'eau froid. That Piston can not yet descend because the Cock is not yet open'd, which admitts the Steam into its antagonist Cylinder. Hence the two Pistons are in equilibrio, being either of them press'd by the atmosphere. Then I say, if the Cock which admitts the Steam into the antagonist Cylinder be open'd gradually and not with a Jerk, that the first mention'd Cylinder will descend gradually and yet not less forceably.

Hence by the management of the steam Cocks the Motion may be accelerated, retarded, destroy'd, revived, instantly and easyly. And if this answers in Practise as it does in theory, the Machine can not fail of Success! Eureka!

The Cocks of cold water may be moved by the great[?] work, but the Steem-Cocks must be managed by the Hand of the Charioteer, who also directs the Rudder-wheel.

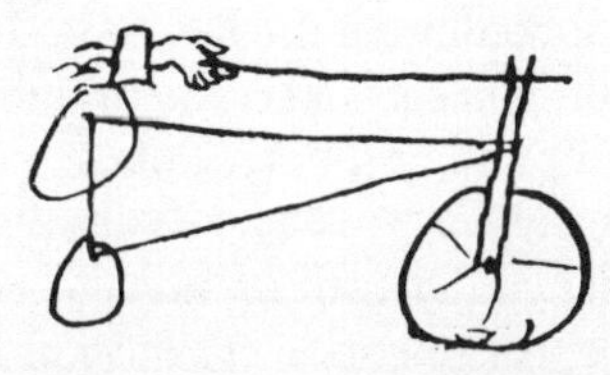

A Plate follows by which this regular and managible force is made rotatory without any loss of Power or Cumber or weight of Machinery.

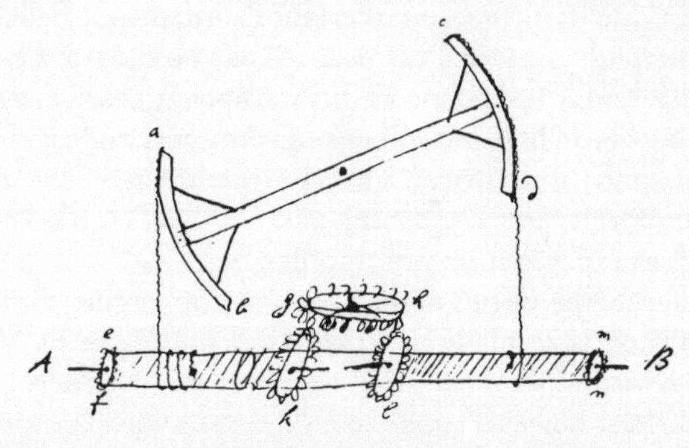

e f g h, is a Roller, and a rachet-wheel like those on which the Cord of a Jack is wound upon. AB are the axis of the two hinder wheels of the fiery Chariot, upon which axis also the Rollers move, when these Rollers move forwards, the Teeth of the Rachets ef and mn, reciprocally carry their neighbouring wheel. Let the arm ab be now elevated, and moves the wheel A forwards, and at the same Time by turning the ingenious wheel with contrart Teeth gh, winds up the Chain upon the roller h.l.m.n. The wheel B all this Time not being acted upon. Then when cd is again elevated the Chain is wound round efgh as at present.

NB. the curves ab and cd are not parts of Circles, they are design'd to vary from circles in order to wrap, and unwrap the Chains on the Rollers, nor will this little deviation from a regular force, by varying the arm of the lever, be at all felt, or may be counteracted by the form of the Roller, on which the Chain wraps. – The contrat wheel gh ought to have been under the Rollers, so that it may be out of the way of the Boyler. The upper surface of the Carriage is in this Form.

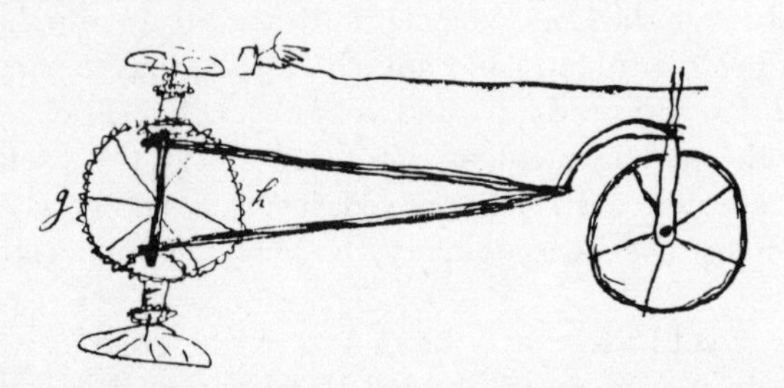

The Size of the Wheel gh is of no consequence, and here lying under the Axle Tree it does no[t] incommode the machine. If the Force required to draw 50 Ton on common wheels is known (to find out which without an Experiment or two I don't know) the Area of each Cylinder may be got to exactness.

I have heard if 5 Horses can just stir a Carriage, six are sufficient to draw it up to London, if this be true one sixth of Power more than what will move a carriage answers our End.

If you could learn the Expence of Coals {of} a common Fire-engine, and the Weight of Water it draws, some certain Estimate may be made if such a Scheme as this would answer. Pray don't shew Wyot this Scheme, for if you think it feasible, and will send me a Critique upon it, I will certainly if I can get somebody to bear half the Expense {with} me, endeavour to build a Fiery Chariot {and} if it answers get a Patent. If you chuse to be Partner with me in the Profit, and Ex{pense} and Trouble, let me know: as I am deter{mined} to execute it, if you approve of it.

Please to minute the Pulses of the common Fire-engine, and say in what manner the Piston is so made as to keep out the Air in its Motion.

By what Way is the Jet d'eau froid let out of the Cylinder? – How full of water is the Boyler? how is it supply'd, and what is the Quantity of its Wast of water? – I sent one Smith to see your Electricity of which He knows nothing, nor is like to know, but it seems He is a retail dealer in your way of Business.

From Sir

Your humble Serv.

Erasmus Darwin

I think four wheels will be better – adieu

Original Birmingham Reference Library, Matthew Boulton Papers, Darwin 6.
Printed Excerpts in *Lunar Society*, pp. 29–31, and *Doctor of Revolution*, pp. 53–5.
Text From original MS. (*Diagrams* From original.)
Notes

(1) This is one of the longest and most important of Darwin's letters, and establishes him as a leading pioneer of the mechanically-propelled car: see L. T. C. Rolt, *Horseless Carriage* (Constable, 1950), pp. 13–16. Darwin's design was never put into practice, because Boulton was already in debt and was not willing to chance his arm on such a speculative venture. The first working steam-carriage was made in 1769, by the French engineer Nicholas-Joseph Cugnot (1725–1804): see S. T. McCloy, *French Inventions of the Eighteenth Century* (University of Kentucky Press, 1952), pp. 36–40.

(2) There is no indication of the date of the letter, but it seems certain to be before 1765, since Dr Small would otherwise have received a mention. 1764 seems the most likely year, but it could be earlier.

(3) A critique of Darwin's scheme would require notes as long as the letter, and could form the subject of a specialized article.

(4) 'Wyot' is probably John Wyatt: see note to letter 64B.

(5) A contrate wheel is a cog-wheel with teeth set at right angles to its plane. Darwin spells it as 'contrart' and 'contrat'.

(6) I have not identified 'one Smith'.

64F. To Matthew Boulton, 3 November 1764

Lichfd. Nov. 3. 1764

Dear Boulton,

I sent to you a few Hours after you was gone to have paid you the Money for the Hay. Please by a Line to tell me who I must pay it to.

I hope whenever it is agreable to you we shall have the Pleasure of your and Mrs Boulton's Company at {our} House: – I must say I am sorry you and Mr Barker could not make up your disagreement, as I believe the affair was misunderstood on both Sides: and am afraid Friend Bleckham was but an unskilful Mediator, and rather widen'd than knit up the Breach between you. – For Instance. "Mrs Boulton did not know that Mr Boulton had such Connections with Mr Barker." – Answ. "He should have inform'd her better." – This is very different from Mr Bleckam's account: etc. etc. etc. – But I am also told, but observe I don't believe it, that Mr Barker ask'd Mr Boulton how He did at the Play, and recived no Answer. – Well! I wish I could reconcile you! – Do come to Lichfield.

Meddleing with other People's Affairs I had nearly forgot my own, viz to remind Mr Boulton of a promise He made me to send me a Piece or two of his Female-Chape-Pieces to make me a Fender with – but this not if it is any Trouble or Inconvenience at all.

Mrs Darwin joins with me in our best Compliments and Wishes to Mrs Boulton and yourself,

from your affect. Friend and Serv.

E Darwin

Addressed For Matth. Boulton Esq. Merch in Birmingham.

Original Birmingham Reference Library, Matthew Boulton Papers, Darwin 9.
Printed Unpublished.
Text From original MS.
Notes
(1) Boulton's mother-in-law Mrs Dorothy Robinson (née Babington) had a sister Catherine who married Ralph Hawkes, and their daughter Anna married Richard Blackham (T. Harwood, *Erdeswick's Survey of Staffordshire*, Nichols, London, 1844, p. 448). So Blackham was Boulton's cousin by marriage, and is probably the 'Friend Bleckham' mentioned in the letter.
(2) I have no information on the disagreement between Boulton and Barker.
(3) This, like many subsequent letters to Boulton, is provoked by the need to pay an account or send in an order.

65A. To Thomas Bentley, [16–21] October 1765

Dear Sir

Sanguine as I am about the Success of this Navigation I can not but be afraid of driving it forwards too fast; if it is thrown out of the House, it is lost for ever! – I have here sent you a hasty Letter from Mr Garbet, whose Knowledge in the trading world is allow'd in this part of the Country, as Mr Bentley's is in another. You will please to return me these Papers as I can not have Time to transcribe them. I can not but still wish, spite of the Expence of reprinting it, for Mr Bentley's Character (who will be conjectured to be the writer), that this Pamphlet had a little more Time pass'd upon it, that it might not discredit the Gentlemen concerned in this affair. – The Octogan Prayer Book, which I believe you refer to in a former Letter, was here thought by me and others to be astonishingly deficient in Language (that is, it being supposed to have had so many criticizers) – as some of the old Prayers were evidently by one additional word or so render'd less strong or less elegant, as if designedly.

I am heartily sorry our opinions are so widely different as I cannot see any advantage the Cause can recieve from the managers appearing bad Writers: if Mr Gilbert advises this, He either has not the feelings of Language, or is not a real Friend to the Cause.

In the Dedication the word "presented" is thought to be improper, as nothing can be presented to the House but in Form. I can't mend the following at present – was not Mr Gilbert's pamphlet dedicated to the House? – do they mind such general dedications?

From, dear Sir, most sincerely your Friend and Serv.

E Darwin

My Heart bleeds for poor Turner.

To the Legislature, The Patrons and Guardians of the public welfare, This Plan, for the Benefit of G. Britain is humbly dedicated.

The Particle 'As' is evidently better omitted.

Ornament is objected to, and all the succeeding part should be interwoven in the Pamphlet.

I shall hope to hear your Opinion of Mr Garbet's design'd letter to the Society of Arts.

Original Not traced. Probably in the possession of Mr W. E. Darwin in 1906.
Printed Wedgwood, *Letters* iii 258–60.
Text From above source.
Notes

(1) During 1765 Darwin became deeply committed to the project for the Trent-and-Mersey or Grand Trunk canal, and began his lifelong friendship with Josiah Wedgwood (1730–1795), who, with Brindley, provided the major driving force behind the promotion of the canal. There are seven long and cordial letters from Wedgwood to Darwin between April and June 1765 printed in the Wedgwood *Letters* iii 227–53, but unfortunately none from Darwin to Wedgwood have survived. In the first of these letters (3 April 1765) Wedgwood tells Darwin, 'you . . . have public spirit enough to be Generalissimo in this affair, and therefore ought to know how matters stand in every quarter, that you may apply your forces accordingly'. Later in April, Wedgwood tells Darwin 'I am quite charm'd with your zeal in this Public spirited scheme'. The other letters go into great detail over the progress of the proposals for the canal. Among Darwin's other contributions is a 60-page pamphlet, 'A Plan for a navigable Canal . . . between Hull and Liverpool' (University College London, Galton Papers, 10).

(2) Thomas Bentley (1730–1780), a merchant in Liverpool, was already a close friend of Wedgwood and in 1768 became his business partner. Bentley had undertaken to write a pamphlet about the canal, to be presented to the House of Commons. Darwin was critical of the wording, and, as this letter shows, his enthusiasm for the project submerged his usual politeness.

(3) Bentley's pamphlet, and Darwin's comments on it, are printed in Wedgwood, *Letters* iii 271–321. Darwin's comments can best be described as cogent but impolite. For example: 'This whole Sentence is formal and parsonic'; 'a very garrulous Sentence'; 'China again! is not this Hobbyhorseycal'; 'sad language'; etc. Bentley wrote to Darwin on 11 and 15 October, quite politely, but defending himself by remarking that he had previous experience of literary criticism in compiling his Octagon prayer book. Darwin then wrote the letter recorded here, rubbing salt in the wound by criticizing the Octagon prayer book too. Bentley replied on 22 October, so Darwin's letter must have been written between 16 and

21 October. Fortunately Bentley's reply was courteous, and this storm-in-a-teacup subsided.

(4) Parliamentary approval for the canal was in fact obtained in April 1766, but Darwin's concern was justified: the Bill for the Macclesfield canal was thrown out in 1766, and the Act was not passed until 1826.

(5) 'Mr Garbet' is Darwin's friend Samuel Garbett (see letter 63B), who was also a supporter of the project. His letter to the Society of Arts is printed in Wedgwood *Letters* iii 261–3.

(6) 'Mr Gilbert' is probably Thomas Gilbert (1720–1798), MP for Newcastle-under-Lyme, 1763–8, who was concerned with canal bills in the House of Commons: see Wedgwood, *Letters* i 19–85; and *DNB*. His brother John Gilbert (1724–1795), the Duke of Bridgewater's agent, had played a major rôle in the promotion of the Bridgewater canal.

(7) 'Poor Turner' is the Liverpool surgeon Matthew Turner (*c.*1730–1791), who had attended Wedgwood when he hurt his leg in 1762 and had introduced him to Bentley. See Meteyard, *Wedgwood* i 299–302 and *The Lancet* (1897) I 905–6. Bentley had mentioned that Turner and his wife were ill.

(8) The dedication of the pamphlet read: 'To the Legislature, as the Patrons and Guardians of the Public Welfare, this Plan for the Benefit and Ornament of Great Britain is humbly presented . . .' The last few lines of Darwin's letter give his suggested amendments to this wording.

(9) For the history of the Canal, see Jean Lindsay, *The Trent-and-Mersey Canal* (David and Charles, 1979).

65B. To Matthew Boulton, 12 December 1765

Lichfd. Dec 12 –65

Dear Boulton

I am undone to know what Observations Dr Franklin supply'd you with about your Steam-Engine, besides giving you his Approbation, and particularly to hear your final Opinion, and Dr Small's on the important Question, whether Evaporation is as [at] the Surface of boiling Water, or not? – or if it be as [at] the Surface of the Vessel, exposed to the Fire, which I rather suspect – <u>For</u> if you boil Water in a Florence-Flask, you see bubbles rising from the Bottom, these are Steam, as they mount they are condensed again by the cold supernatant Fluid, and their Sides clapping together make that Noise call'd <u>Simmerring.</u> <u>2dly</u> When Water boils in the Æolipile or closed Tea-Kettle, what a Quantity of Steam? – is this only as [at] the upper Surface? I humbly concieve not – but as [at] that Surface in contact with the Fire. Now as a Revd. Canon proves the immortality of the Soul from the Steam-Engine, and as the Strength and Activity of Man, or his Momentum, depends on the Quantity of vital Steam rising into the Brain from his boiling Blood; so the Quantity of Steam secreted in the boiler of the Steam-Engine is a Question of the utmost Importance in the Creation of this Animal. And I wish yourself and our ingenious Friend Dr Small will communicate to me your joint Opinion on this Head. And withal your

Thoughts on the best way of constructing the Boiler for the above Purposes.

I find the People of Cheshire are opposing the Navigation being brought to Birmingham and this Country. I hope you will give this Scheme your assistance. Mr Brindley is in a few Days to wait on Mr Garbet for this purpose to survey a Road for it by Lichfield nearer than by Tamworth. I desire you and Dr Small will take this Infection, as you have given me the Infection of Steam-Enginry: for it is well worthy your attention, who are Friends of Mankind, and of the ingenious Arts.

Mrs Darwin with her Compl. desires me to mention to you something about {the} gown She troubled Mrs Boulton with, {as} the Lady it belongs to is shortly going to transmigrate to London, and wishes to have the Bill sent along with it.

Mr Mitchel has been a Day or two with me, and I want to explain Harrison's watch to you – which is very ingenious indeed! – He says Water falling through a Hole in a Vessel has a Velocity, compared to that of Water falling from the Height of its Surface as nearly 7 to 10, is this the Opinion of you Birmingham Philosophers? Hence Bucket Wheels are universally superiour to other kind of Lad[d]les. Pray write me as long a Letter.

adieu
E Darwin

Addressed For Math. Boulton Esqr. at Mr John Motteux's, Merch in Wallbrook, London.

Original Birmingham Reference Library, Matthew Boulton Papers, Darwin 10.
Printed Excerpts in *Lunar Society*, pp. 41, 60, and *Doctor of Revolution*, p. 68.
Text From original MS.
Notes

(1) Boulton's new manufactory at Soho, now nearly complete, relied on a water-mill for its power, but Boulton became alarmed that the mill-pond might dry up in a drought, or if someone diverted the water. So in 1765 he decided to build a steam-engine to pump the water back into the mill-pool when necessary. This new development re-animated Darwin's enthusiasm for steam engines; as this letter shows, he is eager to analyse the modus operandi of the steam – and does so correctly. Boulton continued experiments with this engine until he met James Watt in 1768.

(2) Dr William Small (1734–1775) arrived in Birmingham in May 1765 bearing a complimentary letter of introduction from Franklin, and he quickly became the 'favourite friend' of both Boulton and Darwin. Self-effacing, friendly and accomplished, he served as the 'organizing secretary' of the Lunar group. For seven years previously, he had been Professor of Natural Philosophy at the College of William and Mary at Williamsburg in Virginia, where he had a great and lasting influence on Thomas Jefferson, one of his pupils.

(3) In 1765 Darwin was much occupied with the promotion of the Grand Trunk Canal, in association with Wedgwood, Brindley and Garbett. Now Darwin was asking for Boulton's support, which he readily gave.

(4) 'The gown She troubled Mrs Boulton with' was a black silk negligée belonging to a friend which Mary Darwin sent to Mrs Boulton in October to be painted (Mary Darwin to Anne Boulton, 10 October 1765. Birmingham Reference Library, Matthew Boulton Papers, Darwin 49).

(5) 'Mr Mitchel' is John Michell (1724–1793), a friend of Darwin since his undergraduate days at Cambridge. Michell is now recognized as a founder of modern geology; he also made discoveries in astronomy and magnetism, and was the inventor of the torsion balance. See A. Geikie, *Memoir of John Michell* (CUP, 1918) and R. E. Schofield, *Mechanism and Materialism* (Princeton University Press, Princeton, NJ, 1970), pp. 241–9.

(6) 'You Birmingham philosophers' is the first mention of the embryonic Lunar Society. See *Lunar Society*, p. 115.

(7) In this letter Darwin four times writes 'as' when he appears to mean 'at' – which I have added in square brackets.

(8) John Motteux, 11 Wallbrook, London, was an export–import merchant, who often handled Boulton's wares.

66A. To Peter Templeman, 8 March 1766

Lichfield Mar 8 1766

I am favor'd with your Letter in the Name of the Society for the encouragement of Arts and am sensible of the Honour you are pleased to do me by enquiring about my Carriage.

About seven or eight years ago, I observed that among other lesser defects in the common light Carriages that there were two principle ones, First that the foremost wheels were for the conveniency of turning made too low, by which means the forepart of the Carriage was obliged to rise too suddenly over Stones or other Obstacles in the way; and thence continual Shocks produced, that injured both the Horses and the Carriage, and incommoded the Travellers – Besides the lesser inconvenience of increasing the Friction upon the Axletree. Secondly That in the Time of turning, the Basis on which the Carriage rests, was changed from a Parallellogram to a Triangle

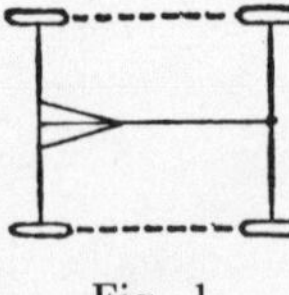

Fig. 1

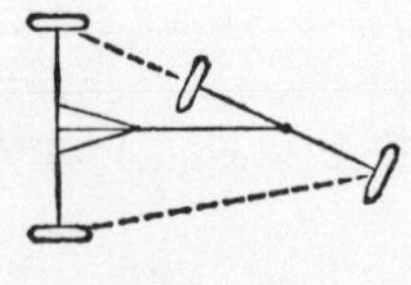

Fig. 2

Thus in Fig 1st the Carriage is a four-footed stool, in the 2nd a three-leg'd Stool. That from this Cause Carriages were most frequently overturn'd, viz at the Time of turning. And that to this Cause was entirely owing the frequent Accidents in Phaetons, where the Body-part with the Driver were placed more forwards, that is, upon the most hazzardous part of this three-leg'd stool in the Act of turning.

I contrived a Method of turning, which gets free of both these inconvenience, viz, it admits the foremost Wheels to be as high as you please, those I made were four feet diameter; and the Carriage stands on the same Basis (that is, is as safe from overturning) in the Act of turning as at other times.

I made a model about 7 or 8 years ago which is perished long since, three years ago I made a Phaeton on this principle, and have rode in it, I dare conjecture, about 10,000 miles and it is yet a good one, and weighs but 500 wgt.

About two years ago I made a Post Chaise on this principle, and have rode in that I dare say 10,000 miles and found no inconvenience from the new manner of turning in either of them.

Each of the foremost Wheels turns not from a Center between them but each on a Center of its own near the Knave. I can not give any drawing or Description that can be at all intelligible, as the parts tho' very simple, lie in different Planes. – But if the Society will bear the Expense will direct a model to be made for them.

Otherwise Mr Greseley of Seal in Leicestershire, to whom I have now given the Phaeton, intends soon after Easter to come to London, and at my desire will shew it to the Society.

I recieved a Letter concerning these Carriages about six weeks ago from Mr Edgeworth of Black Bourton, Witney, Oxon – who I suppose may have been so ingenuous as to mention my name to you.

I had once thoughts of applying for a Patent for making the Axletrees for this new way of turning, which are uncommon shaped Cranks, and require a little Accuracy in making – but I defer'd this till much experience should ascertain there being really useful.

There are some lesser conveniencies in these Carriages, such as a double cross Perch, much lighter, and more springing than the common.

And one Shaft instead of two, by which both Horses can back the Carriage.

And Swingle-Tree Bars to draw by, which prevents the Horses from galling on their Shoulders.

I believe my Carriage has fewer pieces of wood, is lighter, and will last longer than any other Carriage of the Size and Conveniences, besides its being safer from overturning, and easyer to draw.

from, Sir, with the greatest Respect

Your obed. Serv.

Erasmus Darwin

Addressed For Peter Templeman Esq, Secretary to the Society for encouragement of Arts and Sciences, in the Strand, London.

Original Library of the Royal Society of Arts, Guard Book Vol. 12, No. 5.
Printed Excerpts in *Lunar Society* p. 47; longer quotations in *Doctor of Revolution*, pp. 63–4.

Text From original MS. (*Diagrams* Re-drawn.)

Notes

(1) This important letter is the only first-hand account of Darwin's improvements in carriage design, to which he devoted a good deal of time and energy, because he travelled about 10 000 miles a year on his rounds as a doctor and wanted to make his journeys less uncomfortable. Darwin did not seek mechanical fame, because he thought it would injure his medical practice, and he allowed Edgeworth to take the credit for these and further innovations. Darwin later improved the method of steering so that the two front wheels had turning circles with the same centre, which lay on the line of the back axle. This design won Edgeworth a Gold Medal from the Society of Arts in 1769. See *Doctor of Revolution*, pp. 62–7, for further details and discussion. See also letter 67D.

(2) Dr Peter Templeman (1711–1769) was Secretary of the Society for the encouragement of Arts and Sciences. The Society had been founded in 1754 and did much to encourage science and invention in the eighteenth century by offering prizes and awarding medals. For a biography of Templeman, see B. Hutchinson, *Biographica Medica* (Johnson, 1799) ii 437–40, or *DNB*.

(3) 'Mr Greseley of Seal' is probably Rev. Thomas Gresley (1734–1785), Rector of Nether Seal on the border of Leicestershire and Derbyshire. For Gresley family tree, see Nichols, *Leicester* iii 1009–11. There is a biography of Gresley in *Collections for a History of Staffordshire*, Vol. 1(N.S.), Harrison, London, 1898, pp. 113–17.

(4) Richard Lovell Edgeworth (1744–1817) was a talented, wealthy and energetic young man who heard about Darwin's carriage designs in 1765 and enthusiastically began designing improved versions himself. Edgeworth became a prolific inventor and a close friend of Darwin (see letter 66D). For biography of Edgeworth, see D. Clarke, *The Ingenious Mr Edgeworth* (Oldbourne, 1965).

66B. To Matthew Boulton, 11 March 1766

Lichfield Mar.11. 1766

Dear Boulton

I have no design to quarrel with you for not having answer'd my Letters; in respect to my last indeed, desiring you to sign your Consent to the Navigation, your fulfilling my Request was the politest answer you could give me.

Your Model of a Steam Engine I am told gain'd so much approbation in London, that I can not but congratulate you on the mechanical Fame you have acquired by it; which, assure yourself, is as great a Pleasure to me, as it could possibly be to yourself; so much do I interest myself in what belongs to your Reputation. This Model I am so impatient to see, and to hear any Observations you have made or heard on this Subject, that I am determined to spend a Day with you, the first vacant Day that occurs to me: and shall trust to the Stars for meeting with you at Home. – But if the Stars should not shine upon me, pray contrive, that, in your Absence, Mrs Boulton may have it in her Power to shew me this celebrated Model, which I will trust to her good Temper for her complying with.

I should some Months ago have sent you the Impression of Le Brun's Battles of Alexander, but as I was told it is the best Edition, I have determined to endeavour to clean or bleach them, to render them more worthy your Acceptance: but from native Indolence, common to me with the rest of Mankind, I have defer'd doing that long, which it was not necessary to do immediately.

Yesterday I was informed another Physician was arrived in Birmingham, but hope this gives no Uneasyness to our ingenious Friend Dr Small, from whom and from you, when I was last at Birmingham, I recieved Ideas, that for many Days occured to me at the Intervals of the common Business of Life, with inexpressible Pleasure.

Capt. Valentine Gardiner, a Philosopher militant, tells me He heard 3 or 4 Lectures about a Month ago from the Abbè Nollet at Paris, who I find continues to assail Dr Franklin's electrical Theory, but with weak and impotent attacks – and amused his spectators with a floating Electrometer, the Degrees of which were magnified by their Shadows being observed instead of the Scale, and by this means you was not necessitated to approach within the magic Circle of electric Att{ra}ction – but I think this is not new, and no wonderful Discovery.

Don't tease yourself, my dear Friend, about answering this long Letter; indeed I don't expect to hear from you, why then will you let it prey upon your Mind? – But I assure you I intend very shortly, if G-d permit, to pay you a Visit – "Ay, for God's sake, come and welcom," (say you) "any Thing but these damn'd long Letters."

Adieu
E Darwin

Original Birmingham Reference Library, Matthew Boulton Papers, Darwin 11.
Printed Excerpts in *Lunar Society*, pp. 38, 61, and *Doctor of Revolution*, pp. 61, 68.
Text From original MS.
Notes
(1) For 'the Navigation' (the Grand Trunk Canal) and the steam engine, see notes to letter 65B.
(2) Charles Le Brun (1619–1690), the leading French court painter in the reign of Louis XIV, produced a series of paintings of Alexander and his battles.
(3) 'Capt Valentine Gardiner' probably refers to Major Valentine Gardner (1739–1816) whose younger brother Alan became an Admiral, an MP and a Baron. See Burke's *Peerage* (1970), p. 1078, and *Gent. Mag.*, **86**(1) 377 (1816).
(4) Jean-Antoine Nollet (1700–1770), after taking a degree in theology, kept the equivocal title 'abbé' when he withdrew from the religious arena to study science. He worked on anatomy, thermometry, magnetism and electricity (where he was in conflict with Franklin, as this letter shows). His lectures were very popular.
(5) This is one of the most attractive of Darwin's letters, conversational and witty, and full of little kindnesses to Boulton and to Small.

66C. To Josiah Wedgwood, 27 April 1766

* * *

Count Laragaut has been at Birmingham and offer'd the Secret of making the finest old China, as cheap as your Pots. He says the materials are in England. That the Secret has cost £16,000, that He will sell it for £2,000 – He is a Man of Science, dislikes his own Country, was six months in the Bastile for speaking against the Government – loves every thing English. . . . . But I suspect his Scientific Passion is stronger than perfect Sanity. . . . . I congratulate you on the Success of your Act of Parliament. Adieu.

Original Not traced. In the possession of Charles Darwin, 1865.
Printed Meteyard, *Wedgwood* i 436.
Text From above source.
Notes
(1) Meteyard says this letter 'came to hand' on 27 April; so it may be dated a day or two earlier.
(2) 'Count Laragaut' was Louis de Brancas, Comte de Lauraguais (1733–1824), an energetic French nobleman who developed new methods of making porcelain. He took out an English patent and apparently asked Boulton to do the manufacturing. See *Wedgwood Circle*, p. 49 and Meteyard, *Wedgwood* i 437. De Lauraguais was also a prolific writer, of tragedies, autobiography and political pamphlets – for which he was imprisoned, as Darwin remarks. For a biography of de Lauraguais, see *Biographie Universelle* (Paris, 1843) Vol. 23, pp. 357–60.
(3) The House of Commons agreed the Bill for the Trent-and-Mersey canal on 21 April 1766, and it received the Royal assent on 14 May.

66D. To Matthew Boulton, [Summer 1766]

Dear Boulton

I have got with me a mechanical Friend, Mr Edgeworth from Oxfordshire – The greatest Conjurer I ever saw – G-d send fair Weather, and pray come to my assistance, and p{revai}l on Dr Small and Mrs Boulton to attend you, tomorrow Morning, and we will reconvoy you to Birmingham on Monday, if the D——l permit, adieu

E Darwin

He has the principles of Nature in his Palm, and moulds them as He pleases.

Can take away Polarity or give it to the Needle by rubbing it thrice on the Palm of his Hand.

And can see through two solid Oak Boards without Glasses, wonderful! astonishing! diabolical!!! Pray tell Dr Small He must come to see these Miracles.

Addressed For Matth. Boulton Esqr., Snow-Hill, Birmingham. Please to give the Post-Boy six Pence if He delivers it to night (Saturday).

Original Birmingham Reference Library, Matthew Boulton Papers, Darwin 12.
Printed Lunar Society, p. 45; *Doctor of Revolution*, p. 66; and elsewhere.
Text From original MS.
Notes
(1) This typically exuberant letter marks the beginning of Darwin's lifelong friendship with Edgeworth. For the background to the meeting see Edgeworth, *Memoirs* i 160–65 or *Doctor of Revolution*, pp. 65–7.
(2) During the previous year Edgeworth had been occupied in organizing exhibitions of magical feats at Sir Francis Delaval's house in Downing Street, London, and he performed some of these conjuring tricks for Darwin.
(3) The letter has the following endorsement by Boulton: 'Mr and Mrs Boulton depart at 7 on Sunday morning to Litchfd. and Wichnor in order to dine with Mr Levett[:] they should be glad to place Dr Small in the middle of the Chaise and drop him at Litchd. where he may enjoy a philosophical feast and return with Boulton and Wife on Monday Noon to Birmingham'.

67A. To Unknown Man, 7 February 1767

Lichfield Feb. 7 –1767

Dear Sir

I am sorry you should think it necessary to make any excuse for the Letter I this morning recieved from you. The Cause of Humanity needs no Apology to me.

Before the Lungs of animals have been inflated with Air, they are as solid, and as heavy as other animal Flesh, and will sink in Water.

When the Lungs have been in any degree inflated with Air, some portion of that Air remaines in them and prevents them from sinking, by enlarging their Bulk.

Hence the Swimming of the Lungs shews they have been in part distended with Air; but {,you} ask, can not this be done accidentally after the Death of the Child? – I am certain it may, and that much easyer, and more frequently, than, at first View, one should imagine.

The Cavity of the Breast is well resembled by a pair of Bellows, the Bones of the Breast and Ribs are so contrived, that when they are lifted up, the Cavity of the Breast is enlarged; precisely as the hollow part of the common Bellows when the top is lifted upwards; and the Air rushes into this incipient Vacuum, is recieved into an open Chest in the Bellows, and into innumerable minute Pipes in the animal Lungs.

Thus then it appears, that by lifting up the Breast-Bone or Ribs with ones Fingers, the Lungs may became in part inflated, tho' the Child was destroy'd immediately before or in its Birth. As the same Effect would follow, whether the Breast Bone or Ribs were lifted upwards by the power of the living Child, or by the Pressure of the Fingers of another Person.

To ascertain the Truth of what is here said, any one may easyly try the following Experiment – Cut off the Head of a Cock, and by lifting up the Breast-Bone or pressing the Ribs with ones Fingers, it is easy to make him produce a Noise like Crowing, and that as long as you please: which could not be done but by Air passing into his Lungs, and out of them; in the same manner as when the Creature was alive.

This Inflation of the Lungs may be produced by any accidental Handling of a Child lately dead, or by a fall on its side: precisely and accurately in the same manner as a pair of Bellows might be in part open'd by falling with its sides on a Stone or other unequal Surface.

In respect to the particular Case you have mention'd to me, the Child might be kill'd by its fall on the Plaister-Floor, which seems to have been accidental. And its Head appearing black confirms this Supposition.

Secondly [*if*] the Body appearing black or bruised from an accidental fall is no certain Criterion of the Child being born alive: this is further evinced in many parts of the Bodie, in some putrid cases, turning black Hours after Death: and [I] must observe that, in these Cases, as the Bodies of the Children are not decay'd, they are not supposed to have been long dead.

The death of this Child might have been owing to the Loss of Blood from the umbilical Chord, which [it] seems was cut, and not tyed. If it [is] rent or torn off, or bit off, I believe it will cease to bleed spontaneously, as in other Animals.

The Women that have committed this most unnatural Crime, are real Objects of our greatest Pity; their education has produced in them so much Modesty, or Sense of Shame, that this artificial Passion overturns the very Instincts of Nature! – what Struggles must there be in their Minds, what agonies! – and at a Time when, after the Pains of Parturition, Nature has designed them the sweet Consolation of giving Suck to a little helpless Babe, that depends upon them for its hourly Existance!

Hence the Cause of this most horrid Crime is an excess of what is really a Virtue, of the Sense of Shame, or Modesty. Such is the Condition of human Nature!

I have carefully avoided the use of scientific Terms in this Letter, that you might make any use of it you may think proper: and shall only add that I am veryly convinced of the Truth of every part of it,

and am, dear Sir,

Your affect. Friend and Serv.

Erasmus Darwin

Original In the collection of Sir Geoffrey Keynes.

Printed Excerpt (the first and the last three paragraphs) in C. Darwin, *Life of E. Darwin*, pp. 28–9.

Text from original MS.

Notes

(1) There is no indication of the recipient of this letter. He might possibly be Boulton: the ending is the same as 64F.

(2) Evidently a baby was found dead, and its mother was suspected of murder. Darwin's sentiments, and the promptness of his reply, offer a good example of his 'benevolence'.

(3) This one of the very few of Darwin's letters with numerous alterations: however, these are all deletions of scientific terms, as indicated in the last paragraph. One of the more substantial changes is in the third paragraph, where 'remaines in them' originally read 'continues in the Branches of the Wind-pipe, call'd Bronchia'. It is very probable that the manuscript is a draft and that a fair copy of it was sent.

(4) Unusually, Darwin writes *its* as *it's* in this letter.

67B. To Josiah Wedgwood, 2 July 1767

Newcastle Jul. 2 –67

Dear Sir,

It greatly grieves me that I cannot steal an hour to come over to Burslem this evening, to see your curiosities, but more to hear your remarks upon them. I postponed writing you my thanks for those you favoured me with, as I intended to have ascertained the species of fern that were impressed on some of them. The cylindrical stone set round with stellations has taken its form from the roots of gladiolus aquaticus (water-flag) which is a Species of Iris. The bone seems the third vertebra of the back of a camel. The horn is larger than any modern horn I have measured, and must have been that of a Patagonian ox I believe.

If at your leisure you will be at the trouble to tell me the strata they shall penetrate, in the order they penetrate them, and their different thickness, I will in return send you some mineral observations of exactly the same value (weighed nicely) for I must inform you I mean to gain full as much knowledge, from you who can spare it so well, as I return you in exchange.

I have lately travel'd two days journey into the bowels of the earth, with three most able philosophers, and have seen the Goddess of Minerals naked, as she lay in her inmost bowers, and have made such drawings and measurements of her Divinity-ship, as would much **amuse**, I had like to have said **inform**, you – but this in exchange, by Ounus[?]

Adieu E D

I beg my Compliments to Mrs W—— and the fine little Wedgwoodikins.

Original Not traced. There is a MS copy in the Wedgwood Commonplace Book, Wedgwood archive, Keele University Library.

Printed Excerpts in *Doctor of Revolution*, pp. 74–5.

Text From MS copy.

Notes

(1) The greatest engineering work in the construction of the Grand Trunk Canal

was the Harecastle Tunnel, 2880 yards in length. Not only was it the longest tunnel that had been attempted in Britain, but also it had to be cut through very hard, wet, fractured rock. Excavation began in 1766 and was yielding a rich harvest of fossils, including many huge bones. Wedgwood himself describes them in a letter to Bentley quoted in Meteyard, *Wedgwood* i 500–502. Strangely enough, there is no mention of such large fossils in the standard geological texts, such as J. Ward, in *Geology of the North Staffordshire Coalfields* (HMSO, 1905), pp. 285–357, where it is also stated that the 'first recorded discovery of organic remains' there 'dates as far back as the year 1835'. Wedgwood's letter makes it clear that the huge bones were found on the south side of Harecastle 'under a bed of clay, at the depth of five yards from the surface'. They were almost certainly pleistocene, presumably being embedded in the drift deposits brought down by glaciers, and they may have been bones from extinct varieties of elephant (the 'horn' mentioned by Darwin being a tusk). The impressions of ferns and other plants were much older: they were found at the north end of Harecastle in a stratum underlying a bed of coal, and were from the Carboniferous era, 300 million years ago. Darwin's failure to identify the huge bones is not surprising; but in the eighteenth century doctors were expected to be able to identify bones, so Darwin's failure involved 'loss of face', which he tried to cover up with a veneer of humour. Beneath the mask, however, Darwin was deeply impressed by the fossils. They brought him face-to-face with evolution. Within three years he had accepted the idea of evolution and had the motto *E conchis omnia*, or 'everything from shells', painted on his carriage (see *Doctor of Revolution*, pp. 74–5). But it was another twenty years before he published his opinions on the course of evolution, in *Zoonomia* (1794).

(2) The 'three able philosophers' were probably Whitehurst (see letter 67E) and the Tissingtons (see next letter).

(3) 'Ounus' might perhaps be Oeneus, king of Calydon, who gave his name to wine.

(4) 'My Compliments to Mrs W——' is the first of many references in these letters to Wedgwood's wife Sarah (1734–1815). The 'little Wedgwoodikins' were Susannah (born 1765) and John (born 1766). For their biographies, see *Wedgwood Circle*.

67C. To Matthew Boulton, 29 July 1767

Lichfd.

Dear Boulton

I have paid Mr Cob 6–7–0 for your Acct. and desire you will pay Mr Baumgartner when you settle with him, I mean four guineas for one Chest of Florence {and} 43shil for a half Chest. I still owe him for six Buckets but don't know what the[y] amou[n]t to.

I want to see you and Dr Small much, if you will fix a Day, I will send an Express to prevent you if I am absent.

I have been into the Bowels of old Mother Earth, and seen Wonders and learnt much curious Knowlege in the Regions of Darkness – pray tell Dr Small I have taken the Liberty to set a Person to transcribe his Papers, and am much obliged to him for the Lone of them. I have seen the two

Tissingtons, subterranean Genii! And am going to make innumerable Experiments on aqueous, sulphureous, metallic, and saline Vapours. Food for Fire-Engines! – Adieu, the Lord defend you,

E Darwin

Jul. 29 –67

Original Birmingham Reference Library, Matthew Boulton Papers, Darwin 13.
Printed Excerpts in *Lunar Society*, p. 101, and *Doctor of Revolution*, p. 75.
Text From original MS.
Notes

(1) 'Mr Cob' was the Lichfield banker, Francis Cobb, see letter 63D. Mr Baumgartner was John Lewis Baumgartner, a leading Birmingham import–export merchant, principal of Baumgartner and Hoofstetters, 25 New Street, Birmingham, and 35 Little George Street, Aldgate, London. Born in Geneva, Baumgartner came to England in the 1750s and handled much of Boulton's export trade. See E. Robinson, *Univ. of Birmingham Hist. J.*, **7**, 60–79 (1959).

(2) Having received a Chest of Florence wine from Baumgartner, as well as a half-Chest and six Buckets, Darwin presumably had a well-stocked cellar in 1767, though he condemned alcohol in later years.

(3) As in the previous letter, Darwin is very enthusiastic about his geological explorations. Anthony Tissington, FRS (1703–1776), of Swanwick, near Alfreton, Derbyshire was a talented and energetic mine agent, a friend of Whitehurst (see letter 67E), and a leading authority on mineralogy and the science of mining. Elected FRS in 1767, Tissington was from 1760 onwards a friend and correspondent of Benjamin Franklin, who visited him in Derbyshire more than once; six of his letters to Franklin appear in volumes 13–19 of *The Papers of Benjamin Franklin* (Yale Univ. Press, New Haven, 1969–75), and there are other references to him in volumes 9–12. A manuscript biography of Tissington, by his grandson John Tatlow, which I have seen by courtesy of the owner, Dr W. F. Tissington Tatlow, of Hudson, Quebec, Canada, reveals him as able and admirable but also rather quarrelsome, involving himself too much in other people's troubles. Tissington had a younger brother George, who was also employed in mining, at Winster. He is probably the second of 'the two Tissingtons'. (The MS biography states that he died in 1760; but this is wrong, because he wrote to the Duke of Devonshire in March 1764 (Chatsworth, Devonshire Collections, 666.0), and a letter to him dated 2 July 1768 is printed in Whitehurst's *Inquiry into the Original State and formation of the Earth* (1778), pp. 188–9). The second Tissington might (less probably) be Anthony's son, Anthony Tissington the younger (1733?–1815), a mineral agent and a capable businessman, who indulged too freely in drink and 'expensive women', according to the MS biography.

(4) Darwin's proposed experiments probably involved strong heating of the minerals he had collected, to find what gases were given off.

67D. To James Watt, 18 August 1767

Lichfield Augt. 18 –67

Dear Watt,

I have been so continually engaged in travelling about the Country from sick People to sick People, that it was not till this Afternoon in my Power to measure the Iron-work of my Chaise. There is an iron Crank in this Shape, and the Arm of the Axle is on the same Crank, and stands out at (W).

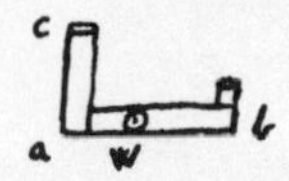

Perhaps it would be better at the angle (a)? But in this place it admits of the Wheels being larger in the act of Turning. ab is 13 In./aW 1½ Inch/ac 12 In/ab is square 1½ Inch/ac is round. The Pearch is a Cross.

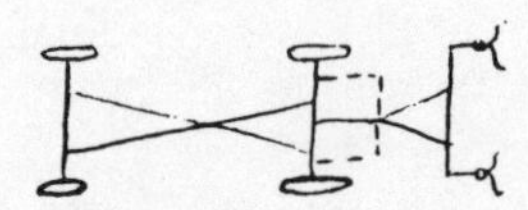

Mr Edgeworth has put the arm of the Axle at the angle a in one He shew'd to the Society of Arts, and the Duke of York has a Phaeton made after it. The two fore-Wheels are 4 ft high.

My Cart is to draw a Ton, by Two Horses, a-breast. The wheels are seven feet Hight. The Body swings under the Axis. Now I think the Body should be fix'd to the Axis, and the Arms of the Axis should be cylindrical, and not the frustrum of a Cone. And 3rd the Axis should, as near as may be, pass through the Center of the Load. 4th the Horses should be link'd to the Body of the Cart, to avoid the additional Weight of the Frame in which my Cart-Body swings. By these means a very slight Pole alone is necessary which may rest on the Back of either of the Horses; and they may back as mine do, by pressing against the Swingle-tree Bar, and have a broad leather behind in the Harness, that their naked Shins may not come against the Woodwork. NB I have not well digested this!

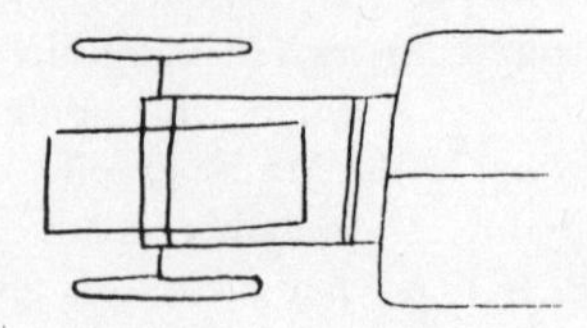

This is my Cart as elegant as I can draw it, the Pole is suspended by a Strick across the Backs of the two Horses.

Enough of Carts and Chaises! Now my dear new Friend, I first hope you are well, and less hypochondriacal; and that Mrs Watt and your Child are well. The Plan of your Steam-Improvements I have religiously kept secret, but begin myself {to} see some difficulties in the Execution that did not strike me when you was here. I have got another and another new Hobby-Horse since I saw you. I wish the Lord would send you to pass a week with me, and Mrs Watt along with you, – a Week! – a Month, a Year.

You promised to send me an Instrument to draw landscapes with. – If you ever move your Place of Residence for any long time from Glasgow, pray acquaint me. – I hope you have heard from Antony Tissington. – Adieu

your Friend E Darwin

Addressed For Mr James Watt, Merch in Glasgow, Scotland. (*Postmark* Lichfield).

Original Privately owned.
Printed Short excerpt published by Muirhead, *James Watt* i 15, and frequently reprinted.
Text From original MS. (*Diagrams* Redrawn.)
Notes
(1) It was in 1765 that James Watt (1736–1819), greatest of British engineers, made his invention of the improved steam engine with separate condenser. Watt's friend Joseph Black (see letter 78H) introduced him to John Roebuck (see letter 63B), whose Carron Iron Works was the leading Scottish manufactory. Roebuck needed a new engine to pump water from his mines at Bo'ness, and agreed to develop Watt's engine. Watt meanwhile took employment as a canal surveyor and engineer. In April 1767 Watt travelled to London, and Roebuck asked him to visit Birmingham and call on Samuel Garbett (see letter 63B), Roebuck's partner. Watt did so, and Garbett introduced him to Darwin, 'almost my most ancient friend and acquaintance in England', as Watt recalled 35 years later. Watt stayed overnight with Darwin: they took to each other immediately, and Watt confided the secret of his engine to both Darwin and Small. Though Boulton was away, Watt looked over the Soho works, and returned in 1768, when he met Boulton. Darwin and Small had complete faith in Watt's invention, and Small constantly urged Watt to move to Birmingham, which he eventually did in 1774. For biography of Watt, see L. T. C. Rolt, *James Watt* (Batsford, 1962) and Muirhead, *James Watt*.
(2) The first paragraphs of the letter, and the diagrams, supply helpful further details about Darwin's carriage and should be read in conjunction with letter 66A.
(3) The Duke of York was King George III's younger brother Edward (1739–1767), noted for his 'immoderate pursuit of pleasure' in Horace Walpole's phrase. Edgeworth had met the Duke through Sir Francis Delaval in 1765.
(4) Mrs Watt was Watt's first wife, his cousin Margaret Miller, whom he married in July 1764. Their first child John (born 1765) died in infancy, and the child Darwin mentions was their daughter Margaret (1767–1791), from whom the present family of Gibson-Watt is descended. Mrs Margaret Watt died in 1773, and it was after her death that Watt moved to Birmingham.

67E. To Josiah Wedgwood, 8 November 1767

Dear Wedgewood,

I have the Pleasure to introduce to your acquaintance Captain James Keir, an old Friend of mine, a successful cultivator of both Arts and Arms. He begs the Favour of seeing your elegant manufactory and hopes to meet our common Friend, the Philosopher, Mr Whitehurst at your House. The Civilities you shew Capt. Keir will be recieved by Dear Sir your affect. humble Serv.

E Darwin

Original Not traced. Seen by Eliza Meteyard in 1865.
Printed Meteyard, *Wedgwood* ii 207.
Text From above source.
Notes
(1) James Keir (1735–1820), Darwin's student friend at the Edinburgh Medical School, had served in the Army for ten years (including the Seven Years' War), and had now retired 'to study chemistry and philosophy more at leisure'. He had decided to settle near Birmingham and was a welcome recruit to the Lunar circle. He was already acquainted with Edgeworth and was now meeting Wedgwood and Whitehurst.
(2) John Whitehurst (1713–1788) of Derby was an unassuming man of little formal education but great natural talent. He gained a high reputation as a clock-maker in the 1750s, when he first met Boulton. In the 1760s he was turning towards geology and was probably Darwin's guide to the Derbyshire caverns in July 1767 (see letter 67B). Whitehurst was the best geologist in the Lunar group and author of the famous book *An enquiry into the original state and formation of the Earth* (1778). In Darwin's view, his 'ingenuity, integrity, and humanity, were rarely equalled in any station of life' (*Economy of Vegetation* II 17, note).

67F. To Josiah Wedgwood, [12–18] November 1767

* * *

Capt. Keir is at present at Birmingham. He desired I would say all the fine Things I could think of to thank Mrs Wedgwood for the Trouble He gave her; of whom indeed he speaks very highly, and was much entertain'd with your manufactory.

* * *

Original Not traced. In possession of Charles Darwin, 1865.
Printed Meteyard, *Wedgwood* ii 207.
Text From above source.
Notes
(1) According to Meteyard, this letter is 'undated but referable to a few days later' – than 8 November. It seems likely to have been written between 12 and 18 November.

(2) 'Keir saw the manufactory, and was shown the plan of the future Etruria Hall, which . . . he greatly admired' (Meteyard, *Wedgwood* ii 207). Presumably Keir saw the Brick-House Works, and perhaps the beginnings of Etruria (which did not open until 1769).

67G. To Josiah Wedgwood, [late 1767]

[Darwin is discussing his horizontal windmill.]

* * *

I think it is peculiarly adapted to your kind of Business, where the motion is slow and horizontal. If the mixture of Clay or Flint and Water should grow stiff by the wind ceasing, I think it may be so contrived with ease as to be put into Motion again gradually, not suddenly: of this I can not judge quite distinctly, without seeing the consistence this mass will acquire by a few Hours rest. I will make you the model if you require it immediately, which I can do for the expense of 3 or 4 guineas, so as to evince its Effects. The advantages are, 1. Its Power may be extended much farther than the common Wind-mill. 2. It has fewer moving Parts. 3. In your Business no Tooth and Pinion-work will be necessary. Plain Countries are preferable to hilly ones for windmils, because the wind acquires eddies in the latter.

* * *

Original Not traced. Seen by Eliza Meteyard in 1865.
Printed Meteyard, *Wedgwood* ii 30.
Text From above source.
Notes
(1) This letter was written between November 1767 and February 1768, and may be early 1768.
(2) Darwin thought that the horizontal windmill he had designed would be suitable for use in Wedgwood's new manufactory (Etruria). It had a vertical shaft and a horizontal sail, and Darwin here goes over its advantages. The windmill was installed at Etruria in 1779 and was busy grinding flints for 13 years.
(3) For further discussion of the windmill, see letters 68A, 69A and 69B, and *Doctor of Revolution*, pp. 79–81, 129–30.

68A. To Josiah Wedgwood, [February] 1768

[This is a letter about the horizontal windmill, and Wedgwood includes a verbatim quotation from it in his letter to Bentley of 3 March 1768.]

* * *

The advantages of the Mill in preference to the common ones are:
1. There are no wheels or Cogs.

2. No moving parts but the Sail and its upright Shaft.

3. Much less friction in this way.

4. The power may be increas'd after the mill is built.

5. It goes with all winds night and day [to which Wedgwood adds the comment, 'What Windmill does not'].

6. There is a large room under it for the convenience of the work. The Principle on which this mill is constructed has not been attended to, viz, that all elastic bodys (as the Air is) moving against an oblique fix'd Plane, change their direction only, but lose nothing of their velocity. Hence in this Mill the wind is bent upwards before it moves against the sail. You will please to send some inginious People over to Lichfield to pass a day and observe this Model, for it is too large for Carr[iage], and my opinions are not demonstrative nor do they satisfy myself.

* * *

[Wedgwood adds: "This Windmill is to grind colours (if it should happen to grind anything) for an intended ornamental Manufacture at Hetruria." He then says it is a good reason for a visit to Lichfield, and goes on to other matters.]

Original Letter from Wedgwood to Bentley, 3 March 1768. Keele University Library, Wedgwood Archive, 18191–25.
Printed Unpublished.
Text From Wedgwood's MS.
Notes
(1) The section of Wedgwood's letter preceding the verbatim quotation reads as follows: 'I purpose leaving home on Sunday, or Monday next, spending a day or two at Birmingham and from thence shall meet Mr Pickford at Lichfield and must spend some Time there in examining a model for a Windmill of Dr Darwin's projection, which he writes me will take 68,000 bricks, 4000 ft of inch deal boards, and a shaft 30 ft long and a foot diameter, having two sails.' The quotation from Darwin's letter follows.
(2) Wedgwood duly made the journey to Birmingham and Lichfield in mid March. At Birmingham he had long discussions with Boulton on combining clay and metal, and at Lichfield he inspected Darwin's windmill. His scepticism about its performance turned to admiration – 'I think [it] a very ingenious invention and have some hopes that it will answer our expectations' (Wedgwood, *Letters* i 207).
(3) The date of Darwin's letter is probably late February. Wedgwood does not mention it in his letters to Bentley of 22 and 25 February.
(4) Darwin's list of the advantages of his windmill is much the same as in his previous letter, though he adds the principle that air striking a plane inclined at a small angle does not lose its velocity. This is basically true, apart from boundary-layer friction losses: he is groping towards aerofoil aerodynamics. His idea of a windmill with a vertical shaft has been much in favour recently.

68B. To Josiah Wedgwood, 14 June 1768

[Dear Wedgwood]

I am heartily vex'd that I could not get to this Place soon enough to day to have come to spend the Evening with you. But I hear you go on well, and that gives me, I assure you, much Satisfaction, and I have an engagement to morrow near Coleshil, which is on the other side of Lichfield, which prevents me of calling of you tomorrow morning. Mr Boulton is gone to Buxton for his Health, and talk'd of coming by Newcastle on his Return to call of you and console you. . . . Mr Boulton has got a new metal which rivals Silver both in Lustre, Whiteness, and endures the Air with as little Tarnish. Capt. Keir is endeavouring to unravel this metal. Mr Edgeworth, a philosophical Friend of mine in Oxfordshire, writes me word He has nearly completed a Waggon drawn by Fire, and a walking Table which will carry 40 men. This is all the news I can think of to amuse you with in the philosophical Arts.

* * *

Original Not traced. In the possession of Charles Darwin, 1865.
Printed Meteyard, *Wedgwood* ii 42–3.
Text From above source.
Notes
(1) Wedgwood had his right leg amputated on 28 May 1768, and Darwin (who was not involved in the operation) visited him more than once in the subsequent weeks. This letter is a substitute for a visit.
(2) 'This place' is Newcastle, where the letter was written.
(3) Boulton's 'new metal' may possibly have been platinum. Keir was already acting as chemical adviser to Boulton, and was making preparations for the glass manufactory which he established at Stourbridge in 1772.
(4) From this letter it appears that Wedgwood had not yet met Edgeworth, though he was soon to do so. The reference to Edgeworth's 'Waggon drawn by Fire' is tantalizing: Dr Small also refers to Edgeworth having 'made considerable progress' in his attempts to 'move land and water carriages by steam'. But it seems that Edgeworth never succeeded in producing a working steam-car or steam-boat. Edgeworth was also developing a 'portable railway', but this seems to have been a carriage with caterpillar tracks instead of wheels, and was apparently not mechanically propelled. The 'walking Table' for 40 men was presumably a primitive omnibus. So Edgeworth was on the brink of inventing the steam locomotive, the steam-car, the steam-boat, the tank and the bus – almost the whole spectrum of modern surface transport: but he never quite succeeded.

68C. To Matthew Boulton, 31 July 1768

Jul. 31 –68

Dear Boulton,

You promised me sometime since to make me a plated Sauce-Pan with the Silver thrice or four times as thick as you usually make it. I should be glad of this as soon {as it} is convenient to you, about {the size} of a Quart or three-Pint Measure.

And Mr Dawkins desires me to bespeak a two Quart Sauce-Pan, a Quart one, and a Pint one, of the same kind, that is with the Silver at least thrice or four Times as thick as you usually make them: and these as soon as may be convenient to you.

Mrs Darwin continues to get better, and desires her Compliments to yourself and Mrs Boulton.

Adieu E Darwin

You will please to send them to Dawkins Esqr at Standlinch near Solsbury with a Note, and Mr Dawkins will take an opportunity to return you the Money.

Addressed For Matt. Boulton Esqr., Soho, near Birmingham.

Original Birmingham Reference Library, Boulton and Watt Collection (Timmins Part I f. 19).
Printed Unpublished.
Text From original MS.
Notes
(1) This is another of the letters of Darwin-as-customer to Boulton-as-merchant, and it is Darwin at his most domesticated, taking an interest in the saucepans.
(2) Mr Dawkins was Henry Dawkins MP (1728–1814), a wealthy landowner, who bought the estate at Standlynch Park, 5 miles south-east of Salisbury, in 1766. He was MP for Southampton and later Chippenham. See Namier & Brooke, *Commons* ii 305. Darwin later became Dawkins's half brother-in-law, as it were, for Dawkins was married to Juliana, daughter of the Earl of Portmore and half-sister of Darwin's second wife Elizabeth.
(3) Darwin's first wife Mary was suffering paroxysms of increasing severity in the late 1760s, described in letter 92A. Here Darwin records a temporary improvement in her condition.

68D. To Thomas Feilde, 26 August 1768

Dear Sir,

If you admit into your account of Staffordshire the wonders of art as well as those of nature, I know no curiosity in this county so worthy your attention as Mr Boulton's works at Soho.

On the other side I have sent you an account of his situation and manufactory; and am, dear Fielde,

Your affectionate humble servant,
E Darwin

August 26, 1768

Soho is the name of a hill in the county of Stafford, about two miles from Birmingham; which a very few years ago, was a barren heath, on the bleak summit of which stood a naked hut, the habitation of a warrener.

The transformation of this place is a recent monument of the effects of trade on population. A beautiful garden, with wood, lawn, and water, now covers one side of this hill; five spacious squares of building, erected on the other side, supply workshops, or houses, for above six hundred people. The extensive pool at the approach to this building is conveyed to a large water-wheel in one of the courts, and communicates motion to a prodigious number of different tools. And the mechanic inventions for this purpose are superior in multitude, variety, and simplicity, to those of any manufactory (I suppose) in the known world.

Toys, and utensils of various kinds, in gold, silver, steel, copper, tortoisshell, enamels, and many vitreous and metallic compositions, with gilded, plated, and inlaid works, are wrought up to the highest elegance of taste, and perfection of execution, in this place.

Mr Boulton, who has established this great work, has joined taste and philosophy with manufacture and commerce; and, from the various branches of chemistry, and the numerous mechanic arts he employs, and his extensive correspondence to every corner of the world, is furnished with the highest entertainment as well as the most lucrative employment.

Original Not traced.

Printed Stebbing Shaw, *The History and Antiquities of Staffordshire*, Vol. II, p. 117 (Nichols, 1801).

Text From above source.

Notes

(1) Stebbing Shaw remarks that he intends to begin his account of the Soho manufactory 'with a letter from a learned and philosophical admirer of the works of art and science, addressed to the Rev. Mr Feilde, of Brewood, then engaged in a History of Staffordshire'. He then gives the letter printed above.

(2) Rev. Thomas Feilde (1729–1781?) entered St John's College, Cambridge, in 1747, and presumably met Darwin there. On entry, his name was spelt Fielde (like that of his father Thomas Fielde, a former Fellow of St John's), but he graduated BA in 1750 as 'Feilde'. He became MA, and was ordained priest, in 1754. Feilde was instituted Rector of Eastwick from 1764 and of Stanstead Abbotts from 1767, both these parishes being in Hertfordshire: see Scott, *Admissions*, Part III, p. 567; and J. E. Cussans, *Hertfordshire* (1870) i 38. However, Feilde lived at Brewood, near Wolverhampton, where he was Vicar and headmaster of the Grammar School from 1762 to 1769: see J. Hicks Smith, *Brewood* (Parke, Wolverhampton, 1874), pp. 12, 28; and C. Dunkley, *Brewood Grammar School* (Cannock, 1936), p. 34. Also *Alum. Oxon.* records his son Thomas Feilde (1766–1847) as 'of Brewood'. There is a pedigree, showing that Feilde the elder had ten children, in R. Clutterbuck, *History . . . of Hertford* (Nichols, 1815) iii 244. Feilde's History of Staffordshire was proposed for publication in 1769, but R. Gough (*British Topography* (1780) ii 231) remarks that this design 'will never be executed'

because of 'the editor's ill-conduct'. Feilde left Brewood in 1769, apparently because of financial problems, and settled in Virginia, where he died (see J. Hay & W. Parke, *Notes and Collections relating to Brewood*, priv. printed, Wolverhampton, 1858, p. 105). The manuscripts of his History were acquired and used by Stebbing Shaw.

(3) Darwin's picture of Soho may seem over-idyllic, but it is in keeping with the splendid engravings of the manufactory printed by Stebbing Shaw.

69A. To Peter Templeman, 4 February 1769

Lichfield Feb. 4 –69

Sir,

Some Time ago you favour'd me with a Letter about a wheel-carriage the Model of which I did not think worth sending you.

I have lately constructed the model of an horizontal Windmill, which appears to have a third more Power than any vertical Windmill, whose Sail is of the same diameter, and is in other respects more manageable and less liable to Repair, as it has less wheel-work, having only an upright Shaft on the top of which the Sail is fix'd, and will move three pairs of Stones. These Mills will be so powerful as to answer nearly as well as water, for the Flour-Trade, and putting this Trade into more Hands will tend to prevent Monopoly in that Business.

This Model is 3 ft diameter, and I should not chuse the Expense of sending it up, unless there was a Chance, that if I demonstrated the above, viz that this Mill is in many respects much preferable to the common one, that the Society might grant me a Premium to assist me to execute it at large.

I should be glad of your Opinion, if I should have this Chance, and must beg of you, and in this Confidence I write to you, not to mention my Name, as I do not court this kind of Reputation, as I believe it might injure me,

From, Sir with all Respect
Your humble Sert.
E Darwin

The mill consists much of Brick-Work, and could not be executed in its greatest Size for less than 300£ or 400£ with Graneries and all conveniences for the Flour Trade.

[on back page] This Mill for a Family, might be executed at a small Expense –

Addressed For Peter Templeman Esqr., Secretary of the Society of Arts, in the Strand, London.

Original Royal Society of Arts Library, Guard Book No. 68.
Printed Excerpts in *Lunar Society*, p. 74 and *Doctor of Revolution*, p. 80.
Text From original MS.

Notes

(1) Darwin's horizontal windmill, illustrated in Plate VII of *Phytologia* and on page 130 of *Doctor of Revolution*, was a fine invention and did good service for 13 years (from 1779 to 1792) in grinding colours at Wedgwood's factory at Etruria. Unfortunately the Society of Arts was not interested in windmills in 1769, though the Society offered a prize for a horizontal windmill in 1786, when Darwin's mill had been operating for 7 years.

(2) The first sentence of Darwin's letter refers back to his letter to Dr Templeman of 8 March 1766 (66A).

(3) Darwin's request that his name should not be mentioned shows that he feared his medical practice might suffer if he was known to dabble in windmill design.

69B. To Josiah Wedgwood, 15 March 1769

* * *

I should long ago have wrote to you, but waited to learn in what forwardness Mr Watt's Fire-Engine was in. He has taken a Partner, and I can make no conjecture how soon you may be accommodated by Him with a Power so much more convenient than that of Wind. I will make packing Boxes and send you my model, that you may consult the Ingenious. I am of opinion it will be a powerful and a convenient Windmill, but would recommend steam to you if you can wait awhile, as it will on many Accounts be preferable I believe for all Purposes.

* * *

Original Not traced. Seen by Eliza Meteyard in 1865.
Printed Meteyard, *Wedgwood* ii 31.
Text From above source.
Notes

(1) Miss Meteyard comments that this letter 'does the utmost credit to Darwin's generosity, candour, and the previsional character of his intellect'. Darwin was certainly keen on his own invention of the horizontal windmill: yet he saw that it would eventually be superseded by steam, and advised Wedgwood accordingly.

(2) Watt's partner in Scotland was John Roebuck (see note to letter 63B), but it was not until Watt joined forces with Boulton in 1774 that his engine progressed. Darwin's windmill was installed at Etruria in 1779 and displaced by steam in 1792.

69C. To Albert Reimarus, [April 1769]

To Dr Albert Hen. Reimarus.
from Dr Darwin.

I recieved very great Pleasure in hearing by Mr Ferber that my old Friend Dr Reimarus is alive and succeeds in the World. – Your Friend Mr

Keir passed some months with me last Winter, and we often talk'd of you, and should have wrote to you, had we been certain of your Existance in this World. – Mr Keir has been in the Army, but has lately quitted the Camp to study Chemistry and Philosophy more at Leizure. – For my own part I practise Medecine in Lichfield, Staffordshire, where I shall hope to hear from you. – I have a good House, a pleasant Situation, a sensible Wife, and three healthful Children, and as much medical Business, as I can do with Ease, and rather more. – Mechanics, and Chemistry are my Hobby-horses, but a Comparison of the Laws of the Mind with those of the Body, has of late been my favorite Study: on this Head I intend to send you my remarks as soon as I have finish'd them.

Mr Keir and myself continue in the Religion you taught us, we hold you to be a great Reformer of the Church.

In respect to medical Knowlege I shall wait till I hear from you, and then shall be glad to exchange Observations with you on medical Subjects, or on any other parts of Philosophy.

If you should wish to send any of your Children to learn the english Language, there is no part of the Kingdom where it is spoke with more Purity than here: and we have a good latin School in the Town, and they shall be welcom {to} live at my House, and I will send you a Son in exchange some other Time, if such a Trafic be agreable to you. –

from dear Sir your affectionate Friend

Erasmus Darwin

at Lichfield in Staffordshire

Addressed For Dr Albert Hen. Johan. Reimarus, Hamburgh.

Original Historical Society of Pennsylvania, Philadelphia.
Printed Excerpts in *Doctor of Revolution*, p. 78.
Text From original MS.
Notes

(1) Dr Albert Reimarus (see letter 56A) was the son of the German philosopher Hermann Reimarus, who was well known in his day as a thoroughgoing deist, a proponent of 'natural' rather than 'revealed' religion. Albert Reimarus was a close friend of Darwin and Keir when they were undergraduates, and it seems from this letter that he may have been influential in diverting them from Christianity towards deism. At the time Darwin wrote, Reimarus was busy campaigning for the Germans to instal lightning conductors.

(2) 'Mr Ferber' was Johann Jacob Ferber (1743–1790), who was born in Sweden and, having been a pupil of Linnaeus, travelled widely to enlarge his ideas. He was most interested in mineralogy and in April 1769 visited Whitehurst and then Darwin. His early life is described in the introduction of J. J. Ferber, *Travels through Italy in the years 1771 and 1772* (*tr.* R. E. Raspe) (Davis, London, 1776), pp. v–ix. Later he travelled to Russia and Switzerland. See *Biographie Universelle* (Paris, 1843) Vol. 13, p. 529.

(3) This letter is particularly interesting because Darwin relaxes his usual 'vein of reserve' about his personal affairs. The 'three healthful children' were Charles (born 1758), Erasmus (born 1759) and Robert (born 1766). Darwin's first wife Mary was by all accounts both 'sensible' and beautiful: 'healthful', however, she was not, and she died in June 1770 at the age of 30.

(4) On the outside of the letter Ferber wrote a note to Dr Reimarus in Latin, asking him 'to receive this letter from the distinguished Dr Darwin of Lichfield in Staffordshire, of whom I enjoyed hospitality for two days and found the most honest of men'. (The rest of Ferber's note does not appear to be relevant and is badly torn.)

69D. To Matthew Boulton, 9 June 1769

Wolsely Bridge, Jun.9 –69

Dear Boulton

WOMAN, says Aristotle, is a two-leg'd animal without Feathers – He certainly meant very young Women! After having given a definition of the Subject of my Letter, the next Thing necessary is a Drawing, here it is, a b c is the said animal, a the Head, b the Body, cc the Legs.

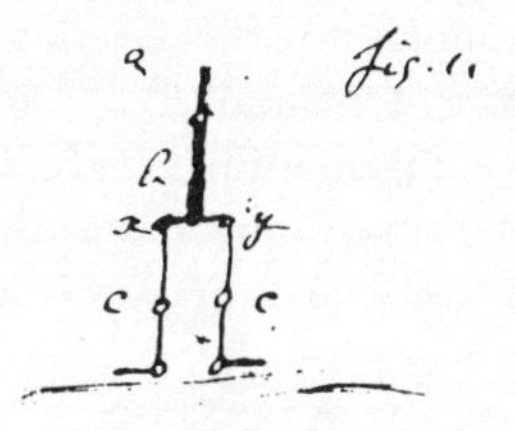

When the animal stands erect, and careless, the Line xy should be horizontal – if it be inclined a few Degrees from the perpendicular, by whatever means this is occasion'd, the whole Figure will gr{adually} undergo the changes express'd in the 2d. f[ig].

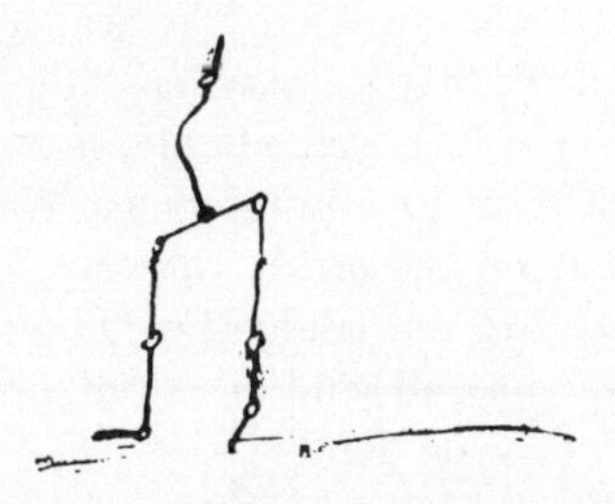

1. The misform'd people, that I have happen'd to inspect, were either distorted Spines, and in these the Disease appear'd to be the Consequence of the unequal growth of some Viscus, as the Lungs, or Heart, or of the Aorta, and then the Malformation of the Spine was a necessary Evil, and

prevented Consumption or Dropsy, or Death in some Shape. These were not lame.

2. Where the gastrocnemii Musculi, which being render'd into the vulgar Tongue, is the Calf of the Leg, were either too strong, of which I have seen two Children, or too Short, the upper part of these muscles having become a hard immoveable Callosity. In this Disease the Patient walks on his Toes, which turn also inwards a little, and the internal Side of the Foot rises, and the Weight of the Body rest[s] on the Balls of the little Toes, and on the external Side or Edge of the Foot. This Figure is not outrès, where the Disease is confirm'd.

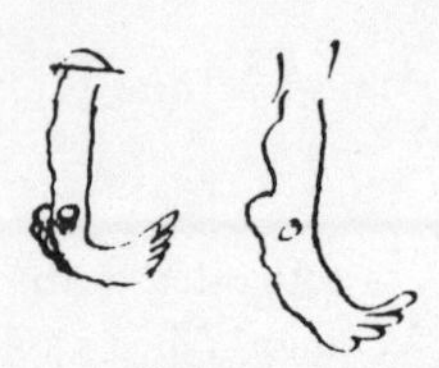

Mr Meynell has this Disease, and the upper part of the Calf of the Leg is quite hard, and feels like an Exostosis.

If this is the Disease there is no other Method but that I described to you, which you will mention to Dr Small.

3. The Thigh Bone is said sometimes by Falls or other Accidents to bend in the Neck, and thus become shorter, a is the natural {bone and} b the diseased one. This Disease I have never examined myself.

Lastly. If you see the Child play sprawling on the Bed, quite naked, and on its feet for some Time, repeatedly, you will see the Defect. If Dr Small and your Honour are not satisfy'd of the Seat of the Disease, pray all of you come over to Lichfield, that we may have a general Consultation, Ladies and Gentlemen. For if it be remediable, it must be NOW. And for this Purpose the Defect must be absolutely ascertain'd – Delay is damnable.

I reciev'd some Time ago of Mr Dawkins for your Use, 3-12-0. – The Candlesticks if they are not sold, Mrs Darwin would rather have back again. – That I beg you will send me a Bill, for I am ashamed of being so long, and so much in your debt.

If you see Mr Baumgartner I shall be glad of a Chest of white florence Wine, if He imports any this Year.

Adieu

Original Birmingham Reference Library, Matthew Boulton Papers, Darwin 14.

Printed Unpublished.
Text From original MS. (*Diagrams* From original.)
Notes
(1) In the late 1760s Dr Small became Boulton's personal physician and closest friend. Small seems to have been puzzled to know how to treat Boulton's one-year-old daughter Anne and referred the case to Darwin.
(2) For Aristotle's views on women, see S. R. L. Clark, *Aristotle's Man* (Clarendon Press, Oxford, 1975), pp. 206–11. To his lasting discredit, Aristotle took the view that 'the female is, as it were, a mutilated male' and regarded the 'female character as being a sort of natural deficiency' (*De Generatione Animalum*, 737[a] and 775[a]). In writing to Darwin, Boulton may perhaps have commented that when Aristotle said the female was deformed, he surely did not mean it literally, as with Anne (who seems to have been crippled by her disease about 20 years later: see H. W. Dickinson, *Matthew Boulton* (CUP, 1937), p. 166).
(3) 'The misform'd people' of Darwin's category 1, the hunchbacks, would most commonly have been victims of Pott's Disease – tuberculosis of the spine – or possibly simple congenital scoliosis. In his category 2, a congenital spastic paralysis could be the cause. Anne Boulton may have suffered a congenital dislocation of the hip, which is very common and leads to deformity and difficulty in walking. The source of such a disorder is not always apparent from external appearances, and Darwin's logical approach to the problem is commendable. The word 'orthopaedics' ('straight children') had only recently been coined, by N. André in *Orthopaedia* (1741). 'Outrès' is the Darwinian spelling of 'outré'.
(4) 'Mr Meynell' may possibly be Hugo Meynell (1735–1808), MP for Lichfield from 1762 to 1768, Master of the Quorn Hunt for 47 years, and 'the real father of the modern English chase', as the Duke of Beaufort called him. See Namier & Brooke, *Commons* iii 134–5.
(5) Darwin's first wife Mary died in June 1770, and this is the last reference to her in these letters until 1792.
(6) Darwin is still ordering Florence wine from Mr Baumgartner (see letter 67C). For Mr Dawkins, see letter 68C.

70A. To John Barker, 7 December 1770

If you think this of Service you will order it to be copied fair over, but it should not be printed at the same Time with that of Mr Barnets, as two Letters will tire the Reader, but as you think best, but don't mention my Name.

Addressed For Mr Barker. (*Endorsed* (in Barker's hand) Darwin 7th Dcemr 1770.)

Original In the possession of the editor.
Printed Unpublished.
Text From original MS.
Notes
(1) The manuscript has the following note: 'Capt. Fernihough vouches for this

being an autograph of the celebrated Dr Darwin. On the back is the signature of the person to whom it was addressed – Mr Barker, Banker, with that Gentleman's memo of its being a note from Darwin, with the date Decr. 7 1770. W.G.B.' Below this is written in another hand 'Mr Bateman's note'. The note is correct, except that Barker's name is written by Darwin, not Barker.

(2) For details of Barker, see letter 63B. For Captain Fernyhough, see letter 64B.

(3) 'Mr Barnet' may possibly be Rev. Richard Barnett, who wrote *Odes* (1761) and *Latin and English Poems* (1809). (See R. Watt, *Bibliotheca Britannica*, Constable, Edinburgh, 1824.)

70B. To Josiah Wedgwood, 9 December 1770

* * *

Mr Watt's Fire-Engine I believe goes on, but I don't know at what rate.

* * *

Original Not traced. In the possession of Charles Darwin, 1865.
Printed Meteyard, *Wedgwood* ii 210.
Text From above source.
Notes

(1) In letter 69B Darwin had recommended that Wedgwood should wait for the development of Watt's steam engine (instead of using Darwin's horizontal windmill) for grinding flints. This excerpt is presumably a reply to a query from Wedgwood about the progress of Watt's engine: the rate of progress was in fact approximately zero, and Darwin might have recommended his own windmill again if he had known the real situation.

(2) Darwin's first wife Mary died in June 1770, and her illness and death deeply depressed him. There are no letters between June 1769 and December 1770, and the first fairly cheerful letter is in February 1771.

71A. To Matthew Boulton, 4 January 1771

Dear Sir

I have just now received a letter from Mr Stuart acquainting me that Mr Anson and himself are obliged to go to London immediately to pass a fortnight, and at their return he will acquaint me. I hope you will receive this letter in time to prevent your journey, and am,

Dear Sir
Jan. 4 –71 Your affect. Friend and Servant
Lichfield E Darwin

I wrote some time ago to bespeak a silver-plated Cross with or without a Lamp; and must now beg you will send two instead of one – better if with

Lamps. – I shall be glad to see you Birmingham-Philosophers-and-Navigators, tho' Mr Anson is gone. You will please to acquaint me if you can and will come on that Day.

Original Birmingham Reference Library, Matthew Boulton Papers, Darwin 16. The first part of the letter, down to 'E Darwin', is written by a scribe. The rest is in Darwin's hand.
Printed Excerpt in *Lunar Society*, p. 85 and *Doctor of Revolution*, p. 87.
Text From original MS.
Notes
(1) Mr Anson was the landowner Thomas Anson of Shugborough Hall (1695–1773), who was MP for Lichfield from 1747 to 1770. See Namier & Brooke, *Commons* ii 23. Anson's younger brother George was Admiral Lord Anson (1697–1762).
(2) Mr Stuart was James Stuart FRS, FSA (1713–1788), known as 'Athenian Stuart', whose careful drawings of Athenian antiquities had a powerful effect on English taste. Stuart designed Lord Anson's house in St James Square, the first in the Grecian style. See *DNB* for Stuart.
(3) On 24 December 1770 Josiah Wedgwood wrote: 'Mr Boulton, Dr Darwin and I are to dine with Mr Anson on New-Year's day'. They were to discuss the construction of a huge tripod to finish the top of the 'Demosthenes lanthorn building' designed for Anson by Stuart. (Wedgwood, *Letters* i 387–8). Darwin's reference to 'Navigators' suggests that he may have wished to discuss new canal projects, including his own idea of a narrow-gauge canal between Lichfield and the Grand Trunk Canal. This letter, though so short, is in effect a microcosm of eighteenth century life, as these notes show.

71B. To Josiah Wedgwood, 15 January 1771

[In this letter Darwin invites Wedgwood, Boulton, Keir]

* * *

and Bentley, if he is in the country,

* * *

[to attend a meeting with Mr Anson at Shugborough.]

Original Not traced. In the possession of Charles Darwin, 1865.
Printed Meteyard, *Wedgwood* ii 223.
Text From above source.
Notes
This fragment is from a follow-up of letter 71A. Wedgwood was unable to accept the invitation because 'he had met with a slight hurt which rendered him unable to move his artificial leg' (Meteyard, *Wedgwood* ii 224). The meeting did take place, for on 24 January Stuart wrote to Wedgwood saying that Boulton and Keir had attended (*Lunar Society*, p. 85). Apparently, nothing came of the meeting, probably because Anson became ill: he died on 30 March 1773.

71C. To Matthew Boulton, 12 February 1771

Dear Boulton

I am quite undone for the Crayons and must beg you will send them to Lichfield as soon as possible. Pray acquaint me if Dr Small be return'd.

I intend to visit you again at Birmingham the first vacant Sunday.

If you don't send the Crayons – – – ! I'll –

and am E Darwin

Feb. 12 –71

Original Birmingham Reference Library, Matthew Boulton Papers, Darwin 17.
Printed Unpublished.
Text From original MS.
Notes

Boulton has endorsed the letter: 'Mr Walker please to enquire if these Crayons are come to London'. The reply was: 'In a letter received from Motteux about a week ago, he says the Vessell on which the Port Crayons are loaded, was not arrived, consequently, it is most probable that she is frostbound[?] in the Elbe and then there is not yet a certainty when they will be in London'. So it looks as though Darwin must have been obliged to – (whatever that may have been). 'Mr Walker' was Zaccheus Walker, Boulton's chief clerk and brother-in-law (see letter 62A). For Motteux, see letter 65B.

71D. To Josiah Wedgwood, October 1771

* * *

Your Windmill sleeps at my house, but shall be sent you, if you wish it, but I should advise you to wait the Wheel-Fire-Engine, which goes on slowly.

* * *

Original Not traced. In the possession of Charles Darwin, 1865.
Printed Meteyard, *Wedgwood* ii 210.
Text From above source.
Notes

This is a continuation of the windmill saga. See Letters 67G, 68A, 69B and 70B. Darwin is still advising Wedgwood against the horizontal windmill, despite its advantages. As a result of this bashfulness of Darwin, the model of the windmill remained 'asleep' for a further eight years before being brought out and improved for use in full-size form at Etruria.

72A. To Benjamin Franklin, 18 July 1772

Lichfield Jul. 18 –72

Dear Sir

I was unfortunate in not being able to go to Birmingham, till a Day after you left it. The apparatus you constructed with the Bladder and Funnel I took into my Pond the next Day, whilst I was bathing, and fill'd the Bladder well with unmix'd Air, that rose from the muddy Bottom, and tying it up, brought it Home, and then pricking the Bladder with a Pin, I apply'd the Flame of a Candle to it at all Distances, but it shew'd no Tendency to catch Fire. I did not try if it was calcareous fixable air.

I shall be glad at your Leizure of any Observations on the Alphabet, and particularly on the number and Formation of Vowels, as these are more intricate than the other Letters. The Welch do not seem to have the W, hence they call Woman, 'oman. I suspect the German W to be a sibilant W, and the same which the Middlesex People or Cockneys use, as in Women and Wine the[y] say 'Vomen and 'Vine, but this Sound I suppose not to be our V, but the german W, or sibilant W.

It is possible the Welch may have consonat Letters form'd in the same part of the Mouth with our H, and the Spanish Ch, but I suspect the Lloyde is no other than the spanish Ch put before the L. The Northumberland People I think make the R a sibilant Letter?

I think there are but four Vowels, their successive Compounds, and their synchronous Compounds. For as they are made by apertures of different parts of the mouth, they may have synchronous, as well as successive Combinations. Aw is made like H or Ch spanish by opening that part of the Fauces a little further. a is made like Sh, or J french, as in ale. e as in Eel is made like s or Z. o as in open, like W. The u as in use, is compounded of eu. i as in while is compounded of aw and e. But what is ah! as in garter? I know not. What the french u?

I have heard of somebody that attempted to make a speaking machine, pray was there any Truth in any such Reports?

I am Sir with all Respect your obliged and obed. Servant

E Darwin

Lichfield Staffordshire

I would return you Dr Priestley's Pamphlet by the Coach but I suppose it is to be purchased at the Booksellers.

My Friend Mr Day who saw you at Lichfield intends himself the pleasure calling of you in London.

Addressed For Dr Benj. Franklin, Craven Street, London.

Original American Philosophical Society, Philadelphia, Franklin Papers III 112.
Printed The Papers of Benjamin Franklin (ed. W. B. Willcox) Yale Univ. Press, New Haven, 1975, Vol. 19, pp. 210–12.

Text From original MS.

Notes

(1) This is one of the untidiest of Darwin's letters, and my text differs slightly from that in *The Papers of Benjamin Franklin*.

(2) Though this is the earliest letter from Darwin to Franklin that has survived, they probably met during Franklin's visit to Birmingham in 1758, when Franklin, then 52, was recognized as the leading man of science of the day, famous for his researches in atmospheric electricity and his invention of the lightning conductor; Darwin, then 26, had already written his paper on atmospheric electricity in the *Philosophical Transactions* of the Royal Society. The two met several times during Franklin's subsequent visits to the Birmingham area: they had much in common, with their wide-ranging intellectual interests and their passion for science.

(3) In 1772 the chemistry of gases (or 'airs', as they were called) was a mystery. But Joseph Priestley (1733–1804) was conducting his epoch-making experiments, some of which Franklin had seen in June (see F. W. Gibbs, *Joseph Priestley*, Nelson, 1965, Chap. 6). Despite the general ignorance, fire-damp (methane) was known, from its presence in mines; and inflammable air (hydrogen) had been prepared by Cavendish in 1766. It was one of these two gases that Darwin hoped to collect from his pond.

(4) In 1771 Darwin had produced his most spectacular invention – a speaking machine that really worked and greatly impressed his Lunar friends. No drawings of it have survived, but Darwin describes it in a long Note to *The Temple of Nature* (Additional Note XV, pp. 107–20). This letter shows that he was still working on the theory, trying to decide how and where each sound is made. ('The Fauces' is the name for the passage between the mouth and the pharynx.)

(5) 'Priestley's Pamphlet' was probably his 'Directions for impregnating water with fixed air'.

(6) The eccentric Thomas Day (1748–1789) had become a friend of Darwin in 1768 and had been living at Stowe House in Lichfield during 1770–71. The best biography of Day is G. W. Gignilliat's *The Author of Sandford and Merton* (Columbia University Press, New York, 1932). See also letter 81C and notes.

(7) Franklin sent a long and interesting reply to Darwin's letter on 1 August 1772, though this was not known until 1977, when the last three pages of the manuscript of Franklin's reply were discovered at the Cambridge University Library. At the beginning of the fragment Franklin is in the middle of commenting on the phonetic questions, and says the vowels are, 'as you observe Compositions, as aui or i, ieu for u; the latter is a barbarous and abominable one: We have however many Words in which each of these Vowels has its natural simple Sound, as inimitable, peruse, etc'. Franklin then describes a speaking clock made by a man in the North of Ireland, in which a little wooden figure dressed like a Watchman popped out at 12 o'clock, pronounced the words *Past twelve a Clock*, and then went in again. Franklin complains that he has suffered from gout, fever and headache for 16 days, for which he blames 'the Amount of Dabbling in and over your Ponds and Ditches and those of Mr Bolton, after Sunset, and Snuffing up too much of their Effluvia. A 1000 Thanks for your Civilities while I had the Pleasure of being with you'. He says he began the letter ten days before and has written 'a little at a time from Day to Day as I was able . . . it is long and too circumstantial, like a sick Man's Letter. . . . I shall be glad to see Mr Day here, for whom I have much

Esteem'. Franklin adds a PS about researches on the Torpedo by Mr Walsh [John Walsh FRS (1725?–1795)]. At the end are three rather illegible lines on phonetics written by Darwin. (I thank Peter Gautrey for making this letter available to me.)

72B. To Josiah Wedgwood, 30 September 1772

Lichfield, Sept. 30 –72

Dear Wedgwood,

I did not return soon enough out of Derbyshire to answer your letter by yesterday's Post. Your second letter gave me great consolation about Mrs Wedgewood, but gave me most sincere grief about Mr Brindley, whom I have always esteemed to be a great Genius, and whose loss is truly a public one. I don't believe he has left his equal. I think the various Navigations should erect him a monument in Westminster Abbey, and hope you will at a proper Time give them this Hint.

Mr Stanier sent me no account of him, except of his death, tho' I so much desir'd it, since if I had understood that he got worse, nothing should have hinder'd me from seeing Him again. If Mr Henshaw took any Journal of his illness or other circumstances after I saw Him, I wish you would ask him for it and enclose it to me. And any Circumstances that you recollect of his Life should be wrote down, and I will some time digest them into an Eulogium. These men should not die, this Nature denys, but their Memories are above her Malice – Enough!

* * *

Original Not traced. In the possession of Charles Darwin, 1879.
Printed C. Darwin, *Life of E. Darwin*, pp. 30–31. Excerpts in Meteyard, *Wedgwood* ii 243.
Text From the two printed sources.
Notes
(1) Wedgwood's wife Sarah had been seriously ill (see next letter), and Darwin, 'her favorite Esculapius', was treating her.
(2) James Brindley (1716–1772), the great canal engineer who had been closely associated with Wedgwood and Darwin in promoting the Grand Trunk Canal, died of diabetes on 27 September 1772. Darwin had diagnosed the disease a few months before, but there was no cure.
(3) Mr Stanier was Brindley's doctor, and he did later send an account of Brindley's illness, which is recorded in three (unnumbered) pages at the end of Darwin's Commonplace Book at Down House.
(4) 'Mr Henshaw' is Hugh Henshall (1733/4–1816), Brindley's brother-in-law, assistant and successor as canal engineer. Henshall managed to complete most of Brindley's unfinished projects, including the Harecastle Tunnel: see A. W. Skempton and E. C. Wright, *Trans. Newcomen Soc.*, **44**, 23–47 (1972); and *Gent. Mag.*, **86**(2), 478 (1816).

(5) Darwin's promise of 'an Eulogium' was generously fulfilled in *The Economy of Vegetation* (III 329–36):

> So with strong arm immortal BRINDLEY leads
> His long canals, and parts the velvet meads;
> Winding in lucid lines, the watery mass
> Mines the firm rock, or loads the deep morass,
> With rising locks a thousand hills alarms,
> Flings o'er a thousand streams its silver arms,
> Feeds the long vale, the nodding woodland laves,
> And Plenty, Arts, and Commerce freight the waves.

And there are another 14 lines about Brindley: of all the scientists and inventors celebrated by Darwin in the poem, only Franklin receives as long a tribute as Brindley. For a biography of Brindley, see C. T. G. Boucher, *James Brindley Engineer* (Goose, Norwich, 1968).

72C. To Josiah Wedgwood, 15 October 1772

* * *

I would advise you to live as high as your constitution will admit of, in respect to both eating and drinking. This advice could be given to very few people! If you were to weigh yourself once a month you would in a few months learn whether this method was of service to you.

* * *

Original Not traced. In the possession of Charles Darwin, 1879.
Printed C. Darwin, *Life of E. Darwin*, p. 57.
Text From above source.
Notes

During 1772 Wedgwood's wife Sarah was dangerously ill several times, and Darwin was often called on to make the thirty-mile journey to Etruria to treat her. These extracts from Wedgwood's letters are typical of many more: 'She is as complete a Cripple as you can easily imagine' (30 March) . . . 'Mrs W is not yet able to dress herself' (12 May) . . . '[She] miscarried this morning' (7 September). . . . '[She] does not seem to have a drop of blood in her body' (10 September). Wedgwood was worn down by the worry, and on 12 October he told Bentley: 'Shewing Dr Darwin by one of my Wastcoats how much I was sunk in 9 or 10 months, he said it was wrong, and I must be very carefull of my health'. This letter is a written reinforcement of the prescription given in conversation. (Incidentally, Sarah Wedgwood outlived her husband by twenty years.) The quotations above are from Wedgwood, *Letters* ii 66, 72, 92, 94 and 105.

74A. To Benjamin Franklin, 24 January 1774

Lichfield Jan. 24 –74

Dear Sir,

I have inclosed a medico-philosophical Paper which I should take it as a Favour if you will communicate to the royal Society, if you think it worthy a Place in their Volum; otherwise must desire you to return it to the Writer.

I have another very curious Paper containing Experiments on the Colours seen in the closed Eye after having gazed some Time on luminous Objects, which is not quite transcribed, but which I will also send you, if you think it is likely to be acceptable to the Society at this Time, but will otherwise let it lie by me another year. I hope you continue to enjoy your Health, and that I shall sometime again have the Pleasure of seeing you in Staffordshire, and am, dear Sir

Your affectionate Friend

Eras. Darwin

NB. If Dr Franklin is not in England, I hope the Person intrusted to read his Letters will return the inclosed Papers to Dr Darwin

at Lichfield

Staffordshire,

which will be gratefully acknowledged.

Addressed For Dr Benj. Franklin, Craven Street, Strand, London.

Original American Philosophical Society, Philadelphia, Pa., Franklin Papers IV 5.

Printed The Papers of Benjamin Franklin, Vol. 21 (Yale Univ. Press, New Haven, 1978), pp. 24–5.

Text From original MS.

Notes

(1) The paper mentioned in the first paragraph was 'Experiments on Animal Fluids in the exhausted Receiver'. Franklin duly transmitted it to the Royal Society on 27 January 1774, and it was published in *Phil. Trans.*, **64**, 344–9 (1774). The manuscript of the paper (Royal Society Letters and Papers 1741–1806, Decade VI No. 39) is not in Darwin's hand. The words 'by Dr Darwin MD' are written across it, thus explaining why the author is given as D. Darwin in the *Phil. Trans.*

(2) The other paper, on 'the Colours seen in the closed Eye after having gazed some Time on luminous Objects', seems to have been an early draft of the paper on Ocular Spectra which appeared in *Phil. Trans.*, **76**, 313–48 (1786) with Darwin's son Robert as author.

(3) See the next letter, which is a postscript to this one.

74B. To [Benjamin Franklin], 3 February 1774

Lichfield Feb. 3 –74

Dear Sir,

I should be much obliged to you, if you could send the following Addition to the Paper I transmitted to you for the royal Society containing "Experiments on animal {fluids in the exhausted Receiver"}

* * *

Original New York Public Library, Erasmus Darwin Misc. MSS.
Printed Unpublished.
Text From original MS.
Notes

This is a postscript to letter 74A, and it is possible that the text given is complete apart from the signature (the last line and signature being cut off the manuscript). The words, 'The writing of the celebrated Dr Darwin', appear on the reverse of the MS.

75A. To William Withering, 25 February 1775

Lichfield, Feb. 25 –75

Dear Doctor,

I am this moment returned from a melancholy scene, the death of a friend who was most dear to me, Dr Small, of Birmingham, whose strength of Reasoning, quickness of Invention, Learning in the Discoveries of other men, and Integrity of Heart (which is worth them all), had no equal. Mr Boulton suffers an inconceivable loss from the Doctor's mechanical as well as medical abilities.

A person at Birmingham desired I would acquaint you with Dr Small's death as soon as I could, but would not permit his name to be mentioned, least he might disoblige some whom he did not wish to disoblige. It was said that Dr Smith, who has been there a few months, had no chance at all of succeeding in that place, from his defect in hearing. Now it occurred to me that if you should choose that situation your philosophical Taste would gain you the Friendship of Mr Boulton, which would operate all that for you which it did for Dr Small. I saw by Dr Small's papers that he had gain'd about £500 a year at an average taking the whole time he had been at Birmingham, and above £600 on the last years. Now as this was chiefly in the town, without the Expense and Fatigue of Travelling and Horse-keeping, and without being troubled with visiting the people, for he lived quite a recluse studious Life, it appears to me a very eligible situation. Add to this that he had increased his Fortune by some other circumstance of manufacture or Schemes which such a town affords. If you should think this Prospect worth your going over to see Mr Boulton at Soho to inquire further into, I will take care to leave at Home a proper letter for you to him if I should not see you.

I was very fortunate in recommending Dr Bates to Aylesbury, and Dr Wright to Newark, but think in my own mind this of the internal Business of Birmingham to be, all put together, the most eligible of any country situation; but I think no one who has not some philosophical acquirements as well as medical is likely to succeed in it.

I shall not mention having wrote this Letter to you, but shall be glad of a line in answer, and please to put private on the internal cover.

Adieu
E Darwin

Addressed For Dr Withering, Private.

Original Not traced.

Printed J. Hill & R. K. Dent, *Memorials of the Old Square* (Taylor, Birmingham, 1897), pp. 111–12. There is a MS copy of this printed version in Birmingham Reference Library, Matthew Boulton Papers. Long extracts are printed in T. W. Peck & K. D. Wilkinson, *William Withering of Birmingham* (Wright, Bristol, 1950), p. 1; *Lunar Society*, p. 121; and *Doctor of Revolution*, pp. 101–2.

Text From J. Hill & R. K. Dent, *Memorials of the Old Square*, pp. 111–12.

Notes

(1) Dr William Withering (1741–1799) had been physician at the Stafford Infirmary since 1766. He had met Darwin in the early 1760s, when Withering was a medical student and in the summer often stayed with his uncle Dr Brooke Hector at Lichfield. Withering was interested in chemistry and botany as well as medicine, and Darwin hoped that he might fill the gap in the Lunar Society created by the death of Dr Small. That hope was not altogether fulfilled, for Dr Small's power to weld the group together was unique and irreplaceable. Withering contributed intellectually, but was deficient in camaraderie.

(2) By all accounts Dr Small was Darwin's 'favourite friend', and Boulton's too. Darwin's tribute to Small was well-merited and sincere. Darwin repeated it two years later (letter 77B) and also wrote an elegy for Small (quoted in Muirhead, *James Watt* i clvi–clviii).

(3) The 'person in Birmingham' may have been either Boulton or Dr Small's partner Dr John Ash (1723–1798), later physician of the General Hospital, Birmingham (see *Lunar Society*, p. 88, and *DNB*).

(4) 'Dr Smith' is probably Thomas Smith, MD Edinburgh 1767, whose doctoral thesis was on the action of the muscles. He is recorded as a Birmingham physician in the *Medical Register* for 1779.

(5) 'Dr Bates' is recorded as a physician of Missenden, near Aylesbury, in the *Medical Register* for 1779.

(6) Dr Wright of Newark was a subscriber to Whitehurst's *Inquiry* in 1778. He is recorded in the *Medical Register* for 1780 as Thomas Wright, MD Edinburgh 1769, whose doctoral thesis was entitled *De Variolis*.

75B. To William Withering, [late February 1775]

Dear Sir,

I have this moment recieved the inclosed, which as I have not yet heard from you, I have determined to transmit to you.

I intend to go to see Mr Boulton tomorrow morning at Soho, and to return at night. The inclosed Letter is encouraging, but I beg leave not to advise you in this affair in the least respect, all I mean is to send you the Facts, and Circumstances. Yourself alone must judge in so important an affair. Let me add that I must beg you will not shew to any Body that part of Mr Boulton's Letter in which my name is mention'd, and that you will please to return me his Letters.

If you are well acquainted with any philosophical Physician, who is at the same Time a Man of good Heart, I should be glad if you would acquaint me for Mr Boulton's sake. I know no other but yourself.

Mr Charles Darwin begs you would acquaint with the Title of your Edition of Linnei Genera, which has the generic Distinctions opposit to the Catalogue of each Class.

I hope Mrs Withering is well, and am, dear Sir your affect. Friend

E Darwin

Addressed For Dr Withering, Stafford. To be deliver'd to Night.

Original Darwin Museum, Down House, Downe, Kent.
Printed Unpublished.
Text From original MS.
Notes

(1) This letter seems to have been written within two or three days of the previous one, but it might belong to early March. See next letter.

(2) Presumably Boulton told Darwin that he would like to see a physician of philosophical inclination and good heart as Dr Small's successor. The letter shows that Darwin was on very good terms with Withering at this time, and was being extremely kind to him. 'Your affect. friend' was a form of words usually reserved for Boulton, Wedgwood, Watt, Keir, Edgeworth and Franklin.

(3) 'Mr Charles Darwin' was Darwin's son, now 16, and studying at Christ Church, Oxford. Disappointed by the torpid classical atmosphere there, he transferred to the Edinburgh Medical School in the autumn.

(4) Mrs Withering was expecting a baby: her daughter Helena was born on 5 March.

75C. To William Withering, [March 1775]

Dear Doctor

I am sorry for your disappointment, which indeed could not be foreseen. I have wrote to Mr Boulton this Post to know if He is concern'd in the

Recommendation of Dr Roebuck's Son, or Dr Pemberton, and will acquaint you as soon as I have his answer. I was told when Dr Small first settled at Birmingham that Dr Ash wished to recommend Dr Cambel, and Dr Groom, but Dr Small by having a philosophical Tast, which suits and is useful to the People of Birmingham, and this with Mr Boulton's Friendship proved a much superiour Interest – and would prove so again I dare say, as they now lament Dr Small's Loss both as a Physician and Philosopher.

May I wish you Joy of an Increase of your Family, from with Compliments to Mrs Withering

Your affect. Friend
and Serv.
E Darwin

Addressed For Dr Withering, Stafford.

Original Darwin Museum, Down House, Downe, Kent.
Printed T. W. Peck and K. D. Wilkinson, *William Withering*, p. 3.
Text From original MS.
Notes
(1) This letter was probably written within a week or two of the previous one. Withering's daughter Helena was born on 5 March 1775, so a date between 6 and 12 March is probable.
(2) Other candidates for Dr Small's practice were apparently being considered. Withering was successful, however, probably because of Darwin's support. Withering moved to Birmingham in May.
(3) For Dr Roebuck (senior), see letter 63B. Dr Pemberton may be Stephen Pemberton (1743–1831), B. Med. Oxford, 1770. For Dr Ash, see letter 75A. For Dr 'Cambel', see letter 75E. Dr Groom may possibly be William Groome (born 1735), who came from Trysull, Staffordshire, entered Balliol College, Oxford, and took his B.Med. in 1764. See *Alum. Oxon.* for Pemberton and Groome.

75D. To James Watt, 29 March 1775

Lichfield Mar. 29 –75

Dear Watt,

I have had no Time for Poetry or Prose, since I was favour'd with your elegant Plan for an Elegy, nor indeed Inclination, as I well know how far short any thing of this kind must fall, when it is written from the Heart, compared with those written from the Head.

I was much grieved to hear from Dr Withering that Mr Boulton has been much indisposed since He was in London, with a Return of the Giddyness of his Head, but hope He is perfectly recover'd.

Your Bill in the House I hope meets with no further Obstruction, pray favour me with a Line to know when yourself and Mr Boulton intend, or expect, to come down to Soho.

Lord, how frighten'd I was, when I heard a Russian Bear had laid hold of you with his great Paw, and was dragging you to Russia. – Pray don't go, if you can help it: Russia is like the Den of Cacus, you see the Footsteps of many Beasts going thither but of few returning. I hope your Fire-Machines will keep you here.

If a Vote or two could be for Service to you, I could write to Mr Coke, Mr Gisburn, and Sir Charles Sedley, and could wait on one or two more if I came up to Town, as Lord George Cavendish, Mr Curzons two of them, Mr Anson, Sir John Wrottesley. If I can serve you by letters with the first three, you must send me their Places of Residence in Town, and if you think the others material to you, I will come up to London for a few Days.

Now how must I direct to you? – Your Letter is only dated London, like other great Men, no Street or House is mention'd. So I shall direct to Mr Boulton.

I want the following Books which I shall be obliged to you to procure for me if you can.

1. I have 27 volums of Bufon's Histoire naturelle in 12mo and wish to have what may have been published since, if they can be had in duodecimo like the 27 Volums which I have.

2. I have all the Commentarii de rebus in {Sc}ientia et Medicina Lipsiae, from the Beginning {to} to 17th Volum inclusive, and also the Supplementum primum decadis secunda, and also the primae decadis Index triplex. Excepting the sixth and seventh volums of the said Leipsic Commentaries. Hence please to observe, that I want the sixth and Seventh Volums of the Leipsic Commentaries, and all that have been publish'd since the 17th Volum printed in 1771 and 1772. And since the Secundae Decadis Supplementum primum printed in 1772.

3. I have the first Tom of Linnei Systema Naturae and want the remainder of that Work.

Pray remember me to Mr and Mrs Boulton. – Charles Darwin is now at Christ-Church Oxford, and would be glad to see you, if you return that Way.

Adieu E Darwin

Addressed For Mr J. Watt. To the care of M. Boulton Esqr. at Mr Matthews, Bush Lane, Canon-Street, London. (*Postmark* Lichfield 31MR.)

Original Privately owned.
Printed Fourth paragraph published by Muirhead, *James Watt* ii 84, and frequently reprinted.
Text From original MS.
Notes
(1) After six years of trying, Small, Boulton and Darwin at last succeeded in persuading Watt to move from Scotland to Birmingham, and he arrived in May 1774. He was released from his agreement with John Roebuck in 1773 after the latter's bankruptcy, and soon afterwards, in September 1773, Watt's wife

Margaret died: 'I am heartsick of this country', he wrote, and took the decision to join Boulton.

(2) The Elegy was of course in memory of William Small. Darwin did eventually write it but, as he foresaw, it fell short of his wishes. It is given in Muirhead, *James Watt* i clvi–clviii.

(3) Boulton was greatly upset by Small's death, and this may have been the origin of his 'giddyness'.

(4) Early in 1775, Watt received a letter from his friend John Robison, who was in Russia, offering him a post there at a salary of £1000 a year. Watt was tempted, and Darwin was alarmed that the new partnership of Boulton and Watt, which promised so much, would collapse before it had really begun. Hence his urgent appeal. In the event, of course, Watt stayed and the Boulton-and-Watt engines were decisive in helping the Industrial Revolution to go full steam ahead.

(5) The reference to 'a Russian bear' may derive from Pope's lines:

> Which made old Ben and surly Dennis swear,
> 'No Lord's anointed, but a Russian bear'.

(*Imitations of Horace,* Book 2, epistle 1, lines 388–9.)

(6) In classical mythology the robber Cacus, son of Vulcan and Medusa, was a three-headed monster who lived in a cave in Italy. He dragged in his victims and slaughtered them. He is supposed to have been killed by Hercules, from whom Cacus stole some cattle during his tenth labour. See *Economy of Vegetation* I 297–334.

(7) Boulton had promoted a Private Bill in Parliament to prolong the life of Watt's steam engine patent for a further 25 years. There was vigorous opposition in Parliament to this unusual procedure, but Boulton succeeded, and the Act received the Royal Assent on 22 May 1775. I do not know whether Darwin's offer was taken up, but, as in the promotion of the Grand Trunk Canal, he was keen to help by canvassing MPs who were or had been patients or friends.

(8) Mr Coke is probably Wenman Coke (1717?–1776) of Longford, Derbyshire, who was MP for Norfolk from 1774–1776, being succeeded by his son T. W. Coke. But Darwin may possibly be referring to Daniel Parker Coke (1745–1825), who contested Derby in 1775 and, although defeated, was reinstated in 1776 after an objection against his opponent.

(9) 'Mr Gisburn' is John Gisborne (1716–1779) of Yoxall, Staffs, elected MP for Derby in 1775 but unseated on appeal in favour of D. P. Coke. His son John later married Darwin's step-daughter Millicent Pole (see letter 00G).

(10) Sir Charles Sedley Bt. (1721?–1778) of Nuthall, near Nottingham, was MP for Nottingham 1774–8.

(11) Lord George Cavendish (1728–1794), uncle of the 5th Duke of Devonshire, had been a schoolfellow of Darwin, and they were blown up together when playing with gunpowder. Cavendish was MP for Derbyshire 1754–1780.

(12) 'Mr Curzons two of them' refers to Assheton Curzon (1730–1820), MP for Clitheroe 1754–80, and Hon. Nathaniel Curzon (1751–1837), MP for Derbyshire 1775–84, later Lord Scarsdale.

(13) 'Mr Anson' refers to George Adams (1731–1789) of Shugborough Hall, MP for Lichfield 1770–1789, nephew of Thomas Anson (see letter 71A). Adams took the name Anson in 1773 on succeeding to the estates of his uncle.

(14) Sir John Wrottesley Bt. (1744–1787) of Wrottesley Hall, Staffs, a nephew of Lord Gower, was MP for Staffordshire 1768–87. For biographies of all these MPs, see Namier and Brooke, *Commons.*

(15) The influential *Histoire Naturelle* of the Comte de Buffon (1707–1788) was published in 44 volumes between 1749 and 1804, and covered the whole of natural knowledge.

(16) Another massive work of reference was the 'Leipzig Commentaries': *Commentarii de rebus in scientia naturali et medicina gestis* (Lipsiae, 1752–1808) 37 volumes, each in 4 parts; plus the *Primae (-tertiae) decadis supplementum*, 3 vols. (1763–85), and *Primae (-tertiae) decadis index triplex*, 3 vols. (1770–93).

(17) The tenth edition of the *Systema Naturae* (1758) of Carl Linnaeus (1707–1778) is the basis for all plant names. Darwin omits the 'e' in both 'Tome' and 'Volume'.

(18) For Charles Darwin, see letter 75B.

75E. To William Withering, 13 May 1775

Dear Doctor,

The Title of your Book should be easily remember'd, and easily distinguish'd from Lee etc. as "The scientific Herbal", "Linnean Herbal", "English Botany"[,] "Botanologia anglica in which the Science of Botany is reduced to English etc.["] But we'll settle all this at Mr Boulton's with the assistance {of} Mr Keir and Mr Watt.

{"}Pedunculis axillaxilus" "Flower-Stems in the angles" or in the viz The Pythagorean Letter y , surely in the angles will be as intelligible as armpit, or Groin, or any other delicate Idea.

Poor Cambel! – I fear the Infirmary will become a Workhouse, and think that it is Hughes Interest as well as Cambel's to unite – unite or perish

Many will withdraw their Subscriptions, who only want an Excuse for so doing. Unless these two great Men agree the Infirmary will be lost – about which perhaps neither of them care much. The last Board should have elected Cambel for the present and have directed Hughes to have taken Care of the patients in Cambel's Absence, which would be always, and have increased his Sallery – at present the Character of the Infirmary must suffer for want of a Physician, and there is an Excuse for withdrawing Subscriptions. The only proper Thing they did was putting your name and Elogium in the Newspaper.

I hope Mrs Withering daily recovers to whom I beg my best Compliments when you see her, and am, dear Dr

your affect. Friend

E Darwin

May 13 –75

P.S. I will send a light Cart for the Books on Tuesday, annuente D.O.M.

Addressed For Dr Withering, Birmingham. (*Postmark* Litchfield.)

Original Darwin Museum, Down House, Downe, Kent.
Printed Excerpt (first paragraph) in *Lunar Society*, p. 125.
Text From original MS. (*Diagram* Re-drawn.)
Notes

(1) Withering was now established as Dr Small's successor, and was about to attend a 'Lunar' meeting, 'at Mr Boulton's with . . . Mr Keir and Mr Watt', probably for the first time, perhaps on Sunday 20 May. He was planning to publish his book on English plants, and apparently asked Darwin's advice about the title. Darwin's hastily-written reply was rather too bluff and hearty for Withering to swallow: 'We'll settle all this', though kindly meant, must have seemed domineering to Withering. He rejected Darwin's advice and adopted a title which runs to 24 lines, beginning *A Botanical Arrangement of all the Vegetables naturally growing in Great Britain* . . . (2 vols., 1776). Darwin's suggestion, *English Botany*, was better, but Withering felt impelled to reject it, and this rejection seems to have marked the beginning of the long feud between them (see letter 88E).

(2) 'Lee' was James Lee (1715–1795), who translated part of Linnaeus's works into English in his *Introduction to the Science of Botany* (1760). See *DNB* for details on Lee.

(3) Withering was much concerned that his book should not include any improper words. He used 'chive' for stamen and 'pointal' for pistil, so that ladies might read the book without being offended. Darwin was rather scathing about this mealy-mouthedness, because the Linnean system's sexual basis cannot be concealed. *Pedunculus* means flower-stem, and *axillus* armpit, so the translation he quotes is reasonable: 'any other delicate Idea' refers back to 'in the angles', of course.

(4) Dr 'Cambel' was Dr Archibald Campbell (1738/9–1805), who graduated MD at Edinburgh in 1765 and is recorded as one of the doctors in Stafford in *Bailey's British Directory* (1784). According to the *Medical Register* for 1779, Campbell was Withering's successor as Physician to the Stafford Infirmary; Hughes was presumably his assistant. Darwin obviously had a poor opinion of both, which his drawing of a trisected worm expresses.

(5) 'Annuente D.O.M.' means 'God willing' – literally, 'annuente Deo optimo maximo', or 'the most good and most great God nodding'.

75F. To Joseph Cradock, 21 November 1775

Lichfield, Nov. 21 –75

Dear Sir,

I am favoured with your book and Epilogue, both which I have read with pleasure, and think the world, as well as myself, obliged to you for them. In your "Village Memoirs" are a great many lessons of good morality, and let me add, of good Christianity, agreeably and yet forcibly impressed. Your satire is strong and yet delicate: in some places you hew indeed at the root of vice with an axe, but in others you dissect the branches with the delicate knife of an anatomist.

What shall I send you in return for these? I who have for twenty years neglected the Muses, and cultivated medicine alone with all my industry! Medical Dissertations I have several finished for the press, but dare not publish them, well knowing the reception a living writer in medicine is sure to meet with from those who wish to raise their own reputation on the ruin of their antagonists. Faults may be found or invented; or at least ridicule may cast blots on a book were it written with a pen from the wings of the angel Gabriel. I lately interceded with a Derbyshire lady to desist from lopping a grove of trees, which has occasioned me, since you saw me (I suppose from inspiration, or rather infection I might catch from you), to try again the long neglected art of verse-making, which I shall inclose to amuse you, promising, at the same time, never to write another verse as long as I live, but to apply my time to finishing a work on some branches of medicine, which I intend for a posthumous publication. But do you go on to study the fine arts of composition and of speaking: so shall you amuse yourself and improve others. Let me add, that I beg you will present my most respectful compliments to Mrs Cradock, whom I look upon to be your inspiring muse; and believe me, dear Sir, your affectionate friend and servant,

E Darwin

Original Not traced.

Printed J. Cradock, *Literary and Miscellaneous Memoirs*. 4 vols. Nichols, London, 1828, Vol. 4, pp. 143–4. Excerpts in C. Darwin, *Life of E. Darwin*, p. 89, and *Doctor of Revolution*, p. 104.

Text From Cradock's *Memoirs*.

Notes

(1) Joseph Cradock (1742–1826) was a country gentleman with a taste for pleasure, anecdotes and literature. Cradock had met Darwin in the early 1760s when Darwin treated him after he caught a fever at the Derby races. Cradock had now sent Darwin a copy of his rather sententious book *Village Memoirs*, in which he encouraged the poor to be virtuous while himself pursuing a life of pleasure. Darwin's polite reply seems to be unsarcastic, so perhaps he accepted Cradock at his face value. See *DNB* for Cradock.

(2) Darwin's reluctance to publish is strongly exposed in this letter, but his resolution 'never to write another verse as long as I live' was the vain hope of a confirmed addict. It was soon after this that he started *The Botanic Garden*, and in this letter he is trying to persuade himself not to. For the verses he mentions, see next letter.

75G. To Elizabeth Pole, [1775?]

[Addressed to a Lady (Mrs Pole) who had given instructions to cut down some branches of trees in Dr Darwin's garden at Lichfield.]

Speech of a Wood-nymph

Hear, bright Eliza! ere thy dread commands
Lop my green arms, my leafy tresses tear,
Relent, sweet Belle, nor with unpitying hands
Wound the sad Nymphs, who dwell and tremble here.

Know, in this Grove there sleeps in every tree
A Nymph, embalmed by some poetic spell,
Who once had beauty, wit and life like thee.
Oh, spare the mansions, where thy sisters dwell!

So shall thy praise in all our echoes flow,
Thy name be carved on every smooth green rind;
For thee, sweet Belle, our earliest fruits shall glow,
For thee our sweetest blossoms scent the wind.

The love-struck swain, when summer's heat invades,
Or winter's blasts perplex the billowy sky,
With folded arms shall walk beneath our shades,
And think on bright Eliza – think, and sigh!

Original Not traced. There is a typescript copy at University College London, Pearson Papers, 577, page 12.
Printed Unpublished.
Text From typescript copy.
Notes
(1) Mrs Elizabeth Pole (1747–1832), who was to become Darwin's second wife in 1781, was the illegitimate daughter of Charles Colyear, second Earl of Portmore (1700–1785) and (probably) Elizabeth Collier (1713–?), governess to the Duchess of Leeds. Elizabeth Pole was brought up at Farnham, Surrey, by Mrs Mainwaring (who remained a friend in later life, see letter 02D). It was in 1769 that Elizabeth married Colonel Edward Sacheverel Pole (1718–1780), of Radburn Hall near Derby. Colonel Pole, a valorous soldier, had fought in eleven battles in Germany and had been left for dead in three of them, including the Battle of Minden, when a bullet entered his left eye and came out at the back of his head. The prescription book at the Darwin Museum, Down House, shows that Colonel Pole was among Darwin's patients in 1771, and presumably his wife and children also became patients of Darwin a year or two before or after 1771 (her first child was born in 1769).
(2) It seems very probable that this is the poem mentioned in the previous letter, written to persuade 'a Derbyshire lady' to 'desist from lopping a grove of trees'.

But it is much more than that: decently disguised as a wood nymph, Darwin succeeds in declaring his admiration for Elizabeth in the first three stanzas; the fourth stanza is more subtle, for only those who knew the trees were in Darwin's garden could confidently identify him as 'the love-struck swain'.

(3) The explanatory note printed here at the beginning of the poem was apparently written by Sir Francis S. Darwin (1786–1859) towards the end of his life, and was presumably based on his mother's recollections. He gives the whole volume of poems the title 'A Poetical Courtship', and a dating as early as 1775 for this poem is consistent with the fact that it comes first in the list of contents of the volume.

(4) Darwin's first wife Mary had died in June 1770, and, by 1771, Mary Parker (1753–1820) had become Darwin's sexual partner. She bore him two daughters, Susan (1772–1856) and Mary (1774–1859); they were brought up in his house and later established a school at Ashbourne (see *Doctor of Revolution*, pp. 234–7). The existence of this poem, and the absence of further children by Mary Parker, suggests that by 1775 Darwin was succumbing to the attractions of Elizabeth Pole. And apparently she was bold enough to order the lopping of trees in his garden.

77A. To Matthew Boulton, [early 1777]

Dear Boulton

I am sorry to acquaint you that I can not wait on you next Sunday, and shall therefore wait, till I recieve another summons from you. I shall write the same to Mr Keir. I hope you will not forget to remember to remind Mr Henderson of his promise of getting me some square and round cylinders in brass or steel. Adieu.

I hope Behemoth has strength in his loines. Belial, and Ashtororth are two other Devils of consequence, and good names for Engines of fire.

E D

Original Birmingham Reference Library, Matthew Boulton Papers, Darwin 21.
Printed Unpublished.
Text From original MS.
Notes

(1) Darwin is apologizing for being unable to attend a Lunar meeting at Boulton's house.

(2) Behemoth was the name given to an engine erected at Hawkesbury Colliery, Bedworth, which began working on 10 March 1777. Though Belial is a well-known devil, Darwin's second choice seems less appropriate, since Ashtaroth (also known as Astarte, Ashtoreth or Ishtar) is the eastern equivalent of Aphrodite.

(3) The letter is undated, but probably not more than a month or two earlier than that of 21 March. There was a Lunar meeting on 3 March, which Boulton (letter to Keir, 1 March) was expecting Darwin to attend. So this may be a letter of apology for the February meeting.

(4) 'Mr Henderson' is probably Lieut. Logan Henderson, who acted as Boulton and Watt's representative in Cornwall. See E. Robinson, *Trans. Newcomen Soc.* **35**, 153–71 (1963).

77B. To The Royal Society, 10 March 1777

Lichfield Mar. 10 1777.

Gentlemen,

The subsequent case of Squinting, as a similar one has not been recorded or explain'd by others, may perhaps merit your attention from its novelty.

[The next 13 pages of the letter are, with minor variations, as printed in *Philosophical Transactions*, Vol. 68, pages 86–94 (1778). Darwin investigates the nature of the blind spot in the retina and at the end mentions one example of a man who had no such blind spot.] One instance I am certain I have seen, as it was in a man capable of the most patient and most accurate observation, and who on numberless repeated trials at different times in my presence could never loose sight of the smallest object with either of his eyes. [Here the printed text ends.] I mean my ever to be lamented friend Doctor Small of Birmingham, whose strength of reasonning, quickness of Invention, patience of investigation, and, what is above them all, whose benevolence of Heart, have scarcely left an equal, not a superiour, in these sublunary realms!

I am, Gentlemen, with the greatest respect,
your most obed. servant
Erasmus Darwin

To the President and Fellows of the royal society.

Supplement to the case of squinting
[This consists of three pages, printed in *Phil. Trans.* **68**, 94–6 (1778), with the following (unprinted) ending:] from
Gentlemen, with the greatest respect,
your most obed. Serv.
Apr. 20 1777 E Darwin

Original Royal Society archives, Letters and papers, Decade VII, No. 2.
Printed Philosophical Transactions, Vol. 68, pages 86–96 (1778), with the exception of the few sentences recorded above.
Text From original MS.
Notes
(1) The Royal Society's editor changed Darwin's title, 'A Case of Squinting', to 'A Case in Squinting', an amendment which scarcely seems worthwhile. Darwin's paper was an excellent one: he reported the case of a five-year-old boy, Master Sandford, who looked at every object with only one eye (the one furthest from the object) by turning his head so that the image fell on the blind spot in the retina of

the nearer eye. Darwin believed the habit arose through his wearing, as a baby, a nightcap which had appendages at each side visible only to the opposite eye. Darwin made experiments which showed that Master Sandford had blind spots four times larger than normal. To decide what was normal, he tried the experiments on his friends, including Dr Small, whose blind spots were apparently so small as to be imperceptible.

(2) It was more than two years since Small's death and Darwin's tribute to him is heartfelt, though not scientific enough to earn a place in the *Phil. Trans.*

77C. To Matthew Boulton, [21 March 1777]

* * *

pray remind Mr Henderson of my order of hollow round and square wire.

E Darwin

Original Birmingham Reference Library, Matthew Boulton Papers, Darwin 22.
Printed Unpublished.
Text From original MS.
Notes

This fragment is undated, and the date given is that in the Birmingham Reference Library's Catalogue. An earlier reminder is in letter 77A.

77D. To Matthew Boulton, 29 March 1777

I want in all haste an elegant Tea-Tray and Waiter of papier machée of about 2 or 3 guineas expence for a young Clergyman in the country. He is in a violent hurry for them, that if you do not deal in such things will you be so good as to promise them for me, and to direct them to be directed and sent

To the revd. Mr Hall at Westbro' to be left at Mr Brooksbys of Newark Nottinghamshire.

E D

Mar. 29 –77

Original Birmingham Reference Library, Matthew Boulton Papers, Darwin 23.
Printed Unpublished.
Text From original MS.
Notes

(1) Boulton told Boswell in 1776 that 'I sell here, Sir, what all the world desires to have – POWER'. This letter shows that he was also still dealing in smaller items.

(2) Darwin's sister Elizabeth (1725–1800) married Rev. Thomas Hall (1717–1775), Rector of Westborough, which is about six miles east of Elston. Their eldest son, who is 'the revd. Mr Hall' of Darwin's letter, was Thomas Hall (1753–1801), who followed his father as Rector of Westborough. The 'elegant tea-tray and waiter' may have been a birthday present for his future wife Catharine Litchford (1754–1841), whom he married in 1778.

(3) 'Mr Brooksby' is probably Samuel Brooksby, JP (1726/7–1795), who became an alderman of Newark in 1772 and mayor in 1785, and was probably connected with 'Brooksby and Sons, mercers and drapers' of Newark, recorded in *Bailey's Directory* of 1784. He knew the Darwins of Elston and was witness of a lease in 1756 with Robert Darwin. See Notts. Record Office MS M.1477; C. Brown, *History of Newark* (1907) ii 265; Nottingham University Library, East Stoke Collection, Sm 250; *Gent. Mag.* **65**, 792 (1795).

77E. To Matthew Boulton, [1777 autumn?]

Dear Boulton,

I trouble you with this line to remind you of procuring me Mrs Kaufin's 4 or 5 prints of graces, and at the same time I must beg you to get them framed for me, along with the Physician and Lover, which I left roll'd up over your book-case, and I should wish to have your own picture framed at the same time, if you think I shall never be likely to have a better picture of your honour's worship.

Now if I could invent any thing witty to have heel-piece'd this letter with, I should have been happy, as it is pray read it, and don't forget to procure me the prints – the Lord take care of you, adieu

E Darwin

Original Birmingham Reference Library, Matthew Boulton Papers, Darwin 24.
Printed Unpublished.
Text From original MS.
Notes
(1) 'Mrs Kaufin' is Angelica Kauffmann (1741–1807) the well-known painter and original member of the Royal Academy. See *DNB*.
(2) There is no indication of the date of this letter, but it is presumably some months before that of 4 February 1778, when Darwin expresses the hope that the Kauffmann prints 'are long since arrived'.

77F. To Matthew Dixie, [14 October 1777?]

Dear Sir

As Mr Webster continues to be troubled with Dizzyness in his Head, it would be of Service to Him to have Issues made on his back; or if He should think this too great a Trouble, He may try a small Blister, kept open for a Fortnight: and should take twice a Day the following Draught and Bolus. I beg my Compliments to Him, and Mrs Webster, and am, dear Mat,

your affect. Friend
and Serv.
E Darwin

Addressed For Mr Dixie Surgn., in Melbourn, near Derby (*Postmark* chfield.)

Original Yale University, Beinecke Rare Book and Manuscript Library, Im. J637 +W791gh.
Printed Unpublished.
Text From original MS.
Notes
(1) This letter is undated, but there is with it a prescription for 'Mrs Ratcliff of Inglebye', dated 14 October 1777. Since Ingleby is only 2 miles from Melbourne, this prescription seems almost certain to have been sent to Dixie, though it is not the prescription referred to in the letter. It is possible that the prescriptions were sent at the same time, and in the absence of any other indications I have used this date.
(2) Matthew Dixie was 'an eminent surgeon and apothecary of Melbourne', who died 'of a paralytic stroke' on 31 July 1778 (*Harrison's Derby and Nottingham Journal*, 6 August 1778). I have no information on Mr Webster.

78A. To Matthew Boulton, 11 January 1778

Dear Boulton

I have polish'd, japan'd, and handed up, the verses of Miss Rogers, which you sent me. This I think the greatest compliment I could pay to the lady: since if I had not thought the contour fine, and several of the figures graceful and vivid, I would not have taken the labour to have touch'd them over again, have preserved the keeping, and put on the varnish.

Miss Seward sais they are the most beautiful lines she ever saw, and longs to be acquainted with the lady-author; and I beg leave to say, that if the fair Poetess should not be displeased with the liberty I have taken with her poem, and at any time wishes my criticism on any of her future productions, I shall be happy with her correspondence.

A Critique on some parts of her poem seems necessary to add to this letter least yourself or Miss Rogers may think some [of] the alterations were not necessary.

1. As <u>Fancy</u> is the principle personage it was necessary to keep the Queen of Melancoly rather in the shade, to give light to the figure of Fancy, and thus preserve what the painters call Keeping. Hence "come pensive maid" tho' more lively than "and as with silent hand" was omitted.

It was necessary to make <u>Melancholy</u> a <u>goddess</u> tho' unseen, in the first stanza, as she was to be so in the second, hence <u>Luxury</u> of Melancholy was changed.

The 4th and 5th stanzas were thought burthensome to the conduct of the poem and the 5th not appropriated to the subject tho' very good lines.

"Sighs resistless" is a better line taken separately than "ceasless sorrow" but <u>sorrow</u> is contrasted to <u>love</u>: whereas <u>sighs</u> should have been contrasted to laughter as the two former are passions, the two latter bodily actions. This is felt by all readers of the stanza, tho' not easily explain'd.

I am tired of writing.

The Lord keep you, adieu!
E Darwin

Lichfield
Jan 11 –78

The inclosed letter is from my Sister, you will please to direct the sundries but don't lose the letter as I have no coppy.

Original Birmingham Reference Library, Matthew Boulton Papers, Darwin 25.
Printed Unpublished.
Text From original MS.
Notes

(1) This letter shows Darwin in his literary role. After his poem on the death of Prince Frederick in 1751, he abandoned literature for twenty years. The revival began in 1775 with an elegy for Small and letter 75G, and continued in 1776 with the 'Swilcar Oak', followed by the beginnings of *The Loves of the Plants* in 1777–1778. The letter shows he is now well versed in matters of style and adept in the nuances of personification.

(2) Miss Ann Rogers (*c.*1755–*c.*1840) was a young lady living at Dronfield, 5 miles north of Chesterfield. In July 1784 she married Dr Jonathan Stokes (1755–1831): see *Derbyshire Parish Registers, Marriages* Vol. XI (Phillimore, 1913) p. 73. Dr Stokes was a knowledgeable botanist and also the youngest member of the Lunar group: see *Journal of Botany*, **52**, 299–306 (1914) and *Lunar Society*, pp. 223–6, 311–17. Dr and Mrs Stokes moved to Shrewsbury in 1786 (like Robert Darwin), and then to Kidderminster in 1787. Anna Seward's wish to meet Miss Rogers was fulfilled: they became friends, and eighteen of Anna's lengthy letters to her (mostly as Mrs Stokes) appear in the *Letters* of Anna Seward. The most informative, in 1781, is in Volume 2, p. 61. Despite Anna's encouragement, Miss Rogers apparently did not continue seriously with writing poetry.

(3) 'My Sister' is Susannah, his housekeeper (see letter 49A).

78B. To Matthew Boulton, 4 February 1778

Dear Boulton,

Not having heard of you for so many moons has induced me to think you are dead – pray send me word that I may begin your epitaph – if you are still alive I design to visit you next Sunday but one (which I believe to be the 15th) if it is agreable and convenient and if Mr Keir will visit you at the same time.

I sent a plate-order, which I hope was recieved. – And I hope you will direct the Mrs Koffman's four prints of graces to be framed for me at your manufactory – which prints I hope are long since arrived. Pray desire Mr

Keir to give me a line, or yourself, to know if you will give us room and dinner as above.

Adieu with best compl. to Mrs Boulton

E Darwin

Feb. 4 –78

Original Birmingham Reference Library, Matthew Boulton Papers, Darwin 26.
Printed Excerpt in *Lunar Society*, p. 143, and *Doctor of Revolution*, p. 120.
Text From original MS.
Notes
(1) Darwin had written to Boulton on 11 January, so it was only in a 'Lunar' sense that he had not heard from his friend. It seems that there were few if any Lunar meetings in the second half of 1777, and Darwin is taking unilateral action to get them started again.
(2) The Kauffmann prints are still giving trouble. Boulton was, not unnaturally, keener on manufacturing Watt's engines than on framing prints.

78C. To Matthew Boulton, 5 April 1778

Dear Boulton

I am sorry the infernal Divinities, who visit mankind with diseases, and are therefore at perpetual war with Doctors, should have prevented my seeing all you great Men at Soho to day -- Lord! what inventions, what wit, what rhetoric, metaphysical, mechanical and pyrotecnical, will be on the wing, bandy'd like a shuttlecock from one to another of your troop of philosophers! while poor I, I by myself I, imprizon'd in a post chaise, am joggled, and jostled, and bump'd, and bruised along the King's high road, to make war upon a pox or a fever!

Now had I pack'd up a print of Prince Pretty-man, for you, and a book on opake colours for the society, in vain!

Now I wish you would do me the honour to let me open a correspondence with one of your clarks, to know if I am ever to have the prints framed of Mrs Kaufman – and direct the said Clark to write to me that he has pack'd them up, and sent them to Lichfield – – far [be] it from me to think of corresponding on these subjects with the great man himself.

When my Brother's Plate is sent, you will also please to order the bill to be sent to me.

There is a small account also between you and me, viz. that last year when you sent me a receipt for 200£ I sent you 205£ not having other bills. This also when I pay for the plate may be settled.

Adieu, the Lord bless you and keep you.

<u>remember my prints</u>,

E D

Apr. 5 –78

Original Birmingham Reference Library, Matthew Boulton Papers, Darwin 27.
Printed The first paragraph has often been published, e.g. *Essential Writings*, p. 33; *Lunar Society*, pp. 143–4; *Doctor of Revolution*, p. 121.
Text From original MS.
Notes
(1) The first paragraph of this letter expresses the very essence of Darwin's personality, as well as being one of the best notes of apology ever written and the most vivid of the few existing descriptions of the Lunar Society in session.
(2) Prince Pretty-man is a character in the farcical comedy *The Rehearsal* (1671) by (probably) the Duke of Buckingham.
(3) From the last paragraphs of the letter it appears that Boulton acted as general factotum whenever Darwin (or his family) wanted any household item, whether a saucepan, a tea-tray, silver plate or picture framing.

78D. To Matthew Boulton, 21 April 1778

Dear Boulton

I am sorry to tease you so much about the prints of Mrs Kaufman, which you was so good as to [promise to] buy for me, and to direct them to be framed. Now as you are going to London the favour I want to beg of {you} is to know whether the said prints are bought? – whether the framing them is directed? And thirdly the name of the workman I may write to every week during your absence?

Pray let me have one line from you on this subject before you go.

If you do not send me word, what workman of yours I may write to about them, I shall not trust to your letter, nor ever expect to have the prints, but shall come over to Soho in your absence, and search out the shop they are laid bye in, and carry them off by force and arms,

Adieu

I beg my best compl. to Mrs Boulton and am,

Your affect. friend
E Darwin
Apr. 21 –78

Pray when and where does Mr Keir intend to reside in your horizon.

Original Birmingham Reference Library, Matthew Boulton Papers, Darwin 28.
Printed Unpublished.
Text From original MS.
Notes
(1) Darwin really is having a hard time getting his Kauffmann prints framed, and this trivial commission has led to what is almost a crisis of trust with Boulton. Their friendship survived, outwardly uninjured, but Darwin never again troubled the giant of manufacturing with so trivial a task.
(2) In 1778 Boulton was trying to persuade Keir to act as manager of the Soho works in his absence. Keir eventually did so in the autumn of 1778 and again in 1779.

(After seeing the state of Boulton's accounts, Keir declined an offer of partnership.)

(3) Boulton replied on 13 May, saying that there was still no news of the prints (Boulton letter books, Birmingham Reference Library).

78E. To MATTHEW BOULTON [April? 1778]

Friend Boulton! take these ingots fine
From rich Potosi's sparkling mine,
With your nice art a Tea-vase mould,
Your art more valued than the gold!

With orient Pearl in letters white
Around it 'To the Fairest' write;
And where proud Radbourne's turrets rise,
To bright Eliza send the prize,

I'll have no bending serpents kiss
The foaming wave, and seem to hiss;
No sprawling dragons gape with ire,
And snort out steam, or vomit fire;
No Naiads weep, no sphinxes stare,
Nor tail-hung dolphins swim in air.

Let leaves of myrtle round the rim,
With rose-buds twisting, shade the brim;
Each side let woodbine stalks descend,
And form the handles as they bend;
While on the foot a Cupid stands,
And twines the wreaths with both his hands.

Perch'd on the rising lid above,
O, place a love-lorn turtle dove,
With hanging wing, and ruffled plume,
And gasping beak, and eye of gloom.

Last, let the swelling bosses shine
With silver-white, and burnish fine;
Bright as the fount, whose banks beside
Narcissus gazed, and loved, and died.

Vase! when Eliza deigns to pour
With snow-white hand thy boiling shower,
And sweetly talks, and smiles, and sips
Thy fragrant steam with ruby lips,

More charms thy polished orb shall shew
Than Titian's glowing pencil drew,
More than *his* chisel soft unfurled,
Whose Heaven-wrought statue charms the world!

Original Not traced. There is a typescript copy at University College London, Pearson Papers, 577, page 13, under the title 'Directions to Mr Boulton for making a Tea-Vase for Mrs Pole'.

Printed Seward, *Darwin*, pp. 105–6; excerpts often reproduced subsequently.

Text From typescript copy, with some amendment of the punctuation and the grouping of the lines.

Notes

(1) Anna Seward says that 'in the spring of 1778' Mrs Pole brought her children to Darwin's house because they were ill, and stayed with them there for a few weeks, 'till, by his art . . . their health was restored'. The poem was sent to Boulton 'soon after she left Lichfield, with her renovated little ones' (Seward, *Darwin*, pp. 104–5). She probably returned home in mid-April, and seems unlikely that Darwin would have written the poem after his son Charles fell ill at the end of April: so the best guess at the date would seem to be 'late April'.

(2) This letter in verse is a reminder that Boulton, though now honoured chiefly as a leading spirit in the Industrial Revolution, was and is celebrated for his superb ornamental work in ormolu: see N. Goodison, *Ormolu: the work of Matthew Boulton* (Phaidon, London, 1974). Darwin was paying a well-deserved tribute when he said Boulton's art was more valuable than the gold he worked with.

(3) Potosí in Bolivia was famous for its mineral wealth, particularly its silver mines: 'as rich as Potosí' was a familiar expression. Darwin's reference to 'Titian's glowing pencil' might apply to several of his paintings, but the 'Venus of Urbino' seems the most likely. 'Heaven-wrought statue' is the Venus of the Medici, and *his* refers to Praxiteles. Today the connection with Praxiteles is thought to be tenuous, and the statue is not admired: see Kenneth Clark, *The Nude* (Penguin Books, 1960), pp. 73–81. (Clark calls it 'no more than a large drawing-room ornament'.)

(4) Writing in verse does not prevent Darwin from giving quite specific instructions for making the vase: apparently Boulton obeyed the orders and Elizabeth Pole duly received a very fine tea-vase.

78F. To Erasmus Darwin junior, 12 May 1778

Dear Erasmus

I was flatter'd into a belief, when I arrived here, that your brother, as ill as he appear'd to me, was considerably better than he had been – alass in 12 hours I saw he was declining fast, and that the case as far as I could judge was nearly hopeless. I fear in two or three days he will cease to live. Perhaps you may see me in a day or two after you recieve this letter – adieu

God bless you,

E Darwin

Edinb.
May 12 –78

I shall return by Newcastle, write to me there if any medical business on that side the country wants me. And say if Lady Alston of Stafford be living.

Pray write an account of this to Sist. Susanna at Elston as soon as you can.

Original Not traced.
Printed Unpublished.
Text From typescript copy in the collection of Sir Geoffrey Keynes.
Notes

(1) Darwin was called to Edinburgh early in May 1778 because his 19-year-old son Charles, a medical student of brilliant promise, was seriously ill. At the end of April, Charles had cut his finger while dissecting the brain of a child who had died of 'hydrocephalus internus', and on the same evening was seized with a severe headache, followed the next day by delirium and other alarming symptoms. The diagnosis in this letter proved correct, and Charles died on 15 May, to the great grief of his father.
(2) Darwin's second son, Erasmus junior, was now 18 and a law student.
(3) 'Newcastle' seems sure to be Newcastle-under-Lyme. Lady Alston of Stafford (*c.*1725–1778) was the widow of Sir Thomas Alston (1724–1774). They had separated in 1752 after two years of marriage, and in the 1760s Lady Alston had two sons by John Wasse of Stafford. See L. Cresswell, *Stemmata Alstoniana* (1905), and *Complete Baronetage* ii 183.
(4) Darwin's sister Susannah, who kept house for him during most of the 1770s, was now apparently at Elston, perhaps temporarily.
(5) The provenance of this letter is unknown, but its style seems to stamp it as authentic.

78G. To Josiah Wedgwood, 17 May 1778

[Darwin wrote to tell Wedgwood about the death of his son Charles, ending his letter with:]

* * *

God Bless you, my dear friend, may your children succeed better.

Original Not traced. In the possession of Charles Darwin, 1879.
Printed C. Darwin, *Life of E. Darwin*, p. 83.
Text From above source.
Notes

See previous letter. This letter too was probably written at Edinburgh.

78H. To James Hutton, 3 July 1778

Dear Sir

I esteem myself highly obliged to you on many accounts. I have inclosed an inscription, which I wish to be put on marble, ornamented so as to cost between five and ten pounds, in such manner as shall be most agreable to Dr Duncan, whom you will please to consult on this matter – I prefer english inscriptions to latin – hope you will see that it is cut deep in the marble, and not simply painted on the marble, as is the practise of the workmen here, if not look'd after. I must beg in your next you will mention the name of the place he is buried in, that I may some years hence direct his brother to find his tomb!

– Now let me add how sorry I am to hear you complain of headach'd[?] and giddyness. The former generally proceed from a decaying tooth, or a tooth about to decay – if it is one temple only that gives pain, it is the last tooth in the upper jaw on the same side. Giddyness frequently proceeds from taking food too seldom – solid food, of the flesh kind often relieves it.

Vinous spirit from small beer to alcohol destroys us all.

I will send you coal full of vegetable seeds, turned to iron, and some Kenal coal. Coal with spar in it, or with pyrites in it, are too common to send you.

And if Don – what's his name – comes this way I shall be glad of his company for some days at my house, and will accompany him to see wonders of Derbyshire.

I wish yourself and Dr Black would come to England. – I shall not send a bill till I recieve another letter from you, with the additional expense.

I intend shortly to publish my poor Charles's treatise on pus and mucus, and his thesis on the retrograde motions of the lymphatics, and will send you a few copies.

You will please to transmit the inclosed to Dr Duncan and to Mr Broughton and believe me dear Doctor

Your much obliged friend
E Darwin

Jul 3 –78

Pray see Dr Duncan about the inscription before you direct it, to whom I am so much indebted.

Addressed Dr Hutton.

Original Library of the Fitzwilliam Museum, Cambridge, S. G. Perceval Bequest, General Series.

Printed Unpublished.

Text From original MS.

Notes

(1) James Hutton (1726–1797) is generally accepted as 'the founder of modern

geology'. His *Theory of the Earth*, developed over thirty years and finally published in 1795, proposed satisfactory mechanisms for geological change that put geology on a sound base. For a critique of Hutton's work, see G. L. Davies, *The Earth in Decay* (Macdonald, 1969), pp. 154–99. Hutton, who lived in Edinburgh, was a close friend of James Watt in the 1760s and visited Watt in Birmingham in 1774, when he also met Darwin: see *Phil. Trans.*, **78**, 43 (1788). Their acquaintance developed into lifelong friendship during those distressing days in Edinburgh in May 1778; but few of Hutton's papers have survived, and this is the only known letter from Darwin. (There is a letter from Hutton to Darwin at Down House.)

(2) Dr Andrew Duncan (1744–1828) was a distinguished medical professor at Edinburgh, who also taught Erasmus's grandson Charles in 1825: 'The venerable professor spoke to me about him [the elder Charles] with the warmest affection forty-seven years after his death, when I was a young medical student in Edinburgh' (C. Darwin, *Life of E. Darwin*, p. 82). Dr Duncan also wrote a 23-page elegiac poem about Charles, *An Elegy on the much lamented death of a most ingenious Young Gentleman* (Robinson, 1778). See *DNB* for Duncan.

(3) Andrew Duncan offered to have Charles buried in the Duncan family tomb, in the chapel of St Cuthbert, and with this letter Darwin is sending Hutton the inscription for the tomb, which read:
'CHARLES DARWIN was born at Lichfield, September 3, 1758; and died at Edinburgh, May 15, 1778. Possessed of uncommon Abilities and Activity, he had acquired Knowledge in every Department of Medical and Philosophical Science, much beyond his Years. He gained the first Medal offered by the Aesculapian Society for a Criterion to distinguish Matter from Mucus; and had prepared a Thesis for his Graduation on 'The Retrograde Motions of the Lymphatic Vessels in some Diseases'. He cultivated with Success the Friendship of ingenious Men, and was buried by Favor of Dr. A. Duncan in this his Family-Vault.

Fame's boastful chissel, Fortune's silver plume,
Mark but the mouldering urn, or deck the tomb!'

(4) Now that Whitehurst had gone to London, none of Darwin's Lunar friends was primarily interested in geology, so he was pleased to talk geology with Hutton. 'Kenal coal' refers to 'cannel coal', a type of bituminous coal with a high proportion of volatile matter which burns with a luminous flame like a candle (whence the name). Darwin refers to 'cannel, or candle coal' in the Note on coal in *The Economy of Vegetation* (Note XXIII). And in a MS dated 1765 he wrote: 'A mile from Rugeley a blazing kind of Coal is procured call'd Canel Coal' (University College London, Galton Papers, No. 10, Part 3, page 5).

(5) Dr Joseph Black (1728–1799) the great Scottish chemist, who discovered the principle of latent heat, was a helpful friend of Watt and Hutton. For an excellent biography of Black, see J. G. Crowther, *Scientists of the Industrial Revolution* (Cresset Press, 1962) pp. 9–92.

(6) Darwin did publish his son's treatise on Pus and Mucus, two years later: C. Darwin, *Experiments establishing a Criterion between mucaginous and purulent Matter* . . . Lichfield: J. Jackson. London: T. Cadell. Edinburgh: W. Creech. 1780.

(7) I have not identified Don what's his name, nor Mr Broughton.

781. To Elizabeth Pole, [autumn? 1778]

Platonic Epistle to a married Lady

Oh, read these lines, Eliza! – may they move
Thy breast to pity, whom it must not love! –
Whilst from despair the sorrowing numbers flow,
Pour all my heart and modulate my woe,
No lawless wish, no dubious words shall rise,
But truth shall tune them, Virtue's self chastise.
Bear them, soft gales! to chaste Eliza's ear,
Herself may read them – or her husband hear.

If brighter scenes from others' hills extend,
If others' vales more fragrant flowers defend,
Each passing swain, with rural charms inflamed,
Eyes the gay scene, and drinks the gale unblamed.
Say! should I gaze o'er thy fair form with bliss,
Or ask the balmy rapture of a kiss,
Pure as the vestal meets her sister guest,
Or holy lips on sainted marble pressed,
Could truth, could innocence or virtue blame?
And frowns Eliza on my guiltless flame?

Ah, who unmoved that radiant brow descrys,
Sweet pouting lips, and blue voluptuous eyes?
The witching airs, which bend your ivory neck,
Heave your white breast, and dimple on your cheek?
O'er all your form the varying grace that plays,
Youth's vermil bloom, and beauty's vivid blaze?
Charms such as thine, which all my soul possess,
Who views unmoved, is more than man – or less.

Then as, simplicity! thy virgin care
Decks her light limbs, and wreaths her shadowy hair;
Calls out the rising group with pencil chaste,
And gives to Beauty all the aid of taste;
Each charm illumined beams celestial powers,
She moves a Goddess, and the World adores!

Yet not the bloom of youth, nor beauty's blaze,
Not o'er that form the varying grace that plays,
With all thy elegance of fancy drest,
Effused such strong enchantment on my breast,
As when sweet Sensibility array'd
Those loving looks in all her light and shade;

And soft Benevolence with aspect kind
Ray'd through those eyes the sunshine of the mind.
– Caught by fair forms, but vanquished by desert,
We gaze on beauty, but we love the heart!

When thy dear babes the bed of sickness pressed,
Stretched the weak limb, and heaved the panting breast;
While their cold lips the fading roses fled,
And life's dim lamp a doubtful glimmering shed;
You watch'd with kindest looks, and ceaseless toil,
The dawning wish, and waked the expiring smile;
Drop'd the still tear unseen, and half-suppres'd
The rising sigh, which struggled in your breast.
Such soft distress the coldest heart might move,
– I thought it sympathy – but found it love!

But when my hand, by learn'd Hygeia taught,
Gave to their lips the medicated draught;
While life rekindling in their eye-balls glow'd,
And to their cheeks the truant purple flow'd,
Till flushed with health in all their bloom of charms
Rush'd the sweet babes to thy enamoured arms,
Gods! from that tongue what tender accents broke,
How all the Mother lighten'd in your looks!
– No! I can ne'er the dear remembrance lose
Of those sweet smiles, which ruin'd my repose!

Vain are the arts of medecine to secure
Life's crimson stream from Love's malignant power;
No opiate lulls the aching mind to rest,
No oil can still the surges of the breast!
Thrill'd at her touch, where'er Eliza treads,
To circling crowds the electric passion spreads;
And, as her soft infectious eye she turns,
Deep heaves the sigh, and fierce the fever burns.
– I have it! – here the dear delirium reigns,
Clings round my heart, and eddies in my veins!
– Feel how it throbs! – oh, let my torments move
Thy tender bosom – give me love for love!

Ah me! – not love! – too fair Eliza, no!
A cold esteem is all thou canst bestow! –
No guilty wish the fond request inspires,
Nor burns my bosom with unholy fires;
But if sometimes thy gentle mind incline
To sooth such cares, and pity pains like mine,

With lenient words a while my woes dismiss,
Oh, cheer me with a smile, and charm me with a kiss!

Slow moves the pilgrim from Loretto's shade,
Leaves the fair shrine, where all his vows were paid,
Safe to his home some sacred relict bears,
Imprints with kisses, and accosts with prayers.
So, step by step, with oft reverted eyes,
Tears unobserved, and unavailing sighs,
To Radbourne's towery groves I bad adieu,
And tore my heart, my bleeding heart, from you.
So at each dawn I clasp with holy love
This pledge of friendship by your fingers wove;
Admire the leaves, and budding flowers, which join
Their twisted stems in beauty's waving line;
The lovely giver in the gift adore,
And thus – and thus – I kiss it, o'er and o'er.

Original Not traced. There is a typescript copy at University College London, Pearson Papers 577, pages 24–6.

Printed Unpublished.

Text From typescript copy in Pearson Papers, with some amendment of punctuation.

Notes

(1) The dating of this 'epistle' is uncertain, but it very probably belongs to 1778, since Darwin gives details of the illness of Elizabeth's children in the spring of 1778. April is a possible date, but autumn seems more likely because of the retrospective sadness.

(2) Though Darwin manages to be quite saucy, he is not yet skilled enough in verse to handle so delicate a subject with complete success. But there are some good passages, and the bad ones have a biographical interest because, by their lack of sophistication, they reveal his feelings more openly than any other of his writings. The overdone sentiment and the loss of actuality through abstraction (e.g. 'stretched the weak limb, and heaved the panting breast') are defects shared with much of eighteenth-century verse – defects which are even more obvious in the fictional digressions in the *Botanic Garden* (e.g. *Loves of the Plants* IV 13–32).

(3) Darwin's reference to the skills he learnt from Hygeia may seem exaggerated. But his success in curing her children, and his ability to write flattering verses, were his only practicable methods of wooing Elizabeth, so he cannot be blamed for exploiting both.

(4) Footnote to *4th paragraph*: in 'the rising group' the word 'group' is probably a euphemism for 'breasts', which simple taste takes care to outline appealingly: cf. (i) 'the rising bosom' in *Loves of the Plants* IV 242; (ii) the modern expression 'pair'; (iii) the original meaning of group, 'round mass' (*OED*); and (iv) the definition in Chambers's *Cyclopaedia* (5th edn 1741), 'an assemblage, or knot, of two or more figures of men, beasts, fruits, or the like, which have some apparent relation to each other'.

(5) Footnote to *5th paragraph*: 'desert' means 'merit'.
(6) Footnote to *10th paragraph*: (i) 'Loretto' refers to the shrine in the town of Loreto, near Ancona, Italy, which is said to be the holy house at Nazareth, carried through the air in 1291 by angels; (ii) Radbourne is of course Elizabeth's home; (iii) the 'pledge of friendship' was a 'pair of worked ruffles' embroidered with a flower pattern by Elizabeth for Darwin, according to a note on the typescript attributed to Sir Francis Darwin.

78J. To Charles F. Greville, 12 December 1778

Lichfield Dec 12 –78

Dear Sir,

I can assure you at my few vacant hours I have incessantly labour'd at the completion of the Polygrapher in its triple capacity, and hope as my time is so much otherwise employ'd, that you will indulge me with a few weeks longer. And I flatter myself I shall be able to send you one that will perform all its parts to considerable perfection.

In the letter I am now writing to you I have quite the free use of one pen, so as to write quite as easily to myself, and as well as I can write without the machine.

My new manner of fixing the paper delights me of all things, as I invented it with great difficulty, and it is so simple and easily removed, and steady.

The new idea of making one pen quite loose in the hand instead of the stiff one you saw is also a great improvement, as any one writes as facile with it the first time of using it as those who have practised it before.

I have also reduced the height of it from 30 inches to 20, and am in doubt about reduding it to 18.

I hope my sending you this specimen which I assure is the first essay of all these improvements, will induce you to grant me a few more weeks of grace. And I will promise not to loose sight of till I have compleated in [it] in respect to its execution of its business, as far as I am able; the beauty of it I do not mean to consider at all.

I am sorry to see by the papers that we are lik[e]ly to have a violent french war – I hate war! – yet if we loose our colonies we loose our power at sea, for as Sea-men can not be <u>made</u>, but must be <u>educated</u> (their business requiring them to hang by their hands and other bodily exertions, which adult men can not well acquire) I suppose the strength of any nation is exactly as the number of seamen they can employ in time of peace.

What an empire would this have been if Lord Chatham's idea had been followed? England to have supply'd America with all manufactured commodities, and America have supplied England with the raw products of the soil. The number of Seamen thus employ'd in times of Peace might at the time of war have been put into Ships of war, and have conquer'd the world!

But the Lord and King George thought otherwise! perhaps it is for better

– Britain would have enslaved mankind, as antient Room [Rome], and have been at length enslaved themselves, adieu,

I never talk'd so much politicks in my life before, I beg to be remember'd to Lord and Lady Warwick, and am, dear Sir

Your much obliged f[d]
E Darwin

Addressed {The Honorable Cha}rles Greville M.P., Portman Square, London.

Original British Library, Hamilton and Greville papers, Vol. IV, Add MS 42071 ff. 48–50 (mechanically duplicated in ff. 51–52).
Printed Short extracts in *Doctor of Revolution*, pp. 127–8.
Text From original MSS.
Notes
(1) This letter is unique (and also non-unique) because it exists in two identical MS copies, one original and one produced by Darwin's polygrapher described in the letter. The copy is as good as a modern photocopy – indeed better because the same ink is used in both versions. It is difficult to tell which is the original. This may well be the earliest document of which a perfect mechanical copy exists.
(2) Hon. Charles Francis Greville, MP, PC, FRS (1749–1809), second son of the Earl of Warwick and nephew of Sir William Hamilton, was one of Darwin's few London friends. Greville was MP for Warwick from 1774 to 1790, and in 1779 held government office as Lord of Trade. Greville had a taste for science and was already amassing his famous collection of about 15 000 geological specimens, widely regarded (when complete) as the finest in the world, because of 'the beauty of the specimens and their immense variety': see *American Mineralogical Journal*, **I**, 55–6 (1810). On his death, the greater part of his collection was purchased by the British Museum, with the aid of a special Parliamentary grant: see *History of the Collections of the British Museum* (*Natural History*), Vol I, p. 424 (1904). Greville was often in financial difficulty, not only through buying minerals but also because of 'his liberality to unfortunates', as one obituary nicely phrases it. He is best known for his rescue of Emma Hart: he became her lover and protector when she was in her teens, and taught her social graces. She became quite devoted to him. But in 1784 he ran into serious financial trouble and made her over to his uncle Sir William Hamilton, who agreed to pay his debts. Hamilton married Emma in 1791, and they met Nelson in 1793. For biography of Greville, see references already cited; Namier & Brooke, *Commons* ii 550–1; and P. J. Weindling, in *Images of the Earth* (Brit. Soc. Hist. Sci., 1979), pp. 248–71.
(3) Darwin says he has completed 'the Polygrapher in its triple capacity'. From later letters it appears that the 'triple' refers to (i) copying writing, (ii) copying drawings and (iii) producing reduced-size copies.
(4) The letter is also interesting in that it offers one of Darwin's rare (and typically far-sighted) comments on contemporary politics.
(5) There are more mistakes than usual in the spelling and grammar, no doubt because Darwin had to concentrate on the operation of the polygrapher as he was writing and could not correct mistakes. His spelling 'Room' for 'Rome' may be phonetic: cf. Shakespeare, *Julius Caesar* I ii 156: 'Now is it Rome indeed, and Roome enough/When there is in it but one only man'.

79A. To Charles F. Greville, 16 May 1779

Lichfield May 16 –79

Dear Sir,

My so long neglecting to send you the machine, I promised, can only be excused by my desire of making it better worth your attention – it remained many months at Birmingham before my mechanical friends return'd from fire-engine building in Cornwall. It is now as much finish'd as I intend to finish it; and I will send you it, when and where you require; but as something may possibly bring you into this climate sometime this summer, it would be more agreable to me, to shew it to you here, before I pack it up to send you. It performs all its purposes pretty well, but is still capable of greater accuracy. I shall not risk the expense of a patent for it, but would sell it, if I could, to any London artist, who might make advantage of it; – which I know I could not, if I had the patent myself.

Pray favour me with a line, and don't scold me much – and tell me if I have any chance of the pleasure of seeing you this summer: and if so pray let me know a day or two before, that I may be at home.

I think I could entertain you, by shewing you all the new experiments upon airs, or gas's, and would accompany you either to Birmingham, or to Caulk. But perhaps the affairs of politics in this critical time may so much engage your attention, as to withdraw you from the more pleasurable pursuits of philosophy.

This machine I would much wish to shew you here, as it can not well be pack'd up without being taken in pieces; and till you have act{ually} seen it, you would find some trouble in putting it together for all its uses. Pray bear me no malice for so long disappointing you, but come and see me; I have a well-air'd and quiet bed for you; and I want much to hear some fossile philosophy, and a hundred other matters, and am, dear Sir,

with true esteem your much obliged
and affect. friend
E Darwin

Addressed The Hon. Charles Greville Esq, M.P., London.

Original British Library Add MS 42071 ff. 63–4.
Printed Unpublished.
Text From original MS.
Notes

(1) Apparently Darwin lent his polygrapher to Boulton or, more probably, Watt, who was very interested in copying machines and wrote to Darwin in the summer of 1779: 'I have fallen on a way of copying writing chemically, which beats your bigrapher hollow'. Since Darwin's method gives a perfect copy, Watt's method was only better because it could be used *after* the document was written, whereas Darwin's method required the copy to be made *as* the document was written.

(2) It seems that Greville did not visit Lichfield: Darwin took the machine to pieces and sent it to him (see next letter).
(3) 'Caulk' is Calke, near Derby, the home of Sir Harry Harpur, Greville's brother-in-law (see next letter).

79B. To Charles F. Greville, 7 June 1779

concerning a Patent etc.

Dear Sir,

I could not get the packing box etc. made sooner. So I hope you will not think me negligent of your kindness to me.

In respect to selling the machine, I don't know what to say – the drawing machine and reducing apparatus alone, I suppose would sell – and the writing part alone. I believe I have been at more than 20 Guineas expence in making experiments – and I should be willing to sell the invention so as to gain 100 Guineas clear, that is for 120, the purchaser being at all expenses about the patent.

Some time ago I put in a caveat about a patent. I am totally ignorant of these things – I sent such a caveat as I believed would best secure the writing part, as I had not then added the drawing and reducing part. See p. 16 [of this letter]. Of this I will send you a copy.

I don't think any body can make advantage of this machine, but one of the trade, since so much would be expected for selling them – otherwise I could manufacture them here – nor would I go into any kind of partnership with one of the trade – nor would I risk the expense of a Patent, suppose 80£ and the expence of making 100 of the machines, suppose for 200£ – since the person who could sell them in London would expect at least one fourth or ⅓, suppose one guinea for each machine – and they would be sold for 5 guineas each. From this would be some few deductions, as advertisements etc.

If Sir Harry Harpur is in your neighbourhood, I wish you to shew him this machine; as I did intend to have taken it to Cork [Calke] had it been done sooner, as Sir Harry has some claim to the invention, as it was began and first plan'd at Caulk. (NB. I must beg you to forward the inclosed letter to him on another subject as I don't know where he is.)

I wish I could get two such partners as yourself and Sir Harry, to advance with me 100£ each! – if you could only find out a proper person to sell them, such as Pinchbeck – I know nobody in London – I could manufacture them cheaper here – and for 300£ the patent might be gain'd and 100 machines made – if the world like them, some advantage might be afterwards made – if not, it is probable only 100£ out of the three could any how be lost by the experiment. Yet I will not run this risk alone – and I think at this season of poverty nobody will buy the invention.

[These are the first four pages of the letter. On *pages 5–8* Darwin gives a detailed description of the polygrapher, and, with diagrams to help, tells Greville how to put it together. *Page 9* begins as follows:]

Dear Sir,

I am return'd from a long round of medical exertion, and am determin'd to pack up the machine tomorrow.

[The rest of *page 9*, *pages 10–14* and most of *page 15* are occupied with instructions on the method of drawing a landscape, reducing drawings, double writing, etc. Towards the end of *page 15* Darwin resumes the letter, as follows:]

I have now only to add that I must here once for all beg your excuse for all this scrawl – and that if nothing can be done with the machine I hope you will sometime or other return it to me; as I have now no other.

[on *page 16:*] Caveat

About the 1st Sept. 1777, I enter'd a caveat at the Attorney and Sollr. general's offices "to prevent any person from gaining a patent for a machine for the purpose of connecting two pens, or pencils, or graving tools of metallic, vegetable or animal substances, so that when one is used in writing, drawing, graving etc the other shall not only perform the same horizontal motions, but shall rise or descend more or less, correspondent with the former".

This general description I thought was preferable to a more particular one – and I suppose nothing need further be done at present; as I shall not be at the expense of a patent, unless I can first sell the invention.

Adieu

from your much obliged

Jun. 7 –79 E Darwin

[This ends *page 16*. On *page 17* (which is unnumbered) there are instructions for 'taking the outline of the profile of a face'. Finally, on *page 18* (unnumbered) are some thoughts about 'drawing', written on the following morning. This final page ends:]

I can not try this, as I sent the machine by a waggon last night (June 7th).

The Carrier, Joseph Hulse, i{nn}s at the ax Aldermanbury and will be there on friday. My servants saw it put into the waggon.

Addressed The Honorable Charles Greville Esq M.P., Portman Square, London. (*Endorsed* Dr Darwin Jany 7th 1779.) [Date incorrect, Darwin's 'Jun. 7' having presumably been read as 'Jan. 7'.]

Original British Library, Hamilton and Greville papers Vol IV (Add MS 42071), ff. 55–62 and 53–4.

Printed Unpublished.

Text From original MS.

Notes

(1) Since Greville showed no sign of visiting Lichfield, Darwin decided to take the polygrapher to pieces and send it to him. In retrospect it seems a rash decision, and it is unlikely that Greville ever succeeded in putting it together again in London: Darwin's splendid invention was destined for oblivion. Even if

Greville had got it to work, it is doubtful whether he would have developed the invention. He was an aristocratic dilettante rather than an engineer, and chronically in debt. When he had to sell something to settle his debts, Emma proved more marketable than Darwin's polygrapher. Still, Greville remained on friendly terms with Darwin, and was the proposer of Darwin's son Robert as a Fellow of the Royal Society in 1787 – a favour for the son to repair the injury to his father?

(2) Sir Harry Harpur (1738/9–1789) of Calke, near Derby, was MP for Derbyshire from 1761 to 1768, and in 1762 he married Lady Frances Elizabeth Greville, Charles Greville's sister. Apparently Harpur was involved in the original idea for the invention, but there is no evidence that he played any part in its development, and Darwin is probably being over-generous to him in the hope of obtaining his financial support. For Harpur, see Glover, *Derby* ii 186, and Namier & Brooke, *Commons* ii 587–8.

(3) Although not appropriate for inclusion here, Darwin's instructional manual for the polygrapher, which is interwoven with the letter, is fascinating in itself and might show how the polygrapher worked, if carefully studied.

(4) 'The ax Aldermanbury' refers to the Axe Inn.

79C. To Josiah Wedgwood, December 1779

Lichfield Dec 1779

Your little boys are very good, and learn french and drawing with avidity and I hope you will let them stay, till we write you word we are tired of them. Josiah is a boy of great abilities and little Thomas has much humour.

Original Not traced. There is a MS copy, made between 1850 and 1900, in the Keele University Library, Wedgwood Archive, 26764–35.

Printed Doctor of Revolution, p. 131.

Text From Keele MS copy.

Notes

(1) 'Your little boys' are John Wedgwood (1766–1844), Josiah Wedgwood II (1769–1843) and Thomas Wedgwood (1771–1805).

(2) Darwin and Wedgwood became dissatisfied with school education in 1779 and in the autumn they began a scheme for home-education of their sons – John, Joss and Tom Wedgwood, and Robert Darwin. The course began at Etruria, and then for most of December the young scholars moved to Lichfield, where Darwin had engaged a young French prisoner to teach them French. For details see Wedgwood, *Letters* ii 532–63, *passim*.

80A. To John Walcott, 3 January 1780

Sir

I am sorry to be acquainted by the Bookseller, that you have discontinued the publication of your useful plates of the indigenous plants of this

Country, a work which I think was likely to advance the knowlege of Botany – both by facilitating the Study of it to those, who are beginners; and by reviving the names of the plants in the minds of those who have made greater progress in it. You will oblige me, who am unknown to you, if you would favor me with a line to know your reasons for discontinuing the work, as I would willingly give you what assistance is in my power, in selecting and getting drawings made of a few plants occasionally as they may occur, if such kind of assistance is wanted.

I am Sir your Obed. Serv.
Jan. 3rd 1780 E Darwin
Phisician at Lichfield, Staffordshire

Original Not traced. There is a MS copy written on the front end-paper of Walcott's *Flora Britannica Indigena* at Central Reference Library, Bristol.
Printed Unpublished.
Text From MS copy.
Notes
(1) John Walcott (1755?–1831) was a naturalist living at Bath and a founder member of the newly-formed Bath Philosophical Society. He was author of *Flora Britannica Indigena; or plates of the indigenous plants of Great Britain* (Bath, 1778–9), which was issued in 14 monthly parts, each with 12 plates. This is the work Darwin mentions: presumably it was abandoned for financial reasons. For further details of Walcott, see *Science and Music in 18th Century Bath* (University of Bath, 1977), pp. 83–94 (notes by H. S. Torrens); and p. 69 (portrait). Also H. S. Torrens, in *Images of the Earth* (Brit. Soc. Hist. Sci., 1979), pp. 215–47.
(2) The letter shows how keen Darwin was to 'advance the knowledge of botany', a task on which he was to spend several years of hard work later in the 1780s. This letter – polite, charming and helpful – is typical of Darwin's letters to strangers on scientific subjects.

80B. To Anna Seward, 6 November 1780

Lichfield.

Dear Miss Pussy,

As I sat the other day carelessly basking myself in the sun in my parlour window, and saw you in the opposit window washing your beautiful round face, and elegant brinded ears with your velvet paw; and whisking about with graceful sinuosity your meandring tail; the treacherous Porcupine, Cupid, concealing himself behind your tabby beauties, shot one of his too well-aim'd quills, and pierced, Oh cruel fate! my fluttering heart.

Ever since that cruel hour have I watched you all night from my balcony, hoping that the stillness of the star-light evening might induce you to take the air on the cupola of your house – but have watch'd in vain!

Many a serenade have I sung under your window, and with the sound of my shrill voice made the whole neighbourhood reecho, through all its winding lanes and dirty alleys – all heard me, but my cruel fair one! – she, wrapt in soft furr, sat purring with contented insensibility, or slept with untroubled dreams!

Tho' I can not boast the shining tortoise-shell that cloaths my fair brinded charmer; – tho' I can not boast those delicate varieties of melody, with which you sometimes ravish the ear of night and stay the listening stars! – tho' you sleep hourly lulled on the lap of your fair mistress and dip with her permission your white whiskers in delicious cream! – – –

Nor am I, sweet charmer, utterly destitute of the advantages of birth and beauty – derived from persian kings my white fur still retains the splendor and softness of their ermine. This morning as I sat upon my master's tea-table, and saw my reflected features in the slop-bason, my long white whiskers, ivory teeth, and topaz-eyes – and sure the slop bason can not flatter me, which shews the azure flowers upon its border less beautiful than they are painted! –

You know not, dear Miss Pussy, the value of the heart you slight; – new milk have I in flowing streams to regale you, and mice pent up in a hundred garrets for your food, or your amusement; oh, permit me this very afternoon to lay at your divine feet the head of an enormous rat, who has even now stain'd my paws with his gore. And if you will honour me so much, as to sing the following song in the evening, I will bring a band of catgut, and catcall, to accompany your mellifluous voice in the chorus

Air 1.

Cats! I scorn, who sleek and fat,
Shivers at a Norway-rat;
Smooths with nice care his silky fur,
Or fawns with soft seductive pur.

Chorus Rough and hardy, bold and free,
Be the cat, that's made for me.

2.

He whose nervous paw can take
My Lady's lapdog by the neck,
With furious hiss assault the hen,
And snatch a chicken from the pen. –

Chorus Rough and hardy, etc.

3.

If the treacherous swain should prove
Rebellious to my tender love,
My scorn the vengeful paw shall dart,
Shall tear his fur, and pierce his heart.

Chorus Soon another good as he
Shall be found the cat for me.

Quow, mow, how, now, quall, wall, pall, gnash, smash, pash! etc.

Deign, most adorable beauty to pur your assent to this request, and believe me with most profound respect your true admirer

Snow Grimalkin
Nov 6 1780

Addressed For Miss Pussy from Grimalkin. [Delivered by hand.]

Original Darwin Museum, Down House, Downe, Kent.

Printed A greatly altered (and weaker) version appears in Seward, *Darwin*, pp. 134–7. This text, and Anna's reply were printed as a tiny book (6 cm × 4 cm) of 34 pages, entitled *The Pussey Cats' Love Letters* (privately printed, W. L. Washburn, Collingswood, N. J., 1934). Extracts from the letter in Anna's version have often appeared in print; there are extracts from the MS version in *Doctor of revolution*, p. 136.

Text From original MS.

Notes

(1) Anna Seward (1742–1809) had known Darwin since he arrived at Lichfield when she was 14. While Darwin was in Lichfield, she lived at the Bishop's Palace in the Cathedral Close, only a hundred yards from Darwin's house, her father Dr Thomas Seward (1708–1790) being Canon Residentiary of the Cathedral. She had talent as a poet, which Darwin encouraged (he was her 'poetic preceptor', as she put it), and in the 1780s she became famous as a poet, being known as 'The Swan of Lichfield'.

(2) This is Darwin's most skittish letter, revealing great delicacy of touch in writing, clumsy though he may have been in person. The letter is amusing in itself and had enough possible double meanings to rouse Anna's curiosity. She entered into its spirit, and made a lengthy reply (Seward, *Darwin*, pp. 138–45). Whatever Anna may have thought, the letter was no more than a jeu d'esprit, a glimpse of the playful author of *The Loves of the Plants*.

(3) Time has gnattered the manuscript, which is not always legible. The typescript copy at University College London (Pearson Papers, 577, pages 80–1), which was made when the MS was more legible, proved useful in deciphering doubtful words. Anna Seward dates the letter 'Sept. 7, 1780', so her text may have been based on an earlier draft. There is a third version of the poem, a MS copy apparently made in December 1781, in the collection of Sir Geoffrey Keynes. The text of this version is approximately midway between the Down House and Seward versions.

80C. To Josiah Wedgwood, 29 November 1780

Lichfield, Nov. 29 –80

Dear Sir,

Your letter communicating to me the death of your friend, and I beg I may call him mine Mr Bentley, gives me very great concern; and a train of very melancholy ideas succeeds in my mind, unconnected indeed with your loss, but which still at times casts a shadow over me, which nothing but exertion in business or in acquiring knowlege can remove. This exertion I must recommend to you, as it for a time dispossesses the disagreable ideas of our loss; and gradually their impression or effect upon us becomes thus weakened, till the traces are scarcely perceptible, and a scar only is left, which reminds us of the past pain of the united wound.

Mr Bentley was possessed of such variety of knowlege, that his loss is a public calamity, as well as to his friends, though they must feel it the most sensibly! Pray pass a day or two with me at Lichfield, if you can spare the time, at your return. I want much to see you; and was truly sorry I was from home as you went up; but I do beg you will always lodge at my house on your road, as I do at yours, whether you meet with me at home or not.

I have searched in vain in Melmoth's translation of Cicero's letters for the famous consolatory letter of Sulpicius to Cicero on the loss of his daughter (as the work has no index), but have found it, the first letter in a small publication called 'Letters on the most common as well as important occasions in Life:' Newberry, St. Paul's, 1758. This letter is a masterly piece of oratory indeed, adapted to the man, the time, and the occasion. I think it contains everything which could be said upon the subject, and if you have not seen it I beg you to send for the book.

For my own part, too sensible of the misfortunes of others for my own happiness, and too pertinacious of the remembrance of my own, I am rather in a situation to demand than to administer consolation. Adieu. God bless you, and believe me, dear Sir, your affectionate friend

E Darwin

Original Not traced. In the possession of Charles Darwin, 1879.
Printed C. Darwin, *Life of E. Darwin*, pp. 31–3.
Text From above source.
Notes

(1) Thomas Bentley, Wedgwood's close friend and business partner for many years, died on 26 November 1780, at the age of 50. It was a serious blow to Wedgwood personally and also commercially, because Bentley was in complete charge of the business in London. Wedgwood went to London immediately, but Darwin was out when he passed through Lichfield, and Wedgwood left a letter; this is Darwin's reply.

(2) This is perhaps the gloomiest of Darwin's letters. He seems to have felt that death was dogging him too assiduously. The tragic death of his son Charles in 1778 still preyed on his mind, as the first paragraph shows; in the summer of 1780 Honora Edgeworth and Mrs Seward had died; and on 27 November, one day after Bentley, Colonel Pole died at Radburn, leaving Darwin with the sad probability that the Colonel's widow would be snatched from him by some wealthy and dashing bachelor.

(3) 'Melmoth's translation' is *The Letters of M. T. Cicero to several of his friends,* translated with remarks by William Melmoth (the younger, 1710–1799), 3 vols.,

London, 1753 (2nd edn 1772). The letter from Sulpicius to Cicero, which Darwin could not find, is in Volume 3, pages 6–13. *Letters on the most common as well as most important Occasions in Life* is one of the library of miniature books, mainly for juveniles, published by John Newbery (1713–1767). The first edition was in 1756, and the seventh in 1767. As Darwin says, it is a 'small publication', about 11 cm × 6 cm, with about 300 pages, and the first letter (pp. 23–30) is that from Sulpicius to Cicero. For Newbery, see S. Roscoe, *John Newbery and his Successors, 1740–1814: a bibliography* (Five Owls Press, 1973). For a modern edition of the letter to Cicero, see *Cicero, the letters to his friends*, trans. by W. G. Williams (Heinemann, 1927) Vol. 1, pp. 268–77.

81A. To James Watt, 6 January 1781

Beau-Desert Jan 6 –81

Dear Mr Watt,

You know there is a perpetual war carried on between the Devil and all holy men. Sometimes one prevails in an odd skirmish or so, and sometimes the other. Now you must know this said Devil has play'd me a slippery trick, and I fear prevented me from coming to join the holy men at your house, by sending the measles with peripneumony amongst nine beautiful children of Lord Paget's. For I must suppose it is a work of the Devil? Surely the Lord could never think of amusing himself by setting nine innocent little animals to cough their hearts up? Pray ask your learned society if this partial evil contributes to any public good? – if this pain is necessary to establish the subordination or [of?] different links in the chain of animation? If one was to be weaker and less perfect than another, must he therefore have pain as a part of his portion? Pray enquire of your Philosophs, and rescue me from Manichaeism.

As to material philosophy, I can tell you some secrets in return for yours, viz., that atmospheric gas is composed of light and the earth of water (and aqueous earth). That water is composed of aqueous gas, which is displaced from its earth by oil of vitriol.

Pray make my best devoirs to all the Phlos, and pray tell Dr Priestly that [I] wish he would try whether a plant insulated in ☿ [mercury] will spoil air.

E Darwin

Original Not traced. There is a photocopy in the Birmingham University Library (L.Add 2657).

Printed Muirhead, *James Watt* ii 124–5.

Text From these sources.

Notes

(1) This is one of the finest of Darwin's letters. He begins with a classic example of Lunar banter – that effervescent essence which kept Lunar meetings bubbling with ideas. He goes on with a deep philosophical question jokily posed, and ends

with an even deeper scientific riddle, which started off the water controversy of the 1780s – and also provided an obscure but largely correct answer. See note 5.

(2) Darwin's letter was written in response to an invitation from Watt on 3 January 1781, also rich in Lunar banter: 'I beg that you would impress on your memory the idea that you promised to dine with sundry men of learning at my house on Monday next. . . . For your encouragement, there is a new book to cut up; and it is to be determined whether or not heat is a compound of phlogiston and empyreal air, and whether a mirror can reflect the heat of the fire . . .' (Muirhead, *James Watt* ii 123). If he could not attend, Darwin was expected to provide the Lunaticks with some bright ideas to discuss.

(3) Beaudesert is near Rugeley and was the seat of Lord Paget (1744–1812), who was made Earl of Uxbridge in 1784. Four of the 'nine beautiful children' suffering 'measles with peripneumony' later had notable careers. The eldest son Henry (1768–1854), a dashing cavalry general, was commander of all the allied cavalry at the battle of Waterloo, after which he was made Marquess of Anglesey and pursued a political career. The second son, Sir Arthur Paget (1771–1840), was a diplomat; the fourth, Sir Edward Paget (1775–1849), was another general, Wellington's second-in-command in the Peninsular War; the fifth, Sir Charles Paget (1778–1839), became an admiral. See *DNB* for their biographies.

(4) Though Darwin pretends to joke about 'the devil and all holy men', his continual exposure to the horrors of eighteenth-century disease had gradually sapped his faith in a benevolent God, and opened his mind to the possibility of biological evolution proceeding naturally, without divine intervention. Manichaeism, which regards the world as a battle between good and evil forces, was officially a heresy; but orthodox Christianity, with its Devil who promoted evil, was always in danger of slipping into Manichaeism, and could never satisfactorily explain why the omnipotent God chose to allow this supposed Devil to operate. Darwin saw the contradictions, and in the evolutionary philosophy which he later developed, evil is a spin-off from the struggle for existence, a struggle in which good – in the form of happiness – triumphs over evil, because it is usually the healthiest and happiest creatures that survive the struggle. Several of Darwin's Lunar friends were religious believers, so he politely wraps his queries in the Great Chain of Being, a popular deistic concept in the eighteenth century: biologically, for example, it offered the interpretation that the Almighty had created a well-designed chain of species.

(5) As well as these philosophical queries, Darwin offers Watt two obscure riddles in 'material philosophy'. The first, that 'atmospheric gas is composed of light . . .', appears nonsensical and may be just a joke, unless there is a wrong word, as Muirhead seems to suggest with his obscure footnote, 'Sic. in orig.' Darwin's second riddle has to be called historic, because it can be read as giving the correct answer to the 'water controversy' – over the relative contributions of Priestley, Cavendish, Watt and Lavoisier in answering Darwin's riddle. In 1781 nearly everyone, both the phlogistians and their opponents, thought water was a 'simple element' that could not be decomposed. Darwin's 'secret' now seems three-headed: first, he was saying water is not an element, but can be decomposed; second, that one of its components is a gas; third that the gas is hydrogen, generated when sulphuric acid (oil of vitriol) acts on a metal (though 'earth' was more often a mineral). By April 1781 Priestley had performed various experi-

ments on exploding air with hydrogen 'to entertain a few philosophical friends, who had formed themselves into a private society' – the Lunar Society. He also fired hydrogen (inflammable air) with oxygen (dephlogisticated air) and produced water, but didn't realise it. Cavendish repeated the experiments quantitatively; Watt offered a partially correct explanation; then Lavoisier was informed of the results, and in November 1783 produced the correct explanation. The controversy over the relative merits of the four contributors continued well into the nineteenth century. Perhaps Darwin also deserves a share. See *Doctor of Revolution*, pp. 141–3, for further details.

(6) Darwin's abbreviation 'all the Phlos' may be another joke, to be taken either straightforwardly as 'all the Philosophers', or obliquely, as 'all the Phlogistians'. All the Lunaticks except Darwin were tenacious in clinging to the erroneous phlogiston theory and Darwin was considered something of a traitor in adopting the 'French heresy' of oxygen in the mid 1780s.

81B. To Thomas Cadell, 5 April 1781

Dr Darwin of Lichfield (who some time ago sent some books to Mr Cadel intitled Experiments on Pus and Mucus) calls on Mr Cadel, and shall be glad to know what number are sold, and shall be glad of an account.

Dr D. is now removed to Radbourn near Derby.

E D

Apr.5 –81

Original William Salt Library, Stafford, S. MS 478.
Printed Unpublished.
Text From original MS.
Notes

(1) Thomas Cadell (1742–1802) was a leading London bookseller and publisher in the 1770s and 1780s. He had a good reputation for looking after his authors, who included Gibbon. He retired in 1793. See *DNB*.

(2) Darwin married Elizabeth Pole on 6 March 1781, four months after the death of her former husband, Colonel Sacheverel Pole. After their marriage, the Darwins spent a few weeks in London before settling at Radburn Hall. While in London, Darwin evidently called at Cadell's shop when Cadell himself was away, leaving this note. See also letters 78H and 85A.

(3) With his second marriage Darwin's life – and letters – enter a new phase. He stayed at home much more, began writing books, and wrote many more letters.

81C. To Thomas Day, 16 May 1781

My Dear Friend,

I was very sorry I had not the pleasure of seeing you in London, and the more so as I understood from Erasmus, that you gave yourself the trouble of coming up on purpose. Now you must know that I was angry at you for giving yourself that trouble, as I do not think any thing like form is necessary between you and me. I know the general benevolence of your heart and your friendly disposition to me from innumerable instances, and, therefore, a visit of form, as I understood your's was to me as a new married man, was by no means necessary. The reason I did not stay at home was your not having fixed the hour, and we were engaged to return numerous visits, which were indispensable, and I thought you as likely to come after noon as before. Enough of this. When you come into this part of the country, I shall hope a visit from you and Mrs Day – not in form.

Mrs Darwin and I had long been acquainted with each other; she is possessed of much inoffensive vivacity, with a clear and distinct understanding, and great active benevolence; like myself, she loves the country and retirement, and makes me as happy as my nature is capable of.

Mrs Darwin begs her compliments to Mr Day; she has read, by the Doctor's permission, the above account of herself, and thinks the Doctor has done her great injustice, as he has left out a principal part of her character, that is, that she loves and esteems her husband. She begs leave to add she was sorry not to have seen Mr Day in London.

Now I have resumed the pen, I am happy to tell you the pleasure I have received from reading a book of Dr Balguy's, "Divine Benevolence asserted" – you may be sure I think it a very *able* performance, as *his sentiments coincide with my own*. He makes it evident that every thing was made by the Lord, with design of producing happiness; and seems to show there is more happiness than the contrary in the world.

Pray, my good friend, why did not you contribute to the *benevolent* designs of Providence, by *buying* a seat in Parliament? Mankind will not be served without being first *pleased* or tickled. They take the present pleasure of *getting drunk* with their candidate, as an *earnest* or proof, that he will contribute to their *future good*; as some men think the goodness of the Lord to us mortals in this world, his temporary goodness, is a proof of his future and eternal goodness to us.

Now you wrap up your talent in a *napkin*, and instead of speaking in the assembly of the nation, and pleading the cause of America and Africa, you are sowing turnips, in which every farmer can equal or excel you. It puts me in mind of Rabelais's Devil. One of the most ingenious devils of all hell chose to turn farmer, but thought best to take an experienced farmer into partnership. – "Mr Clodpole," says the Devil, "the Pope has given me your farm, so if you please we'll go into partnership; you shall plough and sow, I'll take care that no hail or lightning, or storms, shall injure the crop; and that we may not quarrel about the product, I will take all above ground, and you all under ground." The farmer agreed, and set the whole farm with potatoes: when harvest time came, the Devil was much disappointed, and next year he said, "Now, Mr Farmer, it is my turn, you shall have all above ground, and I all under; so Mr Clodpole I shall be even with you." But lo! now the farmer sowed wheat; and Lucifer was again disappointed. "If this is the case," says Old Nick, "I'll e'en go and manage the minds of men, in which nobody excels me, tempting lawyers with gold, lovers with rouge, philosophers with fame, and real statesmen with retirement; and thus, as far as I can, impede the benevolence of Providence." God protect you – pray make my best compliments to Mrs Day, and believe me

Your affectionate Friend
E Darwin

Radbourn, near Derby
May 16 –81

Original Not traced.
Printed T. Lowndes, *Tracts in Prose and Verse* (London, 1827), Vol. II, pp. 332–5.
Text From above source.
Notes
(1) Thomas Day (1748–1789) was the youngest and most eccentric of the original Lunar group. He first met Darwin in 1768 (see Edgeworth, *Memoirs* i 196–7) and they became close friends in 1770–71, when Day came to live at Lichfield, pursuing his plan of educating his ward Sabrina, and then unsuccessfully wooing Honora and Elizabeth Sneyd, who later became the second and third wives of his friend Edgeworth. (See *Doctor of Revolution*, pp. 84–7.) Day himself married Esther Milnes in 1778, and decided to retire from worldly pursuits and devote himself to farming. He kept to this plan, and the rest of his life was spent in farming, first in Essex and then in Surrey, at Ottershaw, near Woking. During the 1780s he wrote his famous book *Sandford and Merton*, which is said to have been the most-read book of the nineteenth century.
(2) From this letter it appears that Day called on the newly-wed Dr and Mrs Darwin during their visit to London in April 1781, but unfortunately missed them. With Darwin's pen-picture of his second wife Elizabeth, and her amendment of it, this is one of the most revealing of Darwin's letters. Erasmus (line 2) is Darwin's son (1759–1799), who was friendly with Day.
(3) Thomas Balguy (1716–1795) was archdeacon of Winchester and a writer of religious books, including *Divine Benevolence asserted* (1781), which was intended to show how benevolently God had arranged things to suit man, providing food to eat, minerals to mine, etc. See *DNB*.
(4) The second half of the letter shows Darwin's concern over Day's career – or rather the lack of it. All Day's friends thought him the most eloquent man they had ever met, and they expected him to have a brilliant career in politics. But he would not compromise his integrity by angling for (or buying) a parliamentary seat, so his brilliant political career remained unstarted. Darwin gently chides him for retiring from the public eye, in a tactful and humorous manner. Darwin's comment about the Lord making everything to produce happiness was itself humorous, as Day must have known; but Darwin *was* keen on promoting happiness, and in *Phytologia* (1800) he sees evolution as maximizing happiness by ensuring that the healthiest and happiest are most likely to survive in the struggle for existence.

(5) Darwin's story from Rabelais (*Gargantua and Pantagruel*, Book 4, chapters 45 and 46) is amazingly apt.

81D. To Joseph Banks, 13 September 1781

Radburn near Derby
Sept 13 –81

Dear Sir

A botanical society at Lichfield have done a great part of a translation of the Genera and Species of Linneus. Of this you will herewith recieve a small specimen, with an account of their design. They hope you will give them leave to prefix the inclosed dedication of their work to you – and should esteem themselves much honour'd if you would favour them through me with any remarks on their design, or if you should think it can be translated on any better plan; and if you think such a translation may be of utility towards propagating the knowlege of Botany; as your opinion will much encourage or retard the progress of the work.

I should be very happy to see you or meet you any where, when you come into this country, and am, dear Sir, with Mrs Darwin's compliments to Lady Banks, your much obliged and obed. serv.

E Darwin

Original Not traced. There is a MS copy at the Botany Library, British Museum (Natural History), London, in the 'Correspondence of the Right Honorable Sir Joseph Banks, Bart', Volume II, p. 31.

Printed Excerpts, forming a summary, in *The Banks Letters*, ed. W. R. Dawson (British Museum, 1958); a short excerpt in *Doctor of Revolution*, p. 144.

Text From MS copy at Botany Library, British Museum (Natural History).

Notes

(1) Sir Joseph Banks (1743–1820) had been President of the Royal Society for three years in 1781, and was to remain President for another 39 years, becoming the unchallenged emperor of British science, a position no one else has since achieved. Botany was his speciality: he had won renown for his work as botanist on Cook's first voyage in 1768–71. So it was natural for Darwin to seek Banks's approval of the project for translating Linnaeus. From the ending of the letter it would appear that Darwin and his wife had met Sir Joseph and Lady Banks during their visit to London in April 1781. This is quite likely, since Greville knew Banks well. See H. C. Cameron, *Sir Joseph Banks* (1952); C. Lyte, *Sir Joseph Banks* (1980).

(2) Darwin had become very interested in botany in the 1770s and created his botanic garden a mile west of Lichfield about 1777. He started the Lichfield Botanical Society in 1778, but it never had more than three members – Darwin himself, his friend Brooke Boothby and William Jackson, a proctor in the Cathedral jurisdiction.

(3) Brooke Boothby (1744–1824), who succeeded to his father's baronetcy in 1789, was a humane and accomplished man of literary tastes. After travelling widely in

Europe in his twenties, when he became a friend of Rousseau, he lived near Lichfield in the 1770s. Though not keen enough on technology to enter the Lunar circle, he was very interested in botany.

(4) For the identification of the third member of the Lichfield Botanical Society, William Jackson (1734/5–1798), I am much indebted to Mrs Jane Hampartumian, archivist at Lichfield Record Office. Jackson, described as a 'yeoman of the City of Lichfield', was married at St Chad's Church on 14 February 1759 to Jane Taylor (1734?–1786?) of the Close, Lichfield, and the register of the Close records the baptism of several of their children in the 1760s, the eldest being a son Taylor (1760–1793). On 16 January 1776 'William Jackson Notary Public', was admitted one of the Procurators General of the Bishop's Consistory Court at Lichfield (Lichfield Record Office B/C/2/108, p. 26), and continued as Proctor until his death on 17 August 1798.

When Darwin left Lichfield in 1781, Jackson took over his botanic garden and 'maintained it on the original plan as long as he lived' (*Gardener's Magazine*, **14**, 345–6 (1838)). Darwin apparently thought well of Jackson, and worked with him for eight years during the writing and publication of the Lichfield Society's translations from Linnaeus. Jackson was also a member of the Derby Philosophical Society, and the Society's loan register (Derby Central Library MS 9230) shows that scientific books were sent regularly to him in 1786–8. A quite different view of Jackson emerges from Anna Seward's venomous pen-portrait: 'Sprung from the lowest possible origin, and wholly uneducated, that man had, by the force of literary ambition and unwearied industry, obtained admittance into the courts of the spiritual law, a profitable share of their emoluments, and had made a tolerable proficiency in the Latin and French languages. His life, which closed at sixty, was probably shortened by late acquired habits of ebriety. He passed through its course a would-be philosopher, a turgid and solemn coxcomb, whose morals were not the best, and who was vain of lancing his pointless sneers at Revealed Religion.' (Seward, *Darwin*, pp. 98–9.) She also says that Jackson admired Boothby, 'worshipped and aped Dr Darwin', and was a 'useful drudge' in the Linnaean translations. The obituary in the *Gentleman's Magazine* (**68**, 730 (1798)) calls Jackson 'a man of literature, and a useful assistant to Dr Darwin in his ingenious publication of the System of Vegetables'. So Jackson's character remains enigmatic.

In the past, confusion has reigned not only about Jackson's character, but also about his name. Most botanical reference books (e.g. R. Desmond, *Dictionary of British and Irish Botanists and Horticulturalists*, Taylor & Francis, 1977) state that the third member of the Lichfield Botanical Society was John Jackson the printer. This is most unlikely, since the *Universal British Directory* (*c.*1793) lists both 'William Jackson, Proctor' and 'John Jackson, Printer' among the Lichfield residents, and Anna Seward would not have confused the proctor she detested with the printer of most of her poems. The error seems to have originated in the *Botanical Commentaries* (1830) of Jonathan Stokes, who had no first-hand knowledge of Lichfield: the idea was plausible because John Jackson was the printer of both the Linnaean translations. A second false scent was laid by R. Simms in his *Bibliotheca Staffordiensis* (1894): his identification of Darwin's collaborator as 'Rev. Joseph Jackson' was followed in *Lunar Society* and *Doctor of Revolution*. Another mistake arose with R. D. Thornton, who says in his *Botanical Extracts* (1810) that

Rev. Dr Richard Jackson, Prebendary of Lichfield, of whom he gives a portrait, was a member of the Botanical Society. Some queries still remain, because William Jackson was such a common name – those born between 1727 and 1737 include the headmaster of Stockport Grammar School (1727–1791) (*Alum. Oxon.*); the musical composer of Exeter (1730–1803) (*DNB*); the suicidal vicar of Pytchley (1735–1785) (*Alum. Cantab.*); and the journalist (1737?–1795) (*DNB*). The proctor was none of these: but he might possibly be the 'William Jackson of Lichfield Close' who wrote the book *Beauties of Nature* (1769), superbly printed by Baskerville, which includes a batch of ribald poems that might have offended Anna.

(5) Darwin formed the Lichfield Botanical Society because he had the ambitious idea of translating several of the works of Linnaeus from Latin into English; but his opening remark in the letter, that 'a great part of a translation' had been done, was probably an exaggeration. Certainly Darwin was only at the beginning of six years of labour on the Linnaean translations, as many subsequent letters show. The books as finally published were the *System of Vegetables* (1783–5) and *The Families of Plants* (1787), running to nearly 2000 pages. For details of the Linnaean works translated, see B. Henrey, *British Botanical and Horticultural Literature before 1800* (OUP, 1975) ii 132. Although the authorship of the translations is attributed to 'A Botanical Society at Lichfield', it seems probable that Darwin did most of the work, but concealed his true role because he was very shy of flaunting himself as an author: he hesitated for several years before publishing *The Loves of the Plants* (anonymously) in 1789. He felt more comfortable sheltering behind the anonymity of the Lichfield Society, and no doubt relished the joke when visitors to Lichfield asked to attend a meeting of the Society (for only Jackson lived there).

(6) Darwin consistently uses the spelling *Linneus*, which was quite usual at that time, and a reasonable adaptation of the author's own version of his name (Carl von Linné). I have used *Linnaeus* and *Linnaean* in my notes, in accordance with modern usage, except when referring to the Linnean Society, which retains the old spelling.

81E. To Carl Linnaeus (the younger), 14 September 1781

Derby Sept 14 –81

Sir,

Being informed of your being in London by my friend Mr Wedgewood, I have taken the liberty to send you through his hands a specimen of a translation of the genera and species plantarum, about to be publish'd by a botanical society at Lichfield. Being informed that you are well acquainted with the english language, and wishing to do every thing in their power to render their translation, as worthy of the original of your illustrious father as they can, they would esteem themselves much obliged to you, for any remarks on their plan of translating, and should be happy to adopt any improve[me]nt, <u>if you think it can be translated on a better plan</u>.

I shall be happy to be honour'd with a line from you on this subject, after you have compared the specimens with the original, and if any thing should bring you into Derbyshire shall hope to have the pleasure of your company at Radbourn near Derby

and am, Sir,
Your obed. serv.
E Darwin

Please to direct to
Dr Darwin
Derby

Addressed Dr Linneus, London.

Original Linnean Society, London: Linnaeus's Correspondence, Vol. III, ff. 205–6.
Printed Unpublished. A microfiche facsimile of the original MS is available in: 'Linnaeus, C. (Correspondence)'. Inter Documentation Company, Zug, 1968. Order No. 3–6490/2.
Text From original MS.
Notes
(1) Dr Carl Linnaeus the younger (1741–1783), son and disciple of the great Linnaeus, succeeded his father as Professor of Botany at the University of Uppsala in 1778, and also inherited his library and natural history collection. These, together with his herbarium and manuscripts, were bought by James Edward Smith in 1784 and after his death in 1828 were bought by the Linnean Society, of which Smith was founder, and President for 40 years.
(2) The specimen of Darwin's translation sent to Linnaeus was presumably the same as that sent to Banks (letter 81D) and 40 other botanists (see letter 81F), and may perhaps be identified with one of two 7-page specimens bound in at the back of a copy of Linnaeus's *Fundamenta Botanica* in the Linnean Society Library.

81F. To Joseph Banks, 29 September 1781

Radbourn near Derby
Sept 29 –81

Sir,

I transmitted to you a little time ago a specimen of a translation of the Genera and Species plantarum of Linneus, by a botanical society at Lichfield; with a dedication, which they hope you will permit them to prefix to their work. A similar specimen has been sent to about forty botanists, many of whom have sent answers with their approbation in general. Mr Lightfoot advises the botanical society to introduce the plants in the Mantissa in their proper places, and also the plants mention'd in the Edition of the regnum vegetabile by Murray. This last book they have not been yet able to procure, but beg to have your opinion on these and other circumstances, which will be decisive.

Mr Crowe of Tuckswood near Norwich has very politely sent me specimens of two rare grasses, the Agrostis littoralis, and the Phalaris phleoides;

he advises a character to be introduced to express the plants, which are native of Britain. And also to mention the habitations of them, and is so obliging as to say he would supply us with many new habitations; this last circumstance I fear by adding a new line or two would prevent the english from being printed on exactly correspondent pages with the latin, which was the design. Of all these we beg your opinion.

Mr Crowe proceeds in the following words "there is a new edition of the species plantarum publishing abroad, many mistakes in the former are in this corrected, some plants placed to different genera, and many other alterations made [;] it is esteemed to be executed in so masterly a manner, that Sir Joseph Banks is altering his Herbarium to it, as far as it is printed. About seven weeks ago He had recieved by the post as far as Hexandria. His noble spirit would give anything, so you might with ease borrow his copy; and he will recieve the remainder, I hope, time enough for the publishing the translation, as it is proposed to be deliver'd in numbers".

You see, Sir, by this extract from Mr Crowe's letter, I am encouraged to ask further from you, not only information but assistance on all these circumstances.

The design of the translators is to form such a language from the english, as Linneus has done from the latin; and as the english language bears compound words better than the latin, they hope to express the meaning of Linneus (as explain'd by himself in the philosophia botanica, or by Elmgreen in a thesis published under Linneus's eye, in Vol. 6 of the amoenitates academicae) with more precision in english, than could be done in latin; so the words awl-pointed, for acuminatum; and bristle-pointed for cuspidatum are more expressive than the latin-words. And so end-hollowed for retusum; end-notch'd for emarginatum, edge-hollow'd for sinuatum; scollop'd for repandum; wire-creeping for sarmentosus; and many others, express the meaning according to his own definitions, better than the latin language was capable of.

Mr Lightfoot wish'd not to have the botanic terms have english terminations, but some of them as Petalum, are already anglicised; and the botanic language in english could not be made as concise and distinct, as the botanic latin language without it; because both the participles, diminutives, and compounds of these words are often wanted; thus petal, five-petaled, petal-like; corol, corollet, corol-like; corymb, corymbed; fascicle, fascicled, fascicular; these few terms will soon be familiarized to the ear.

Some of our friends think the word eggshape should be written eggshaped – we thought sword-form and eggshape and halbert-shape etc might be used adjectively better – as a sisser-formed leafe, would be a leaf cut out by sissers, whereas a sisser-form leaf would mean one in the form of sissers. So thread-formed styles, would be styles formed of thread; and thread-form style, one of the form of a thread. The same in the words compounded with shape – as we never say three-folded nor uniformed –

and were hence induced to think the words signifying general measures (as size, shape, form, weight, price) when compounded, would bear better to be used as adjectives than as participles. As in common conversation we say, it is a large size, a fine shape, elegant form, great weight, high price; instead of it is of a large size etc.

Of these niceties of language I will endeavour to get the opinion of Dr Johnson, and write it eggshaped, and thread-formed, if it be esteemed better.

I have no excuse to offer for giving you this second trouble, but the knowlege of your general love of science, and your philanthropy to wish that science to be propagated amongst your countrymen.

Mrs Darwin begs her best compliments to Lady Banks, and I am, dear Sir, with true respect

Your much obliged and most obed. serv.

E Darwin

Original American Philosophical Society, Philadelphia, Pa, Misc. MSS.
Printed Unpublished, apart from a few phrases in *Doctor of Revolution*, p. 145.
Text From original MS.
Notes

(1) This lengthy letter was sent to Banks before he had replied to the previous letter. Darwin was obviously working on the Linnaean translation project with great energy. He had already sent off the specimens of the translations to forty botanists (see letter 81G for an example) and was corresponding with several of them. It was nothing new for Darwin to display energy: what *is* new in this letter is the scholarly tone and attention to detail. Previously, Darwin had been deeply involved in the Lunar circle, and his Lunar friends, though men of genius, were men of the world rather than scholars. And his medical practice, though arduous, was far from scholarly. The isolation at Radburn after his second marriage was just what he needed to bring out his scholarly talents: without this change of life, he might never have written the books for which he won fame. He might have remained the life and soul of Lunar meetings but 'left not a wrack behind'. Another unusual feature of the letter is the touch of flattery towards Banks, which sprang from a genuine admiration of Banks's work, but was accentuated by the hope that Banks might lend Darwin copies of the books he needed.

(2) John Lightfoot, FRS (1735–1788) was a country clergyman and naturalist, author of *Flora Scotica* (1778). Darwin accepted his advice and translated Linnaeus's *Systema Vegetabilium*, as edited by J. A. Murray (13th edn, Gottingen, 1774) – Darwin sometimes calls it the *Regnum Vegetabile*. So Darwin abandoned (or at least postponed) his original plan of translating the *Genera Plantarum*.

(3) James Crowe (1750–1807) was a Norfolk surgeon and 'a most excellent' botanist, according to Sir J. E. Smith, President of the Linnean Society. Crowe specialized in mosses, fungi and willows. The thesis by 'Elmgreen' is Johannes Elmgren's *Termini Botanici* (Uppsala, 1762), which Darwin translated in October (see letter 81H). Linnaeus published 10 volumes of the *Amoenitates Academicae*: Darwin refers to Vol. 6, published in 1763.

(4) In support of his claim that English is better than Latin for detailed description of the shape of leaves, Darwin goes into surprising detail: it is strange to find him arguing the case of *eggshape* versus *eggshaped*. He did obtain the opinion of Dr Johnson, whose help is acknowledged in the preface.

(5) *Glossary*. Mantissa: addendum (though Linnaeus's *Mantissa Plantarum* (1767) ran to a whole volume). *Regnum vegetabile*: vegetable kingdom. *Systema Vegetabilium*: system of vegetables. *Agrostis littoralis*: bent grass of the sea shore. *Phalaris phleoides*: a type of canary grass. *Hexandria*: plants with six stamens. Awl-pointed, etc: an awl is a flat blade ending in a point, whereas a bristle is cylindrical; hence Darwin's distinction. For the seven Latin terms *acuminatum* . . . *sarmentosus*, it is difficult to improve on Darwin's own translations. Corol (usually corolla): coloured 'coronet' of petals forming inner envelope of a flower. Corolet: floret in aggregate flower (first used 1794, according to *OED*). Corymb: kind of cluster of flowers (or berries). Fascicle: bunch of leaves, flowers, etc (first used 1794, according to *OED*). Halbert (= halberd): spear with axe-like blade. Sissers: scissors. Style: bristle carrying stigma.

81G. To Richard Pulteney, 10 October 1781

Derby Oct 10 1781

Sir,

Enclosed you will receive a Specimen of a Translation of the Genera and Species Plantarum of Linneus by a botanical Society at Litchfield, who have printed a Sheet to transmit to the ingenious, hoping to be improved by their remarks, or encouraged by their approbation. They wou'd be much obliged to you if you woud honor them with a line of your Sentiments, whether it can be translated on a better plan? and if it merits your approbation, they hope you will recommend it to such of your botanical friends, as you think it may be agreable to.

I am Sir
With great respect
Your obedient serv.
Erasmus Darwin

Original Not traced. MS copy at Library of Fitzwilliam Museum, Cambridge, S. G. Perceval Bequest, General Series.
Printed Unpublished.
Text From MS copy at Fitzwilliam Museum Library.
Notes

(1) This is one of the letters 'sent to 40 botanists', which Darwin mentions in letter 81F. Dr Richard Pulteney FRS (1730–1801) lived at Blandford, Dorset, and had a flourishing medical practice in the surrounding counties. He was very keen on botany and was a continual contributor to the *Gentleman's Magazine*, mainly on botanical subjects, and author of *A General View of the Writings of Linnaeus* (1781), which greatly helped to publicize knowledge of the Linnaean system. For further information on Pulteney, see *DNB*.

(2) Pulteney's reply (MS in Fitzwilliam Museum Library) is dated 5 January 1782: he says he did not receive Darwin's letter until 29 December. Pulteney mentions that a new edition of the *Genera* and *Species Plantarum* is expected, gives his approval to the translation project and offers his best wishes for its success. The imminence of a new edition of the *Genera Plantarum* was probably Darwin's main reason for postponing his translation of it.

81H. To Joseph Banks, 24 October 1781

Derby Oct. 24 –81

Dear Sir,

I am much indebted to you for the observations you have favour'd me with, and the assistance you promise in the prosecution of our work.

We have nearly finished a translation of Elmgreen's termini, which we think of prefixing to the first number; as you seem to think some sort of dictionary necessary, and perhaps will approve of this. This translation of the termini, if you will give me leave, I will transmit to you in a week or two, and hope you will just cast your eye over it; and acquaint us with any remarks which may occur to you.

But as I have only the edition of it printed before Hudson's book, which has many typographical errors (witness No. 515, where avidis sororis is printed I suppose for aridis sonoris) I shall be much obliged to you to lend me the sixth volume of the amoenitates academicae, which books I have in vain attempted to buy, along with a Murray, if you can spare one, and the supplementum; both which shall on the very day you fix be returned.

The greatest care shall be observed to avoid any ridiculous terms, particularly in those bordering on obscenity. The word grinning I should imagine can not well be changed; masked in our language is often only covered, as in a masked battery – mask-like might be used, but perhaps most people would prefer grinning as a more literal translation.

If you could put us in a way how to procure the new edition of the species plantarum and would be so good as to direct a bookseller to send it me, I should also be much obliged, but am sorry to give you any trouble about things I may perhaps procure without troubling you.

Mr Boothby writes me word, he has had the pleasure of waiting upon you – and has favour'd me with some remarks you made on a poem, in which I have written several of the notes, and corrected some of the verse, a part of which was written by Miss Seward, and a part by a Mr Sayle. The history you mentioned to Mr Boothby of the Tremella is truly curious: yet I have certainly seen parts of fungi, particularly growing on an old rail, turned to a white gelatinous substance, as I supposed by the frost. I shall beg leave to add your account of it in the note, if it be agreeable to you: otherwise not. The other circumstances that Mr Boothby informs me you criticised, I shall endeavour to have altered – the design of the poem was to induce ladies and other unemploy'd scholars to study Botany, by putting many of the agreeable botanical facts into the notes.

I hope our society will be able to get one number out by Christmas, but shall advise not to print one till they have a few others in advance. We are advised to print about 112 pages in a two-shilling number every month.

I hope you will give me leave to write again to you after we have had a meeting; as we have received letters with different opinions about English names etc from above 20 botanical Gentlemen and Ladies, but all of them much applauding the design.

I am Sir with true respect
your much obliged and obed
E Darwin

If you can spare the books above mention'd, you will please to direct them by the coach to Dr Darwin to be left at Mr Halls Sadler in Derby. Our coach inns at the Swan with two necks in Lad-lane.

Original Not traced. There is a MS copy at the Botany Library, British Museum (Natural History), London, in the 'Correspondence of the Right Honorable Sir Joseph Banks, Bart', Volume II, pp. 52–4.

Printed Summary in *The Banks Letters*, ed. W. R. Dawson (British Museum, 1958); short excerpt in *Doctor of Revolution*, p. 145.

Text From MS copy at Botany Library.

Notes

(1) Apparently Banks made a favourable reply to Darwin's previous letters and offered to lend books – an offer which Darwin quickly took up.

(2) Darwin had obviously been hard at work translating Elmgren's *Termini*, and the translation duly appeared in the published *System of Vegetables* (pp. xv–xl).

(3) 'Hudson' was William Hudson, FRS (1734–1793) and 'Hudson's book' was probably his *Flora Anglica* (1762). The misprints mentioned are in Elmgren, not in Hudson, but do not occur in the 1762 edition of the *Termini*.

(4) Darwin was still unable to obtain Murray's edition of the *Systema Vegetabilium*, or Volume 6 of the *Amoenitates Academicae*, and he succeeded in borrowing copies from Banks. The 'supplementum' is the *Supplementum Plantarum* (Brunswick, 1781) by Carl Linnaeus the younger, and it lists many further plants. Darwin borrowed the *Supplementum* from Banks and included a translation of it in the *System of Vegetables*.

(5) This letter shows Brooke Boothby making a contribution to the project by discussing it personally with Banks. I have not traced the botanical poem mentioned. Perhaps it was based on Anna Seward's verses about Darwin's botanic garden (see Seward, *Darwin*, pp. 125–32), the notes being a first draft of those for *The Loves of the Plants*. Mr Sayle may be William Sayle: see letter 50A, note 5.

(6) Tremella, and its curious properties, are described in *Loves of the Plants* I 373 note. See also letter 93F.

(7) The idea of publishing the translation in monthly parts is pursued in Letter 81J.

(8) 'The Swan with two necks in Lad Lane' was a London coaching inn.

81I. To Joseph Banks, 1 November 1781

Derby Nov 1 –81

Dear Sir

With this I trouble you with a translation of Dr Elmsgreen's termini botanici, and shall be much obliged to you at a leizure hour to run your eye over it, and favour me with what you think may be translated more intelligibly or more elegantly – and with any other words, which Linneus has affix'd peculiar meanings to, which are not there mention'd – as we wish to make this translation, as perfect as our language and the nature of the subject will allow of. If Dr Solander and Dr Linneus, are with you, and will do me the same favour, I shall esteem it a particular obligation, from both of whom I hope at their leizure to be honoured with a line.

We have recieved letters of general approbation and encouragement, with some particular remarks from near thirty of reputed botanists – and shall begin the work as soon as we can be favour'd with the books, you are so kind to lend us – and which will be returned on the day you require them.

Some terms appear to be more distinctly explain'd in the philosophia botanica – in which Linneus has also given instances of particular plants. These I am thinking of adding to Elmsgreen, and printing them with the first number of the work, if you think that a good plan.

If there is any particular bookseller, that you would recommend to sell the work, if you please to mention his name, I will write to him, otherwise my design is to write to Cadel, meaning to print it at our own expense.

I am, Sir, with true respect
Your much obliged
and obed. serv.
E Darwin

We shall wait for your observations on the botanic terms, before we proceed any further.

Addressed Sir Joseph Banks Bar[t].

Original Library of Fitzwilliam Museum, Cambridge, S. G. Perceval Bequest, H 232.
Printed Unpublished.
Text From original MS.
Notes

(1) Darwin has honoured his promise to send the translation of Elmgren's *Termini* 'in a week or two' (letter 81H): this letter is only eight days later.

(2) Dr Daniel Solander (1736–1782) was a Swedish botanist who became Linnaeus's 'favourite pupil': his knowledge of the Linnaean system was unrivalled. He came to England in 1760 and was chief assistant to Banks on the voyage of the *Endeavour*. On his return he became secretary and librarian to Banks and then Keeper of the Natural History Department of the British Museum. Banks kept open house at Soho Square for scientists: Solander was a close friend of his, and

both Solander and Dr Linnaeus may have been staying there, as Darwin surmised.

(3) Linnaeus the elder described his methods in his *Philosophia botanica* (1751), the book Darwin mentions. Cadell was the London bookseller who handled the pamphlet by Darwin's son Charles (see letters 81B and 81J).

81J. To THOMAS CADELL, 2 December 1781

Radburn near Derby
Dec. 2 –81

Sir,

You will recieve this by the hands of my son, Mr Darwin, to whom you will please to pay the bill due to me. I should also be glad to know, if any more of the pamphlets are likely to be wanted, as I have yet 2 or 300 undisposed of.

I am tempted to join in another greater publication, a translation of the botanical works of Linneus by a botanical society at Lichfield. The plan has the approbation of many botanists, and is therefore likely to sell. Our design is to publish it in monthly numbers, 112 pages in a number, at 2/- each with a few prints. Mr Griffiths authour of the monthly review sais He supposes any principle bookseller, whom I may apply to, would give 1/6 each for them, if he has the sole vending them to all other booksellers.

The work will consist of between two and three thousand pages, I suppose about 2500. Our design is to print 1000 copies, the correction of the press will be so difficult, as there will be so many italic and two or 3 kinds of capital characters, besides astronomic and chemical marks, that it can not be printed, but in the country, where we can ourselves correct the press.

I have inclosed you a specimen which has been sent, and gain'd the approbation in general, with remarks for its improvement, of at least thirty eminent botanists.

The prints will be about 11. or 12. one in each number as far as they go.

The Printer in the country, who did my other pamphlet, asks us 25/ per sheet for 1000 coppies, and sais the paper will cost 14/ a ream; and that two quires in each ream are bad, and not to be reckon'd.

The reason of our intending to print it in numbers is that we can not well risk the expense of the whole copy at once – and by this means we shall feel our way in regard to the sale of it.

If you think it worth while to engage with us, pray acquaint me of your terms – with all other expenses which may attend the publication – and whether the above is a proper price of printing and paper, and any other intelligence which may occur to you about this affair. And also what bookseller you would recommend in Edinburgh, and Dublin, and whether these copies should likewise go through your hands.

The work will be dedicated to Sir Joseph Banks, who has likewise approved the design.

I shall hope to hear from you in a few posts, and am Sir

Your humble serv.

E Darwin

Please to direct Dr Darwin, Derby.

Addressed Mr Cadell.

Original Library of Fitzwilliam Museum, Cambridge. S. G. Perceval Bequest, H. 234.

Printed Unpublished, apart from a short excerpt in *Doctor of Revolution*, p. 145.

Text From original MS.

Notes

(1) Darwin's son Erasmus seems to have visited London quite frequently in the course of his legal business, and often carried letters and manuscripts of books to his father's publisher – or bookseller, in the usual phraseology of the day. 'The pamphlet' is the medical thesis of Darwin's son Charles (see letter 78H). For notes on Cadell, see letter 81B.

(2) Darwin was setting himself a hard task in proposing to issue the Linnaean translations at the rate of 112 pages a month, especially in view of the difficult problems of layout and different founts of type. In the end it appears that the translation was issued in four parts between 1783 and 1785, and then reissued as a two-volume book (with the date 1783) under a lengthy title, '*A System of Vegetables, according to their classes, orders* ... translated from the *Systema Vegetabilium* of Linneus by a Botanical Society at Lichfield'. The London booksellers were Leigh and Sotheby, so apparently Cadell did not accept Darwin's offer. Darwin's later books were all published by Johnson (see letter 84G), though Cadell does figure again in Darwin's bibliography as publisher of the posthumous *Beauties of the Botanic Garden* (1805).

(3) 'Mr Griffiths' was Ralph Griffiths (1720–1803), editor of the *Monthly Review*, and a close friend of Wedgwood's late partner, Thomas Bentley (see letter 80C).

82A. To Matthew Boulton, 27 January 1782

Radburn Jan. 27 –82

Dear Boulton,

Whether you are dead, and breathing inflammable air below; or dephogisticated air above; or whether you continue to crawl upon this miry globe, measuring it's surface with your legs instead of compasses, and boring long galleries, as you pass along, through it's dense heterogeneous atmosphere; – – as I am alive, now, I can not recollect how I ment to finish this long period, so here we'l leave it; and pray tell me how you do, and your wife and children and fire-engines, and if you[r] philosophical society have discover'd the longitude and philosopher's stone. And above all if you keep yourself and friends in good spirits, and happyness.

Now a second purport of this letter is to tell you, that I have bought a large house at Derby, and if it be not inconvenient to you, should be glad if you could repay me the 200£ pounds you have of mine. As I shall want money to compleat this purchase. If you could fix a time for this purpose, I will either send Erasmus, or come over to Birmingham myself; and see what all you philosophers are about.

Mr Erasmus has got the recieversip of the Pole-Estate, and is to settle immediately in Derby, in a part of the house I have purchased. I told him and Mr Robt. I was writing to you, who both beg to be remember'd to you. If you and Mrs Boulton should come into Derbyshire I hope you will not pass us. Mrs Darwin begs her Compliments to yourself and Mrs Boulton.

from your affect. friend
E Darwin

Original Birmingham Reference Library, Matthew Boulton Papers, Darwin 29.
Printed Excerpt in *Doctor of Revolution*, p. 148.
Text From original MS.
Notes
(1) There is probably more than meets the eye in the first paragraph: Darwin often wrapped serious speculations in a jokey cover. In 1782 Priestley and the Lunar Society were exercising themselves over the composition of water (or steam). After giving them the 'secret' in January 1781 (letter 81A), Darwin is possibly reiterating it: can it be just by chance that the components of water – hydrogen (inflammable air) and oxygen (dephlogisticated air) – appear in the first two lines of his letter? The idea of Boulton 'boring long galleries' through the air refers obliquely to the boring of the cylinders for his steam engines; and perhaps the 'globe' refers to the globes in which Priestley was exploding hydrogen, as well as to the earth? Hence also the 'heterogeneous atmosphere', and the need for 'compasses', to draw a diagram of the globe? Maybe. But I have no idea why the Lunar Society was trying to discover the longitude (of what?).
(2) Though Darwin bought his house in Derby early in 1782, and sometimes stayed there during 1782–3, his family continued to live at Radburn Hall until the autumn of 1783. 'Mr Erasmus', Darwin's second son, was now 22 and had presumably completed his legal studies. He moved to Derby and was very successful as a solicitor there. 'Mr Robt.' was now 15. Boulton's wife Anne is here mentioned for the last time in these letters: she was drowned in the ornamental pool at Soho in July 1783.

82B. To Joseph Banks, 23 February 1782

Radburn, near Derby
Feb 23 –82

Dear Sir,

By the advise of many of our botanical correspondents the botanical society at Lichfield have determined to translate Murray's edition of the regnum vegetabile, and to insert in their places the essential characters of the plants from Dr Linneus's supplementum. They have at length after much advise and consideration determined the botanical language; which is much more similar to the original than that they first proposed. A number is now in the press, and will, it is hoped, be out in a month or six weeks, and we hope will meet with your approbation.

We have been able to procure but one copy of Murray, and as the members of our society are not all at Lichfield, we can not proceed with this work without your permission to keep your copy of Murray, till new ones are imported, and we hope you will procure for us the remainder of the supplementum; and another copy also of the supplementum, as for the reasons mention'd above, we can not well proceed without two copies.

The volumn of the amaenitates acadomicae I will return by an opportunity of a safe person next week – but must beg, beside the above favours, that you will lend me Du Hamel's Physique de plantes, and Bonnet's usage de feuilles, if you have them, as I can not procure them, and wish to see if the physiology of plants can be brought to anything like science – or be compared with that of animals.

Your known liberality of sentiment, and desire of encourageing every kind of science, is my excuse for asking these favours of {you.}

Mrs Darwin joins in compliments to Lady Banks with Sir

Your much obliged
and obed. serv.
E Darwin

Addressed Sir Joseph Banks Bart., PRS. Soho Square, London.

Original Library of Fitzwilliam Museum, Cambridge, S. G. Perceval Bequest, H 233.

Printed Unpublished, apart from short excerpt in *Doctor of Revolution*, p. 145.

Text From original MS.

Notes

(1) This letter brings out very clearly the generosity of Sir Joseph Banks to his fellow botanists: Darwin was in some danger of treating Banks as though he were the London Library. However, Darwin can scarcely be blamed, because he seems to have been unable to obtain the books elsewhere, and he was polite in his requests. Still, the translations might never have come to fruition without Banks's generous help.

(2) The changes in language decided during the winter are summarized in the preface of the *System of Vegetables* (p. iii): 'We have retained the words calyx for flower-cup; corol for blossom; stamen for chive; pistil for pointal; pericarp for fruit vessel'. Darwin was particularly scornful about the words 'chive' and 'pointal', used by Withering in an attempt to avoid sexual indelicacy.

(3) From this letter it appears that at least one of the other two members of the Lichfield Botanical Society helped with the translation. I would guess – and it can be no more than guess-work – that Darwin did the translation and then passed it on to either Boothby or Jackson (or both) for checking (Boothby?) and fair-copying (Jackson?).

(4) H. L. Duhamel (1700–1782) was a well-known French writer on botany and agriculture: *La Physique des Arbres* (1758) is reckoned his finest work. Charles Bonnet (1720–1793) was an influential Swiss naturalist who was a leading exponent of the 'preformation' theory of development, arguing that every being contained the 'germs' of all subsequent generations. Bonnet's *Recherches sur l'Usage des Feuilles des Plantes*, the book requested by Darwin, was published in 1754. Darwin's confidence that he can unravel the mysteries of plant physiology is rather breath-taking, but typical of his optimistic 'get-things-done' approach.

82C. To Joseph Banks, 17 March 1782

Radburn Mar 17 –82

Dear Sir,

I return'd your sixth volume of the amoen: academ: and thank you for the loan of it. I should have sooner sent it, but hoped to have received another copy of Murray, and also that Dr Linneus's supplementum would have been procured from abroad, and then meant to have returned them together.

Mrs Blackburn favour'd us with a copy of Murray, but desired it to be returned in three months, which it was to a day; and as I could procure but one other, and our society was not all resident at Lichfield, we were distressed on this account, but are still flatter'd with daily hopes of more being imported. I am sorry you say the remainder of the supplementum is not likely soon to be had.

On looking over Malpighi and Grew and Hales, the phisiology of plants appear'd to me not to have hitherto been under the attention of any one perfectly acquainted with the animal oeconomy: last summer I contrived to inspect the absorbent system of the Picris with a colour'd liquor; and as the blood of that plant is white, these two systems were beautifully apparent to the eye. On reading a manuscript translation of Mr ——, a Swedish naturalist, I found the authors I mention'd to you in my last, had made a set of similar experiments; and I had designed to have investigated this subject, so little understood at present, further during the summer.

This, however, I have now laid aside for, perhaps, more important, tho' less ingenious occupations; and shall therefore decline giving you the trouble of sending me the books you are so kind as to offer, both in your last, and in a former letter of yours, and am Sir

With great respect
Your obed. serv.
E Darwin

Original Not traced. There is a MS copy at the Botany Library, British Museum (Natural History), London, in the 'Correspondence of the Right Honorable Sir Joseph Banks, Bart', Volume II, pp. 106–7.

Printed Summary in *The Banks Letters*, ed. W. R. Dawson (British Museum, 1958).

Text From MS copy at Botany Library, British Museum (Natural History).
Notes
(1) See Notes to letters 81E and 81G for explanations of *amoen: academ:,* Murray and *supplementum.*
(2) 'Mrs Blackburn' is Anna Blackburne (1726–1793). She was the daughter of the botanist John Blackburne (1693–1786), of Orford near Warrington, who maintained a famous botanic garden, with many exotic species. Anna was herself a skilled botanist, and she corresponded with Linnaeus and many other naturalists. She also assembled a museum of natural history. Though unmarried, she was known as Mrs Blackburne because she was in charge of her father's household (her mother died in 1740). For a biographical sketch of Anna, see V. P. Wystrach, *J. Soc. Bibliog. Nat. Hist.*, **8**, 148–68 (1977).
(3) Marcello Malpighi (1628–1694) was a great Italian biologist, discoverer of the capillaries and author of a classic book on the anatomy of plants (1679). Nehemiah Grew (1641–1712) was a physician, plant anatomist and early Fellow of the Royal Society, best known for his book *The Anatomy of Plants* (1682). Stephen Hales (1677–1761) was a chemist and a pioneer of plant physiology, who in his classic *Vegetable Staticks* (1727) applied numerical methods to plants, to investigate the behaviour of the sap and plant respiration.
(4) Darwin discusses his experiments on the absorbent system of Picris and other plants in *Phytologia* (pp. 11–12), and also mentions there (pp. 12–19) the work of Malpighi, Grew and Hales. These three are the authors mentioned 'in my last' – i.e. 'in my last sentence'. I have not identified 'Mr ——, a Swedish naturalist'.
(5) The 'more important tho' less ingenious' occupation was presumably the translations of Linnaeus.

82D. To The Gentlemen Proprietors of the Water-Works at Nottingham, 31 August 1782

Derby Augt. 31 –82

Gentlemen,

I have pleasure of recommending to your notice an ingenious man, Mr Hancock, who intends to offer himself a candidate as Engineer for your water works. And have no doubt but you will find him on examination well skill'd in that branch of natural philosophy, and well acquainted with the practical part of it. And in every respect worthy your patronage.

I am Gentlemen, with great respect your obed. serv.

E Darwin

Addressed To the Gentlemen Proprietors of the Water-works at Nottingham.

Original Birmingham University Library MS L.Add 14.
Printed Unpublished.
Text From original MS.
Notes
(1) The Water-works at Nottingham were a fine example of civic enterprise. The

Nottingham Council first considered a proposal for supplying the town with piped water in 1693, and in 1696 the Corporation went ahead with a project for furnishing the inhabitants with 'convenient clean sweet and wholesome water by aqueducts and pipes' from the River Leen, with provision for a pumping engine and reservoirs. The Proprietors seem to have been mainly Councillors. See Freda M. Wilkins, 'Taking the waters', *The Nottinghamshire Historian*, No. 16, pp. 9–10 (1975). For further details, see the numerous references to the Water-works in *Records of the Borough of Nottingham*, Vol. V (Quaritch, 1900); Vol. VI (Forman, Nottingham, 1914); and Vol. VII (ibid, 1947).

(2) Thomas Hancock (*c.*1750–*c.*1810) was duly appointed Civil Engineer to the Water-works Company, probably thanks to Darwin's strong recommendations, and he filled the post from 1782 until 1805, most successfully, it seems: 'It was not until the appointment of Thomas Hancock as Engineer to the Water Works in 1782 that a satisfactory general supply to the town began to be organized' (D. Gray, *Nottingham through 500 years*, 2nd edn, Nottingham, 1960, p. 126). E. Willoughby's *Nottingham Directory* of 1799 records 'Thomas Hancock, Engineer' at 'Engine-house, Leen side'; this entry is repeated in the 1815 edition of C. Sutton's *Nottingham Directory*, but not in the 1818 edition. See also *Records of the Borough of Nottingham*, Vol. VII, p. 404.

82E. To Charles Upton, 11 September 1782

11th September 1782

Dear Sir,

In Answer to your Letter, I have Reason to believe that most, if not all the Windows of my House looking into your Yard occupied by Mr Bruckfield, are ancient Lights, as such you must know I have a Right to their Continuance. With regard to putting out new Lights, Edgar How, the late Owner of Mr Bruckfield's House, by a Deed dated 24th June, 1750, granted "to John Heath, his Heirs and Assigns, full and free Liberty, Power, and Authority, to put out or make any Windows or Lights, in the Wall of the said John Heath's House, to look into the said Edgar How's Yard, etc and Mr Edgar How covenants, that himself nor his Assigns shall disturb or interrupt Mr Heath etc in the Enjoyment of the said Liberty and Privilege etc".

After having stopped up five of the old Windows, viz. two Chamber Windows, two Closet Windows, and one Garret Window, all looking into your House and Yard, I did not expect any Objection from you to my enlarging one Cellar Window; and all I mean to do is to enlarge another, for the better lighting my Cellar Stairs; but had I not been assured that my Authority was perfectly valid in Law, I should not have done this without previously making Application to you.

I am, etc
E Darwin

Addressed To Mr Upton.

Original Not traced.

Printed In a pamphlet entitled 'A State of Facts, relative to the dispute between Dr Darwin, of Derby, and Mr Upton, Attorney in Derby, and Mr Upton's reasons for declining to bring the question to a second trial', by Charles Upton, dated 6 November 1786.

Text From the copy of the pamphlet in Derby Central Library, MS 9283.

Notes

(1) Charles Upton (1752–1814) was for many years an attorney in Derby. In 1779 he married Sarah, daughter of Thomas Evans (1723–1814), the Darley mill-owner and Derby banker, whose children were much intermarried with the Strutts (see letter 01D). Upton was Sheriff of the County in 1809, and purchased several estates near Allestree, where he is buried. See Glover, *Derby* ii 17–18. As the next note shows, he was evidently somewhat pernickety.

(2) Upton's pamphlet begins: 'In the Year 1782 I purchased the house next adjoining Dr Darwin's in the Full Street, in Derby; and the Doctor soon afterwards opened a Window which looked into my Yard . . . I thought it but reasonable to ask Dr Darwin by what Authority he claimed the Power of opening Windows upon me. . . . Dr Darwin favor'd me with an Answer, of which the following is a true copy'. Darwin's letter follows. Upton then goes on to say that after 'a diligent search', he discovered the deed mentioned by Darwin, and found that Darwin had omitted some phrases to the effect that the lights were not to be opposite the parlour windows 'of the said Edgar How's house'. Upton was astonished at 'such a Conduct, in so near a Neighbour': he thought Darwin ought to have quoted the deed verbatim. However, he talked the matter over with Darwin and they drew up a complicated agreement, which Upton quotes in full. For reasons of structural safety Darwin had to leave one window 3 feet 9 inches larger than Upton wished, the measurement having been left blank in the agreement. Upton refused to sign the agreement, and in 1785 heard that as a result Darwin was speaking ill of him to his relatives (presumably the Strutts). Upton therefore consulted an attorney in Leicester, who advised him to obstruct some of Darwin's windows. Darwin brought an action against him, tried at the Derby Summer Assizes 1785. Darwin won, by proving that 'four of the obstructed Lights had been windows between 25 and 26 years'. After exploring various legal avenues and finding little prospect of redress, Upton decided to state his case in a pamphlet.

(3) Darwin cannot have been pleased to be asked to block up one new window (and several old ones) in the house he had recently bought, but his letter is polite. Though we do not know Darwin's side of the story, the pamphlet shows that Upton was legalistic and contentious, and his action in blocking Darwin's 'ancient lights' was foolish as well as unfriendly. Darwin is presumably referring to Upton in letter 95E when he speaks of a 'professed spy', an attorney who 'shoulders us on the right'.

82F. To Josiah Wedgwood, 17 October 1782

* * *

I hope Dr Franklin will live to see peace, to see America recline under her own vine and fig-tree, turning her swords into plough-shares, etc.

* * *

Original Not traced. In the possession of Charles Darwin, 1879.
Printed C. Darwin, *Life of E. Darwin*, p. 45.
Text From above source.
Notes

Although the American War of Independence on land virtually ended in October 1781 with the surrender of Cornwallis at Yorktown, the war at sea was continued (chiefly by the French and Spaniards). The preliminary articles of peace were not signed until November 1782. This is the event Darwin is looking forward to, hoping that his old friend Franklin, whom King George III called the evil genius behind the American Revolution, would be able to enjoy peace in his old age. However, Darwin says it in Biblical clichés (*Isaiah* ii 4 and *Micah* iv 4), instead of with his usual originality of phrasing.

82G. To Matthew Boulton, 26 December 1782

Derby Dec.26 –82

Dear Boulton,

I have purchased a house here and furnished it, which has rather distress'd me for money; that if you could repay me the 200£, it would be a great convenience to me, and I hope not a great inconvenience to you. For this purpose I will at any time send over a proper receipt to you at Soho, or should be happy to see you here, as may be most agreable to you – or will meet you at Lichfield any day you may fix.

I am here cut of[f] from the milk of science, which flows in such redundant streams from your learned lunations; which, I can assure you, is a very great regret to me. I hope Master Matthew rises in growth and abilities, and will rival the activity of his father. I assure you I wish much to make a visit to Soho, which if the Lord permit I intend to perform next summer. Pray if you think of it, make my devoirs to the learned Insane of your society, and believe me,

yours sincerely
E Darwin

Original Birmingham Reference Library, Matthew Boulton Papers, Darwin 30.
Printed Excerpts from the second paragraph have frequently been published, e.g. *Lunar Society*, p. 204, and *Doctor of Revolution*, p. 150.
Text From original MS.

Notes

(1) Darwin's first request to Boulton to repay the £200 (letter 82A) obviously had no effect, so he was asking again. About this time Thomas Day was also asking Boulton to repay a loan, of £1000. But neither Darwin nor Day could persuade Boulton to pay up: his steam engines were still not showing a profit, and he needed the capital borrowed from his friends.

(2) It seems that in the winter of 1782–1783 Darwin operated his medical practice from the new house in Full Street, Derby, which was more convenient than the isolated Radburn Hall.

(3) Boulton's son Matthew was now 12 years old (see letter 89C).

83A. To Matthew Boulton, 4 March 1783

Derby Mar. 4 –83

Dear Sir

I am favour'd with your letter, and the note for 30£. In respect to the principle I should not have written to you for it, but that I supposed the income from your Engins would have render'd it not inconvenient to you to return it. When you can do it therefore conveniently, do so.

We intend to pass this Summer at Radburn, where Mrs D. and myself shall be very happy to see you, and yours; in the Autumn we design to return to Derby to reside for good (I hope) as they say.

We have establish'd an infant philosophical Society at Derby, but do not presume to compare it to you well-grown gigantic philosophers at Birmingham. Perhaps like the free-Mason societies, we may sometime make your society a visit, our number at present amounts to seven, and we meet hebdomidally.

I have repeatedly spoke of your Engins to Arkwright's friends, and hope you may still be employ'd, but know very little at present of the matter. I suppose you use your reciprocating engine for working corn-mills, and not your circular one?

Pray bring your Son to see us at Radburn sometime this Summer, that we may renew our early acquaintance. I wish you would bring a party of your society and hold one Moon at our house. NB. our Society intend to eclipse the Moon on the 18 of this month, pray don't you counteract our conjurations. I beg to be remember'd to all the Insane at your next meeting, and am, dear Boulton,

your affect. friend
and obed. serv. E Darwin

Original Birmingham Reference Library, Matthew Boulton Papers, Darwin 31.

Printed Excerpts frequently published, e.g. *Lunar Society*, p. 236, and *Doctor of Revolution*, pp. 153, 163.

Text From original MS.

Notes

(1) Darwin is now almost having to apologize for having asked for his money.

Boulton's image of easy affluence was excellent for promoting business confidence, but easily shattered by an actual request for money.

(2) The infant Derby Philosophical Society did not really thrive until after Darwin started living in Derby permanently. He did not give his inaugural presidential address until July 1784, and the meetings probably lapsed in the summer of 1783.

(3) The famous textile mill built by Richard Arkwright (1732–1792) at Cromford, near Matlock, was powered by water, and Boulton wanted to sell him a steam engine. Unfortunately Boulton was not on good terms with Arkwright, and that is why Darwin is acting as intermediary, no doubt through the Strutts of Derby, his fellow members of the Derby Philosophical Society, who were close friends of Arkwright. See letter 85B and the article on Arkwright in *DNB*.

83B. To Josiah and Sarah Wedgwood, 18 April 1783

Radburn

Dr Darwin and Mrs Darwin present their best Compliments to Mr and Mrs Wedgewood, and are truly sorry it will not be in their power to wait on them tomorrow morning at Radburn, as Mrs is confined to her room, tho' not yet brought to bed, and expects every hour, and the Dr is obliged to go to Ipstoc in Leicestershire early tomorrow morning unless prevented by Mrs Darwin.

They hope to have the pleasure of the Company of themselves and family as soon as Mrs D. is about again.

Apr. 18 –83

Original Keele University Library, Wedgwood archive, 237–2.
Printed Excerpt in *Doctor of Revolution*, p. 154.
Text From original MS.
Notes

(1) The child about to be born was Violetta, the eldest daughter of Darwin and his second wife, Elizabeth. Violetta was born on 23 April and lived to be ninety. She married Samuel Tertius Galton, and was the mother of Sir Francis Galton.

(2) 'Ipstoc' is Ibstock, about 20 miles south of Derby.

(3) 'Radburn' at the top is written upside-down in the MS.

83C. To The Printer of 'The Weekly Entertainer', 3 September 1783

Derby, September 3 1783

Sir,

The mortality among the horned cattle in this neighbourhood having spread a general alarm through the country, I shall be obliged to you to give the following account of the circumstances of it a place in your Weekly Entertainer, with the subsequent observations.

Sir, your's, etc
E Darwin

Original Not traced.
Printed Derby *Weekly Entertainer*, 29 September 1783, page 301.
Text From above source.
Notes

(1) Darwin's account of the disease tells us that it broke out in August at Melbourne, 7 miles south of Derby and spread to several villages nearby: at Calke (Caulk in Darwin's spelling), Sir Harry Harpur lost 16 cows. (*Weekly Ent.* pp. 301–2).

(2) Darwin's observations on the disease are confident and sensible. He advises that cattle affected by the disease should be destroyed and buried four feet deep to prevent the spread of the infection. He goes on: 'As all putrid diseases are attended with great debility, whatever contributes to decrease the strength of the cattle, makes them more liable to the infection: hence, blood-letting and purging are of the worst consequence; and on the contrary, whatever contributes to encrease their strength, makes them less liable to the infection' (*Weekly Ent.* pp. 302–3). In *Zoonomia* (ii 249) he advises that, if such an infection should recur, a government order should be obtained, to prevent cattle being moved, and all within five miles of the outbreak should be slaughtered – a policy adopted in the twentieth century in Britain.

83D. To The Duke of Devonshire, 20 November 1783

Derby Nov. 20 –83

My Lord Duke,

Inflammations of the internal, and more vital parts, are often by a wonderful process of the animal machine removed from thence upon some external part less necessary to life. The laws and the manner, by which this is brought about, are yet not discover'd. In this country the liver is subject to more frequent inflammation than any other viscus, since those liquors which are the product of fermentation, viz. spirits, wine, and beer, are drank by all ranks of people, and seem to exert a specific influence on the liver. In some constitutions this inflammation of the liver, occasion'd originally by drinking much spirituous or fermented liquor, is removed upon the joints, and causes the gout; which is so far salutary as the inflammation of the liver, a part more essential to life, is relieved by it. In other constitutions it is removed upon the face, and either a number of pustles are produced, call'd gutta rosacea, or a permanent redness, as was the case with your Grace.

Now if by the use of solutions of Lead, or other drugs, apply'd externally so as to harden these parts, or render them less sensible, the inflammation of the liver ceases to have crisis this way, the efforts of nature either act in vain to remove it from the liver, which therefore becomes permanently diseased; or the inflammation is removed to some other part – when this happens the common medical language would say, that the humour is repell'd from one part and is fallen upon another. And as I have twice seen a more extensive paralysis induced by using repellants (as they are term'd) to faces, which were cover'd with the gutta rosacea; I am of opinion the use of the solution of sugar of lead was the remote cause of the disability of the muscles of your face.

What then is to be done? As nature is disapointed in Her usual habit of producing an erysipelas of the face, if a new crisis could be taught her, by inflaming the joints of the feet, this might prevent any other injury of the constitution. That is, if your Grace could acquire the Gout, it would be so far salutary, as to prevent any further attack upon your face. On this account it would be adviseable to try the use of the Bath-water, and also to take some gently stimulant medecine, as the volatile tincture of Guaiacum, two teaspoonfuls twice a day, in a little gruel.

Another method would be perhaps equally useful, viz. to reinstate, if it could be done, the sensibility of the skin of that part of the face, which used to be inflamed most frequently, or most permanently, before the unfortunate use of the solution of Lead. This might be attempted by moistening it once in 3 or 4 days with a very small quantity of tincture of Cantharides.

The third method, which is the safest and most certain to establish and prolong your Grace's health, is to diminish the original cause of this kind of diseases; that is to prevent the frequent returns of slight inflammations of the liver; and thence to have no occasion for the repeated crisis, or removal of those inflammations to the joints or to the face.

This is to be done by taking 4 or 5 grains of calomel once a fortnight for 2 months or 3: and by taking care to have a stool every day, by acquiring a bodily habit of going to the garden at a certain hour, or by taking occasionally the size of a nutmeg of laxitive electurary.

And by obstinately persevering in drinking daily but a certain quantity of spirituous or fermented liquor, and this quantity should be about half that, which you have been generally accustomed to. I express myself strongly upon this head, as I know in this kind of disease Health is not to be bought at any other price.

It is generally said that the leaving off the habitual quantity of vinous liquors is not always prudent. It is true, where people have lived to great excess in this respect, more caution is necessary to be used in the discontinuance of them – but I have discover'd a certain rule to know how far the quantity of spirituous potation may be discontinued safely in all people: and that is by first advising them to omit one half, or one third, of their usual quantity; and then to observe in 4 or 5 days, if their appetite was as good or better than before: if this was the case, they may be certain of recieving no injury, but the greatest good from this omission; and may if they please after some weeks, gradually decrease it further. About 7 or 8 years ago, I had three fits of the Gout in three successive years; and gradually left off drinking ale and wine; and at length small beer also; and have since had not the least attack of Gout.

When spirituous potation is diminished as above, I have generally advised people to take spice after their meals, for a month or two; that they might the less feel the want of the spirit. As ten black-pepper corns cut into 3 or 4 pieces each. And to eat as much flesh-meat, as their palates seem'd to chuse, both to dinner and to supper.

There is a popular outcry against what are call'd high dishes, but as these consist of spice and gravy chiefly, with anchovey, and burnt fat, I don't see much harm in them, unless acids or wine are boil'd in them, which are always render'd unwholesome by having dissolved a part of the lead or copper of the vessels, in which they have been boil'd. Broth, which has long stood in tin'd or copper vessels, is also liable to become impregnated with lead or copper: as the tin-lining of vessels is composed of more than half lead.

There is another popular error, that people are more injured in their health by gluttonous eating, than by inebriate drinking. But besides the deleterious effects of inebriation, which all of our profession daily witness and lament, the temporary effects of great intoxication are such as to induce one to believe the daily ingurgitation of a small quantity of so potent a drug must be hazzardous to the constitution. Give a man unused to vinous fluids a bottle of port wine, or 3 pints of ale – what is the consequence? He looses his understanding, and becomes for a time an Ideot: and falling down on the floor without the use of his limbs, has a temporary paralysis! From both which he recovers by slow degrees, is left stupid in mind and much debilit[at]ed in body for several days. Now stall'd oxen, or indolent men may be cram'd up the throat with **natural** food, without imparing their health, or other great inconveniency, except the inactivity of corpulency.

Wine and Spirits are the product of art, of a nice chemical process – and therefore should not be class'd amongst our natural food, but properly belong to medecine; and should only be found, as formerly, in the shops of the apothecaries. In all mealy roots and seeds, whether esculent ones or poisonous ones, the starch, which is found in them, is the same (as in potatoes or white bryony root). This starch during the growth of the plant is converted into sugar (as in the process of making malt, in the sugar-cane, and in ripe fruits) by a natural process: hence the starch and the sugar are both wholesome nourishment; but when this sugar is converted into spirit by fermentation, which is a process of chemical art, our food is converted into poison; and becomes the greatest curse of the christian world, destroying more than the sword, and wisely interdicted by the disciples of Mahomet and Confucius.

I know not whether your Grace will think it worth while to read so long an account of the method of treating your complaint. If you should read it, how much so ever it may differ from the opinions or language of others of the faculty, I hope your Grace will see in it an earness desire to be of service to you. From, my Lord Duke,

Your much obliged
and obed. serv.
E Darwin

Derby Nov. 20 –1783

Addressed [No address – probably sent by hand.]

Original Chatsworth House, Devonshire Collection, No.566.
Printed Unpublished.
Text From original MS.
Notes
(1) William Cavendish, fifth Duke of Devonshire (1748–1811), succeeded to the title in 1764, after the early death of his father the fourth Duke (1720–1764), who had been Prime Minister in 1756–1757. The fifth Duke was generally considered to be good natured, intelligent and sound in judgement, but rather inactive. He was Lord High Treasurer of Ireland (1766–1793) and Lord Lieutenant of Derbyshire (1782–1811), but took little part in national politics, unlike his wife Georgiana (see letter 00H), who was a strong supporter of the Whig cause. After Georgiana's death in 1806, the Duke married Lady Elizabeth Foster. His own death was attributed to 'water on the chest', and not disease of the liver, as Darwin would presumably have predicted.
(2) Darwin was very familiar with liver disease caused by over-indulgence in alcohol, especially among his wealthy patients, and he was warning the Duke of the dangers. His letter is admirably clear and he defines most of the terms used, but a short glossary (including spelling corrections) may be useful: viscus, soft internal organ; pustle [pustule], pimple; sugar of lead, lead acetate; repellant [repellent], medicine supposed to 'repel morbid humours'; the Bath-water, the waters at Bath; Guaiacum, tropical tree, the tincture being a drug made from its resin; Cantharides, a type of beetle, the tincture obtained being used to induce irritation and blisters on the skin; electurary [electuary], medicine in powder form, to be mixed with honey, etc; esculent, eatable.
(3) The 'three fits of the Gout' which Darwin remembered as 'about 7 or 8 years ago' were probably those recorded in his Commonplace Book in 1779 and 1780 (see *Doctor of Revolution*, pp. 132–3). The history of his gout is recorded in *Zoonomia* ii 452–3. He seems to have given up alcohol by degrees, renouncing hard liquor in his twenties, but continuing to drink wine and ale until about 1780. After that he regarded alcohol, as he says here, as 'the greatest curse of the christian world'. The 'golden rule' of gradual withdrawal is described in *Zoonomia* i 244–52.
(4) At the end of the letter, Darwin probably mis-spelt 'earnest' because it was a word he never used – except to a Duke. The tone of the letter suggests that this was the first time Darwin had been consulted by the Duke, although he had known the Duke's uncle Lord George Cavendish (1728–1794) since his schooldays.

84A. To Matthew Boulton, 17 January 1784

Derby Jan. 17 –84

Dear Boulton,

If it be not very inconvenient to you to pay me the 200£ (as you seem'd to say when I had the pleasure of seeing you), it would be very acceptable to me at this time, as I have a sum of money to pay, to compleat a purchase I made here some time ago.

I hope philosophy and fire-engines continue to go on well. You heard we sent your society an air-balloon, which was calculated to have fallen in your garden at Soho; but the wicked wind carried to Sir Edward Littleton's.

Pray give my Compliments to your learned society, and believe me, dear Sir

Your affect. friend
and obed. serv. E Darwin

The note is due Feb. 7.

Original Birmingham Reference Library, Matthew Boulton Papers, Darwin 32.
Printed Excerpts frequently published, e.g. *Lunar Society*, p. 251 and *Doctor of Revolution*, p. 160.
Text From original MS.
Notes
(1) In the autumn of 1783, Darwin and his family moved from Radburn Hall to their town house in Full Street, Derby – the 'purchase' Darwin mentions. It was two years since Darwin had first asked Boulton for the £200: see letter 82A.
(2) 'Balloon mania' developed after the flights of the Montgolfiers in the summer of 1783, and the Derby Philosophical Society's attempt at 'balloon post' is one manifestation of it.
(3) Sir Edward Littleton (1726/7–1812) was MP for Staffordshire in six Parliaments, from 1784 to 1812. He lived at Teddesley Park, near Penkridge (and previously at Pillaton Hall). He married Frances Horton of Catton, sister-in-law of the Mrs Horton of letter 91A. See Namier & Brooke, *Commons* iii 46.

84B. To Josiah Wedgwood, 11 March 1784

Dear Sir,

I admire the way in which you support your new theory of freezing steam. You say, "Will not vapour freeze with a less degree of cold than water in the mass? instances hoar-frost, etc." Now this same et caetera, my dear friend, seems to me to be a gentleman of such consequence to your theory, that I wish he would unfold himself a little more.

I sent an account of your experiment to Mr Robert, and desired him to show it to Dr Black, so that I shall hope some time to hear his opinion on the very curious fact you mention, of a part of ice (during a thaw) freezing whilst you applied a heated body to another part of it. Now in spite of your et caetera, I know no fact to ascertain that vapour will freeze with less cold than water. I can in no way understand why, during the time you apply a heated body to one part of a piece of ice, when the air of your room was at 50° and the ice had for a day or two been in a thawing state, that a congelation should be formed on another part of the same ice, but from the following circumstances. There is great analogy between the laws of the propagation of heat, and those of electricity, such as the same bodies communicate them easily, as metals, and the same bodies with more difficulty, as glass, wax, air: they are both excitable by friction, both give light, fuse metals, et caetera. <u>Therefore I suppose that atmospheres of heat of different densities, like atmospheres of electricity, will repel each other at certain distances</u>, like globules of quicksilver pressed against each other, and that hence by applying a heated body near one end of a cold body, the more distant end may immediately become colder than the end nearest to the heated body.

<u>March 11, 1784.</u> Since I wrote the above I have reconsidered the matter, and am of opinion that steam, as it contains more of the element of heat than water, must require more absolute cold to turn it into ice, though the same sensible cold, as is necessary to freeze water, and that the phenomenon you have observed, depends on a circumstance which has not been attended to. When water is cooled down to freezing point, its particles come so near together, as to be within the sphere of their reciprocal attractions; – what then happens? – they accede with violence to each other and become a solid, at the same time pressing out from between them some air which is seen to form bubbles in ice and renders the whole mass lighter than water (on which it will swim) by this air having regained its elasticity; and pressing out any saline matters, as sea-salt, or blue vitriol, which have become dissolved in it; and lastly by thus forcibly acceding together, the particles of water press out also some more heat, as is seen by the rising of the thermometer immersed in such freezing water. This last circumstance demands your nice attention, as it explains the curious fact you have observed. When the heat is so far taken away from water, that the particles attract each other, they run together with violence, and press out some remaining heat, which existed in their interstices. Then the contrary must also take place when you add heat to ice, so as to remove the particles into their reciprocal spheres of repulsion: they recede from each other violently, and thence attract more heat into their interstices; and if your piece of hot silver is become cold, and has no more heat to give, or if this thawing water in this its expansile state is in contact with other water which is saturated with heat, it will rob it of a part, or produce freezing if that water was but a little above 32°.

I don't know if I have expressed myself intelligibly. I shall relate an experiment I made twenty-five years ago, which confirms your fact. I filled a brewing-copper, which held about a hogshead and half, with snow; and immersed about half-an-ounce of water at the bottom of a glass tube in this snow, as near the centre as I could guess, and then making a brisk and hasty wood-fire under it, and letting the water run off by a cock as fast as it melted, I found in a few minutes on taking out the tube that the water in it was frozen. This experiment coincides with yours, and I think can only be explained on the above principle. In support of the above theory I can prove from some experiments, that air when it is mechanically expanded always attracts heat from the bodies in its vicinity, and therefore water when expanded should do the same. But this would lengthen out my letter another sheet; I shall therefore defer it till I have the pleasure of a personal conference with you. Thus ice in freezing gives out heat suddenly, and in thawing gives out cold suddenly; but this last fact had not been observed (except in chemical mixtures) because when heat has been applied to thaw ice, it has been applied in too great quantities.

When shall we meet? Our little boy has got the ague, and will not take bark, and Mrs Darwin is therefore unwilling to leave him, and begs to defer her journey to Etruria till later in the season. Pray come this way to London or from London. Our best compliments to all yours.

Adieu,

E Darwin

PS. – Water cooled beneath 32°, becomes instantly ice on any small agitation, or pouring out of one vessel into another, because that the accession of the particles to each other, and the pressing out of the air, or saline matters, and of heat is facilitated.

Original Not traced. In the possession of Charles Darwin, 1879.
Printed (i) Excerpts in Meteyard, *Wedgwood* ii 470; (ii) Complete in C. Darwin, *Life of E. Darwin*, pp. 98–101.
Text From (ii) above.
Notes

(1) Darwin is slightly embarrassed to find Wedgwood putting forward the erroneous idea that water vapour freezes more easily than liquid water. But Darwin has to make the best he can of it, because Wedgwood's conclusion was based on observations which it would be impolite to doubt. So Darwin tries to provide a theory to explain the alleged phenomena, and in doing so greatly overestimates the importance of bubbles of air 'squeezed out'. In the midst of his explanations he throws in the pregnant sentence: 'I can prove from some experiments, that air when it is mechanically expanded always attracts heat from the bodies in its vicinity'. This is a preview of his exposition of the principle of adiabatic expansion, which he propounded and then applied to explain the formation of clouds in *Phil. Trans.*, **78**, 43–52 (1788).

(2) 'Mr Robert' is Darwin's son Robert, now 17 and a medical student at Edinburgh, where the great chemist Joseph Black, discoverer of latent heat, was a professor (see letter 78H). Darwin was no doubt relieved to be able to refer the problem so easily to so eminent an authority.

(3) 'Our little boy' is his son Edward, now two years old.

84C. To Josiah Wedgwood, 13 March 1784

* * *

Mrs Darwin says she hears your whole family are going to town in a body, like a caravan going to Mecca; and we therefore hope you will make Derby a resting-place, and recruit yourselves and your camels for a few days, after having travell'd over the burning sands of Cheadle and Uttoxeter.

* * *

Original Not traced. In the possession of Charles Darwin, 1865.
Printed Meteyard, *Wedgwood* ii 559.
Text From above source.
Notes
Some or all of Wedgwood's family visited the Darwins quite frequently in the 1780s, and this invitation, rich in metaphor, is one of many.

84D. To Josiah Wedgwood, [23 March 1784?]

* * *

Your medallion I have not yet seen; it is covered over with so many strata of caps and ruffles and Miss Wedgwood is whirled off to a card-party – seen and vanished like a shooting-star.

* * *

Original Not traced. In the possession of Charles Darwin, 1865.
Printed Meteyard, *Wedgwood* ii 559.
Text From above source.
Notes
(1) 'Miss Wedgwood' is Josiah's eldest daughter Susannah (1765–1817), who was a great favourite with Erasmus and his family. She married Robert Darwin in 1796 and was the mother of Charles Darwin.
(2) The dating of this letter is doubtful. Meteyard (*Wedgwood* ii 470) refers to letters from Darwin to Wedgwood on 11, 13 and 23 March and 1 April 1784. The first is letter 84B, the second is 84C with the invitation to visit Derby. The excerpt here implies that Miss Wedgwood (and possibly others?) have just arrived at Derby. That is plausible ten days after the invitation, but the letter could refer to a different visit, possibly not in 1784 at all.

84E. To Josiah Wedgwood, [1 April 1784?]

* * *

The detension or expulsion of parts of icy mountains I always attributed to the water from thawing ice in the daytime, filling valleys in the remaining ice, and by its freezing there cracking the old ice by its expansion at freezing, and not by its expansion by an increase of cold.

* * *

Original Not traced. In the possession of Charles Darwin, 1865.
Printed Meteyard, *Wedgwood* ii 470.
Text From above source.
Notes
(1) The dating is doubtful. Meteyard (*Wedgwood* ii 470) refers to letters from Darwin to Wedgwood on 11, 13 and 23 March and 1 April 1784. The first is letter 84B, the second is 84C, and if 84D is the third, this one is most probably the letter of 1 April. The letter appears to be a reply to Wedgwood's reply to 84B, so a 21-day interval is reasonable.
(2) Some of Wedgwood's ideas on the freezing of water were erroneous, and Darwin is here gently putting him right on the fissuring of mountain ice.

84F. To Thomas Henry, 6 May 1784

Sir

I esteem myself greatly obliged to the Litterary and Philosophical Society at Manchester for the honor they have done me, in electing me one of their members; and beg you will present my grateful acknowledgements to them, and accept yourself my thanks for your polite letter to me on this occasion.

I am, Sir, with great respect
Your obed. serv.
E Darwin

Derby
May.6 –1784

Addressed Thos. Henry Esqr., Secretary to the litterary Society, Manchester.

Original The Houghton Library, Harvard University, Cambridge, Massachusetts.
Printed Unpublished.
Text From original MS.
Notes
(1) The Manchester Literary and Philosophical Society was founded in 1781 by Thomas Percival FRS (1740–1804), an energetic Manchester doctor, unitarian and reformer, who was President of the Society till his death. Darwin's Derby Philosophical Society was on similar lines, but smaller, and his election to honorary membership of the Manchester Society was a fitting prelude to his inaugural Presidential Address to the Derby Society on 18 July 1784 (see *Doctor of Revolution*, pp. 159–61). For Percival, see *DNB*.

(2) Thomas Henry FRS (1734–1816) was a surgeon-apothecary in Manchester and a noted chemist, who was the first to discover that carbonic acid in air promotes the growth of plants. He was Secretary of the Manchester Literary and Philosophical Society from its foundation, and President after Percival's death. Many of his papers were published in the *Memoirs* of the Society. For Henry, see *Ambix*, **20**, 183–208 (1973), and *DNB*.

84G. To Joseph Johnson, 23 May 1784

Sir,

I have the favour of your note in a letter from Mr Fuseli, and from the ingenuous manner of your proposal to me, and the character from Mr Fuseli and others, have no doubt of relying on your honour, and of putting intire confidence in you.

1. I would not have my name affix'd to this work on any account, as I think it would be injurious to me in my medical practise, as it has been to all other physicians who have published poetry.

2. I would not wish to part with the intire coppy-right, because that would preclude me from entrenching or altering, or adding to it in any future edition, if such should be worth while to publish.

What I would propose to you then is that we publish one edition consisting of 500 copies, at our joint equal expense, and equal profits, and afterwards, if you think proper, to publish another edition of 500 copies at our joint equal expense and profits, but that after the publication of this 1,000 copies, viz., in the space of one year after such publication, that I shall again be at liberty to do what I please with the work.

The above refers to the work only, which you have seen, call'd the second part of the botanical garden; for I would not yet bind mystelf to publish the first part, which I believe will consist of but 400 lines, but which will have 3 or 4 times the quantity of notes, and those of more learned, and newer matter, but half of which are not yet done.

If you accept of these proposals, you will please to acquaint Mr Fuseli, who is so kind as to promise some ornament for the work, and also to remit the book by the coach, because I have about 60 lines more to add to it, and two or three more notes.

Also pray send me any general or particular criticisms which may have occur'd to you – and particularly if you think it too long, and would wish any part to be omitted, as I am unacquainted with what is like to make a book sell.

Lastly, if you should think equally well of it, I could print it accurately and well at Litchfield by Mr Jackson, who has just now printed "Louisa, a poetical novel by Miss Seward." I could then correct the press myself, and by thus revising the work might perhaps in some places improve it, or the notes.

Adieu from Sir
Your obed. serv.
E Darwin

Derby, May 23 –84.

Original Not traced. Probably in possession of Mr R. H. Thornton, 224, 24th Street, Portland, Oregon, in 1923.

Printed in *Notes and Queries*, 12th series, Vol. 12, p. 449 (9 June 1923). Excerpts in *Lunar Society*, pp. 204–5, and *Doctor of Revolution*, pp. 162–3.

Text From *Notes and Queries*, as specified above.

Notes

(1) Joseph Johnson (1738–1809) was the leading radical bookseller and publisher of the day; he subsequently published all Darwin's books, as well as many books by Priestley, Cowper and others. See *DNB*. Darwin had submitted a first draft of *The Loves of the Plants* to him and had apparently received quite a favourable response.

(2) 'Mr Fuseli' is Henry Fuseli (1741–1825) the well-known painter and author, and a friend of Blake. This letter suggests that Darwin knew Fuseli quite well, but his name does not appear in any other of Darwin's letters. However, Fuseli provided two illustrations for *The Botanic Garden*: the frontispiece, 'Flora attired by the Elements'; and the famous 'Fertilization of Egypt', which Blake engraved with some touches of his own. Fuseli's 'Nightmare' also appears in some editions of *The Botanic Garden*.

(3) The letter shows Darwin's extreme reluctance to publish his poem, because he feared it would injure his medical practice. The publication arrangements he suggests seem designed to allow him to suppress the book if he found it was damaging his medical reputation. The 'first part' mentioned by Darwin eventually ran to 2448 lines, rather than the 400 he predicts. *The Loves of the Plants*, the 'second part', was published in 1789, and the 'first part', *The Economy of Vegetation*, in 1792. The contract for the complete *Botanic Garden*, dated 20 February 1790, is in the Liverpool City Libraries (Hornby Library). It specifies that Johnson is to pay Darwin £300 immediately for the copyright of the second part (*Loves of the Plants*) and £400 on publication of the first part (*Economy of Vegetation*).

(4) Mr Jackson was the excellent Lichfield printer John Jackson, who had just printed Anna Seward's *Louisa* (1783). As suggested here by Darwin, the first edition of *The Loves of the Plants* (1789) was printed by Jackson.

85A. To [Thomas Cadell], 13 January 1785

Sir

I should take it as a favour if you would acquaint me at your leizure, whether you have sold any more of my pamphlet "Experiments establishing a criterion between mucaginous and purulent matter", since if you have disposed of those, I could send you another hundred of them.

I am Sir your obed. serv.

E Darwin

Derby
Jan. 13 1785

Addressed {*name cut off*} London.

Original Bodleian Library, Oxford, MS Montagu d.2.
Printed Unpublished.
Text From original MS.
Notes
The pamphlet (see letter 78H, note 6) was based on the prize essay of Darwin's eldest son Charles, and Cadell sold it in London. See also letter 81B.

85B. To Matthew Boulton, 26 January 1785

Derby Jan 26 –85

Dear Sir,

I was desired lately to look at the specification of Mr Arkwright's patent, about which He had a trial last year; and it was in my opinion unjustly given against him. He has made many improvements in the machine since, all which I am master of, and could make more improvements myself in respect to diminishing the friction etc – but the essential parts of the machine are described in the specification in my opinion.

Now if patents are to be so easily overthrown – does your patent or act of parlament stand on firm foundation? – Suppose one should say, you have not mention'd your working your piston in oil; and if any one should copy your machine, and working their piston with oil, should say, your specification was not similar to your machine; and therefore would infringe it? Would this be just?

I believe some shyness has existed between you and Mr Arkwright, but is it not your interest to assist him with your evidence on his trial? which also may make good humour between you, and you make fire-engines for cotton-works in future; which I understand from spinning wool also, will increase tenfold.

If yourself or Mr Watt think as I do on this affair, and that your own interest is much concern'd in establishing the operation of patents, pray give me a line that I may advise Mr Arkwright to apply to you, and send you a drawing and specification and model.

If Mr Watt be in this part of the country, pray tell him, how glad I shall be to hear that he is amongst the living – more so if any thing should bring him or yourself to Derby.

Adieu, the Lord keep you both in fine preservation,
from your affect. friend
E Darwin

I think it hard, that an inventor of machines has not his property in the invention secured to him for 14 years without paying 100£ for a patent; as a writer of books has; much more so, that his right to his invention should be so easily infringed for want of an intelligent or mechanic jury.

Tho' your inventions, by draining Cornwal, have supply'd work to ten thousands – Mr Arkwright has employ'd his thousands – I think you should defend each other from the ingratitude of mankind. His c[*e*]ase comes on the 14th of February, [so] that the time is urgent. Pray let me soon hear from yourself or Mr Watt on this affair.

Original Birmingham Reference Library, Matthew Boulton Papers, Darwin 33.
Printed Short excerpts in *Doctor of Revolution*, pp. 166, 167.
Text From original MS.
Notes
(1) This letter illuminates Darwin's own personality and the progress of the Industrial Revolution. Injustice always roused Darwin, and because he thought Arkwright had been unjustly treated, he went to great pains to help Arkwright: he made himself master of all the intricacies of Arkwright's machinery, which he celebrated in *The Loves of the Plants* (II 93–104, and Note); he devised improvements in the machinery (his Commonplace Book has 13 pages of designs for cotton spinning machines); and later in the year he abandoned his patients and travelled twice to London for the patent trials. Darwin saw how the new inventions were bringing a surge of prosperity to the country, so he thought the inventors deserved to be protected. He also wanted the entrepreneurs to sink their personal differences and stand together on this issue, whereas Boulton and Watt had been at loggerheads with Arkwright for years. Only three months earlier, Watt had written: 'as to Mr Arkwright, he is to say no worse one of the most self sufficient Ignorant men I have ever met with'. Darwin was the only person able to act as peacemaker, and he did so most effectively in this letter. As a result, Watt agreed to go to London with Darwin in February to appear as a witness for Arkwright at the trial. They had to travel to London again in June for a further trial. Annoying though the journeys were, they greatly strengthened the friendship between Darwin and Watt.
(2) There is an excellent discussion of Arkwright's patents by Schofield, *Lunar Society*, pp. 349–55.

85C. To Richard Dixon, 18 March 1785

Derby 18 March 1785

My dear Friend,

I am glad you find yourself better by losing 7£ – you may say with the Irishman "you have gained a loss" – but I should not advise you to sink yourself any further, but keep where you are. The reason I advised you to emaciate yourself so far was, because I believed your shortness of breath to be owing to some fat about the lungs or heart – so that you should occasionally weigh yourself? fast and pray only when it is necessary. If you could get false teeth, you would find that another consolation as you would speak easier; and if you could get it (for it is but one piece cut to look like 2 or 3 teeth) made of ivory instead of the bone of the seahorse, it would become dusky and look like your other teeth. I should recommend Beardmore to you in Poel[?]-Court Fleetstreet – I advised my Brother at Elston to get . . . . . . tooth but I believe he thought it a sin and would not at all listen to me about it.

Mrs Darwin and Mr Erasmus, and all the branches and twigs are well and beg to be remembered. I sent your compliments to Mrs Day, she is got into her new house at No 21 Prospect Row Birmingham and has a good tempered man to her husband and is very happy I believe – I hope your Son's wife is better and beg my compliments to all of the name of Dixon – pray tell him we have not dared to eat the cheese he was so kind as to send us, as your Brother said it must not be cut for two or three months.

Adieu God bless you if it be possible
from your affect. friend
E Darwin

Addressed Mr Richard Dixon, Hartfordend Mill, Felsted, Essex.

Original Not traced.
Printed Short excerpts in *Doctor of Revolution* p. 171.
Text From MS copy made by R. W. Dixon, Richard Dixon's great-grandson, and sent by R. W. Dixon to Charles Darwin in 1879. This MS is in the Cambridge University Library (Dar. 218).
Notes
(1) Richard Dixon (1731?–1797) was born at Elston, and was a lifelong friend of Darwin. Evidently Dixon knew Mary Parker (1753–1820), the mother of Darwin's daughters Susan and Mary Parker (see letter 75G). Mary Parker (the elder) married Joseph Day (1745–1811) in 1782, and they had now moved to the house in Prospect Row, Birmingham, where Day conducted his business. He is variously described in directories as a toy-maker, a victualler and a stamper. For information on the Day and Parker families, I am indebted to Mrs Kathleen Archer.
(2) 'My Brother at Elston' was probably Robert the naturalist (see letter 99C). John the clergyman was rector of Elston and of Carlton Scroop, and seems to have lived at Carlton.
(3) I have not identified 'Beardmore'. There are five Beardmores in London in *Bailey's British Directory* (1784), but none has an address resembling that given by Darwin (which may be Powel's Court, Queen Street).

85D. To [George?] Goodwin [15 June 1785?]

Dr Darwin's compliments to Mr Goodwin and Mr Ince – He is just arrived at Mr Wedgewood's N° Great George street. – And will wait on them at any hour to-morrow, they shall appoint.

Wednesd. night.

Addressed Mr Goodwin or Mr Ince, N° 40 Norfolk Street, Strand.

Original Cambridge University Library, MS Add 8090.
Printed Unpublished.
Text From original MS.
Notes
(1) Darwin and Watt twice travelled to London, in February and in June 1785, to appear as witnesses in the Arkwright patent cases (see letter 85B). From its similarity to the next letter, this letter could have been written the day before. But this dating is speculative.
(2) Goodwin and Ince seem likely to have been solicitors, but there is no trace of them in the list of London lawyers in *Bailey's British Directory* of 1784 (Vol. IV, pp. 929–69). So they were probably Arkwright's local lawyers, who travelled to London for the case. The *Universal British Directory* of 1793 (Vol. 4, pp. 795 and 799) records 'George Goodwin, Attorney' as a resident of Winster, and 'Richard Ince, Attorney' as a resident of Wirksworth. The notice of George Goodwin's death in the *Derby Mercury* (20 July 1796) refers to him as 'a very eminent attorney-at-law'; and the *Derby Mercury* of 1 July 1795 records the death at Wirksworth of 'Mr Ince, attorney-at-law'. Since Winster and Wirksworth are so close to Cromford, it seems very probable that these are Darwin's 'Mr Goodwin and Mr Ince'. (However, there was also an attorney Richard Goodwin at Ashbourne, so the identification is not certain.)

85E. To Thomas Dixon, 16 June 1785

Dr Darwin's compliments to Mr Thos. Dixon. He is come to Mr Wedgwood's and stays till Tuesday mng. If Mrs Dixon be not yet recovered from her indisposition the Dr will be glad to see her at Mr Wedgwood's at ½ past seven any morning. If Mr Richard Dixon sen^r^. be in London he shall hope to see him – Thursday Night.

Original Not traced. There is a MS copy in the Cambridge University Library (Dar. 218), made by R. W. Dixon and sent by him to Charles Darwin in 1879.
Printed Unpublished.
Text From MS copy.
Notes
(1) R. W. Dixon added the following note to his MS copy: 'a copy of the above letter was forwarded to Mr Richard Dixon Sen. by his Son from London to Harfordend,

Felsted, that he might hasten his journey to London in order to meet Dr Darwin, and the letter is dated Friday (June) 17th 1785'.

(2) As is clear from the letter, Thomas Dixon was a son of Darwin's friend Richard Dixon.

(3) This note was written while Darwin was in London with Watt for the second time in 1785, to appear at the King's Bench as a witness in defence of Arkwright's patent. See Schofield, *Lunar Society*, pp. 349–51 and *Doctor of Revolution*, p. 167.

85F. To Charles Blagden, 17 August 1785

Derby Augt. 17 –85

Sir

In Reichard's edition of the systema plantarum frequent reference is made to a new edition of the genera plantarum – this edition I suppose to have been published since that of Stockholm in 1764, but I can not yet meet with it.

I hope you will do me the favor of a line to acquaint me if such an edition is come over, or how to procure it, as the Lichfield Society intend shortly to publish a translation of the genera plantarum, encouraged by the favorable reception of their translation of the systema vegetabilium. I have no excuse for taking the liberty to trouble you with this letter, but your known attatchment to science, and desire to forward its progress.

And am, Sir with great respect
Your obed. ser.
E Darwin

Addressed Dr Blagden, Gower Street, Bedford Square, London.

Original Royal Society archives, Blagden Letters, BLA.d.8.
Printed Short extract in *Doctor of Revolution*, p. 170.
Text From original MS.
Notes

(1) Dr Charles Blagden, FRS (1748–1820) acted as assistant to Henry Cavendish in the 1780s and was deeply involved in the water controversy (see *Doctor of Revolution*, pp. 141–3). Blagden, a close friend of Banks, was Secretary of the Royal Society from 1784 to 1797 and was knighted in 1792. His own researches were in chemistry, on the depression of freezing points of solutions: see *Biographical Dictionary of Scientists* (Black, 1969).

(2) The new edition of Linnaeus's *Genera Plantarum*, which was used in the Lichfield Society's translation, was the 7th, published in 1778. (See also letters 86J and 87A.)

86A. To Josiah Wedgwood, 29 January 1786.

* * *

I shall be glad to hear that you have purchased at the price you wished the famous vase.

* * *

Original Not traced. In the possession of Charles Darwin, 1865.
Printed Meteyard, *Wedgwood* ii 576.
Text From above source.
Notes
Darwin is referring to the famous Portland Vase, which was brought to England in December 1784 by Sir William Hamilton. It was known as the Barberini Vase at that time. The Duchess of Portland bought it from Sir William Hamilton, but she died in July 1785. Wedgwood was apparently negotiating to buy the vase early in 1786, but in fact it was included in the auction of the whole of the Duchess's museum, which took place between 24 April and 7 June 1786. The vase, which was the last of the 4155 lots in the auction, was bought in (for £1029) by the Duke of Portland and, by prior arrangement, was loaned to Wedgwood three days later, on 10 June 1786. See Meteyard, *Wedgwood* ii 576–7, and *Wedgwood Circle*, p. 90.

86B. To Unknown Doctor, 11 March 1786

[This letter refers to the case of Miss Hollis, who suffered from fever and delirium: Darwin offers advice and mentions his theory of sympathetic fever (later expounded in *Zoonomia* ii 537–625).]

The letter is not available for publication. The owner, Dr Ralph Colp, Jr., of New York, intends to publish it soon in a Medical History Journal.

86C. To Josiah Wedgwood, 22 March 1786

Appleby, Leicestershire.

Dear Sir,

I have not been at home since about an hour after you left [for] Lichfield, having been a Wanderer or a sojourner ever since in this part of the Country – have therefore not yet fabricated another lamp but only thought over it.

1. I think in the specification it should be said the wick is made to rise or sink by a rachet and pinion as in Fig 1 or by a Stud sliding in a nick as in Fig 2 or the like, since the essential circumstance is the wick being placed

within the circumference of the flame not in the manner of elevating or sinking it.

2. The easyest way of putting on the external tube, so as to have no wick in it, will be to make it in two pieces, one to fit upon the other and the stud to be at the joining a b.

3. This joining may be as low down as you please. The air-hole may be round the rim or on the top of the Can at w w.

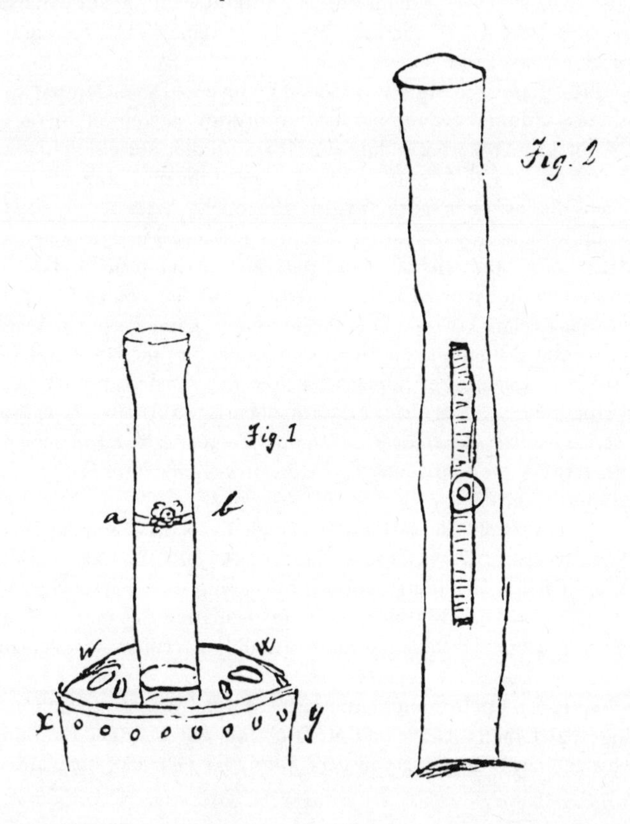

The animal gut should be first blown up with air and dry'd, 2. cut into proper pieces, 3. the ends of each piece made wet with warm water and attached to the two ends, and the wires put round, 4. and then dry'd before the oil is put into it.

I am interrupted and the post is about going.

Adieu, I fear'd the snow would much incommode you, when I found so much more than I expected myself.

From yours

Mar 22 –86 E Darwin

Original Not traced. There is a MS copy at University College London, Galton Papers, 14.
Printed Unpublished.
Text From MS copy. (*Diagrams* From MS copy.)
Notes

(1) This is the first of eleven letters to Wedgwood concerned mainly with oil lamps. Although the text is a manuscript copy, not in Darwin's writing and probably made about 1870, some of the diagrams look as though they are drawn by Darwin – perhaps they were traced carefully from the originals. These diagrams have been reproduced as if they were original.

(2) It was in 1782 that Aimé Argand (1750–1803) invented an oil lamp with tubular wick and glass chimney, which induced an upward current of air on both the inside and outside of the wick and produced a continuous bright light almost free of smoke and smell. The Argand lamp was 'one of the most important domestic inventions of the late eighteenth century' (Schofield, *Lunar Society*, p. 347). The Lunar group were keenly interested: Boulton and Small had designed a similar lamp in the 1770s, and when Argand patented his lamp in England in 1784, Boulton obtained the exclusive rights to manufacture the metal parts. But in 1786 the patent was declared invalid, and Argand never gained any financial reward from his invention. Boulton continued to manufacture the lamp and Watt also took an interest. For further information, see *Lunar Society*, pp. 347–9. For biography of Argand see *Biographie Universelle* (Paris, 1843) Vol. 2, p. 185 and E. Delieb, *The Great Silver Manufactory (1971)*, pp. 107–10. An excellent account of late eighteenth-century oil lamps may be found in Volume 20 of Rees's *Cyclopaedia* (1819), under 'Lamp'.

(3) Darwin's first drawing of an oil lamp is on page 154 of his Commonplace Book at the Darwin Museum, Down House, dated 'Nov. 30 1785', and it is sketched opposite, as it helps in understanding the ensuing letters. Darwin's improvements in design are mainly concerned with regulating the supply of oil, but it is beyond the scope of these notes to attempt a critical assessment of his efforts.

(4) Wedgwood replied briefly to this letter on 29 March (MS copy in Galton Papers, 14), saying 'that I shall observe and profit by the contents, that I am preparing Models, etc . . .' and 'thanking you very kindly for your late hospitality'.

Rough sketch of Darwin's drawing of his lamp.
(From Commonplace Book, page 154.)

Nov. 30 1785

Two calves bladders may be put one within the other, and glue and honey mixed put between to make them hold oil.
Capillary tubes made of a bundle of wire . . . may prevent the oil from throwing up a jet d'oel on setting the lamp down hastily.

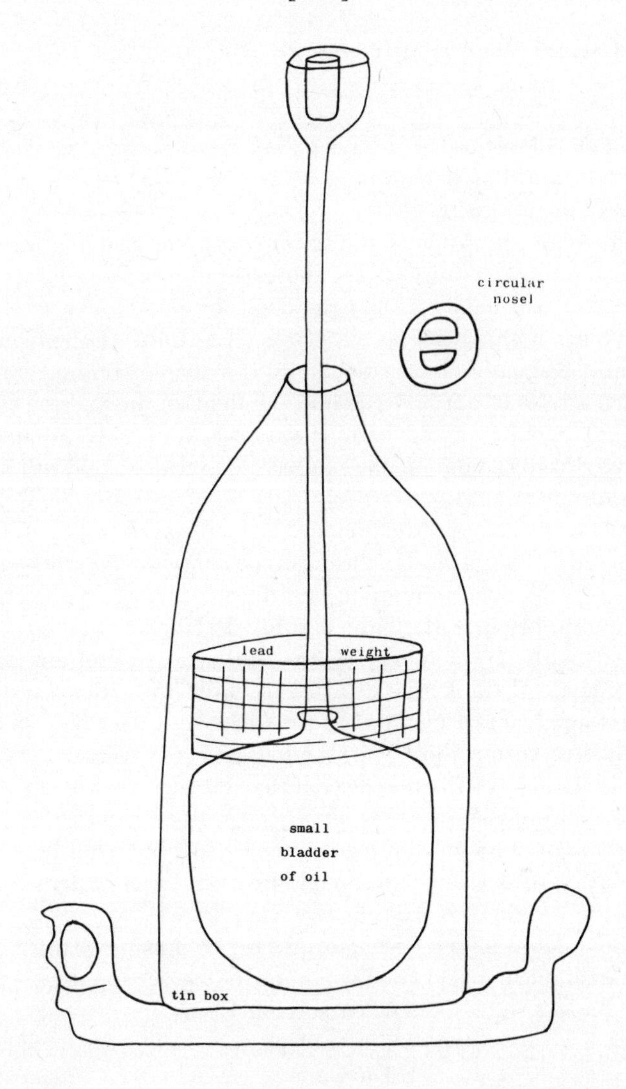

In this ingenious lamp, the lead weight pressing on the bladder raises the oil, and the lamp itself sinks, as the oil wastes.

[*Note:*
jet d'oel is Darwinian French for 'jet of oil']

86D. To JOSIAH WEDGWOOD, 4 April 1786

Apr 4 –86

Dear Sir

I was set off from Drayton to have passed Sunday night last with you at Etruria, but Fate drag'd me another way, or the D——l as the Manichaeans would express it.

I should advise you to permit the ague to continue, as it may introduce a new habit, and thence nature (or the above-mentioned D——l, as you like) may forget her bad habit. If the ague does not weaken her very much I should give her no medicine at all (not even a vomit), except she wants some natural evacuation, as a stool daily. If it should continue a month I should then advise to put an issue into the thigh of the affected side; and then unless the child should become very weak, or have a tendency to jaundice, or a hard tumour on one side of its bowels, I should still leave the ague to continue.

In respect to the lamp, if the pressure be made on the oil by weights, as you propose, the column of oil in the top of the lamp becomes shortened as it wastes both by its own diminution, and by the diminished height of the counter column of oil at the base. I study'd (I believe) all the possible means, both with ☿ [mercury], water, and springs, and air, as in the fountain of Hiero (I think it is called). I try'd the last with great diligence many years ago before I understood the theory of the error, I believe no means will succeed but those with the bag and lead weight.

As lead is about 12 or 13 times heavier than oil, an inch of lead will raise a column 12 or 13 inches.

I have thought of no improvement since I wrote but what I believe was mentioned, that the socket teeth should be turned the contrary way, which will bring the button into the centre of the pillar.

I have not made another, fearing to give my workman two distinct ideas, but if you wish me to make one, I will immediately set about it – I fear you found the Snow troublesome when you left Derby.

Shall we hope to see you all in your passage to London? Adieu

E Darwin

Mrs D. desires with me to be remembered to Mrs and Miss W.

Original Not traced. There is a MS copy at University College London, Galton Papers, 14.

Printed Unpublished.

Text From MS copy. (*Diagram* From MS copy.)

Notes

(1) Market Drayton is 12 miles south-west of Etruria.

(2) For Manichaeism, see letter 81A.

(3) The child with the ague was presumably Wedgwood's youngest daughter Mary Anne, who had been unhealthy from her birth and had received treatment by

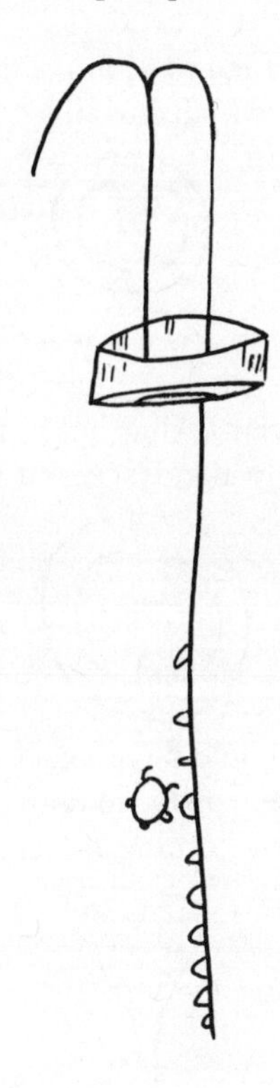

electricity and opium from Darwin in 1779 (see *Doctor of Revolution*, pp. 131–2). She was now eight, and it seems that Darwin had given up hope of her recovery. She died on 21 April.

(4) 'Hiero' is now usually known as Hero (or Heron) of Alexandria. He lived in the first century A.D. and devised many ingenious mechanisms, including syphons, a steam wheel and 'the fountain of Hiero'. These are described in his treatise *Pneumatics*. The mode of operation of the fountain is shown in the diagram in the next letter. Darwin also discusses it in his paper on the formation of clouds, *Phil. Trans.*, **78**, 43–52 (1788).

86E. To Josiah Wedgwood, 7 April 1786

Derby Apr 7 –86

Dear Sir,

I was streighten'd for time when I wrote last, [which] prevented me from saying more about the Lamp.

1. In respect to filling it, the place where the end of the funnel is inserted must be made larger a little (which will not incommode the wick), so that the end of the funnel may stick easily in this part of the wick-holder (the

wick being then taken out), then a section of the funnel would be to fit the part of the wick-holder which is dotted.

2. Another way would be to leave a place on purpose to fill it without taking out the wick at x, in which the funnel with a common form'd pipe might stick upright, and then taking hold of the candlestick part would distend the bladder and help the descent of the oil.

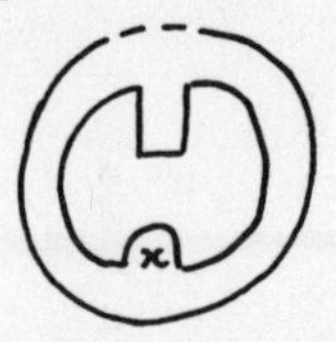

3. The pipe of the funnel should have a cock in it, which would be neither difficult to understand nor expensive to make. Then when the oil becomes visible in the wick-holder, by turning the cock none is spilt over.

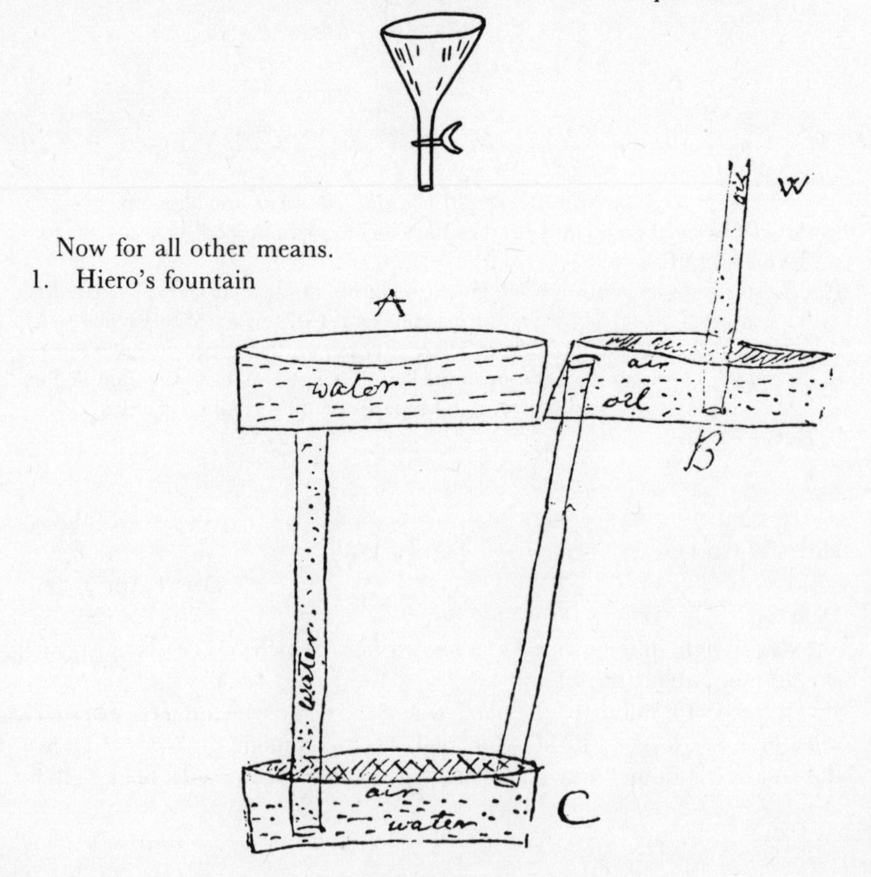

Now for all other means.

1. Hiero's fountain

The Vessel B was placed over the vessel A in the model I made. But this labours under such variety of evils that it cannot succeed, 1. because as the oil at W wastes, the surface of the water both sinks at A and rises in C and hence the error is doubled, 2. the variations of heat and cold much affect it.
2. With ☿ [mercury] and oil here the column of ☿ as it sinks at M, rises xx, hence the error is doubled and the oil subsides in the tube y 16 times as much as the perpendicular height of this double error.

3. Your scheme is I think less objectionable than the above as the error is not doubled, if Oil alone is used. Thus when the Oil in the reservoir sinks an inch, the oil in the wick sinks only an inch. This however is a great defect and the machinery less simple. But the Patent I think should be so expressed as to include this too – which is on the same principle, viz the oil raised by a weight from a reservoir beneath the wick.

If you wish me to construct one here, let me know and I will set about it.

I hope the shadow of the Gent has left.

If a fixed pillar is quite necessary which I doubt it may be done by putting the pillar through the centre of the reservoir, by putting it into a chicken's gut, at a.b.c.d. But I much prefer the other.

Your fair Secretary made no explanation of her elegant copper plate design.

Adieu

best compliments

E Darwin

P.S. I think the best way of filling the lamp is by directing (by a printed direction) the person to sink the wick as low as it will go into the wick-holder by means of the Button; and then the top of that part of the wick-holder, being made rather larger, will contain the nose of the funnel which should be a cock as above.

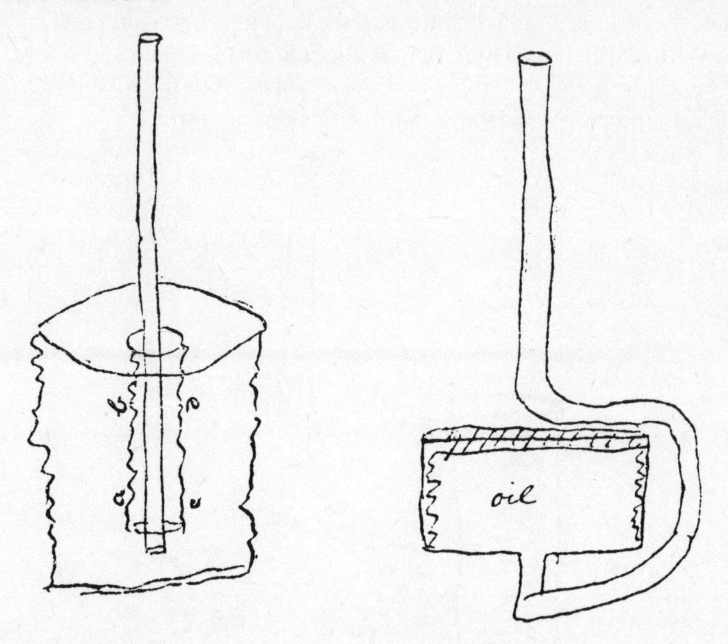

Mr French is endeavouring to forward a navigation to bring coals to this Town, and said he understands you have a plan, or design or survey formerly made by Mr Brindly or others – if you have such a thing, perhaps you would favour him with a sight of it.

Mrs Darwin joins in Compliments and hopes of seeing you in your road to Town.

Original Not traced. There is a MS copy at University College London, Galton Papers, 14.

Printed Unpublished.

Text From MS copy. (*Diagrams*: From MS copy.)

Notes

(1) 'The Gent' was presumably an unwelcome visitor of Wedgwood; or perhaps a ghost?

(2) 'Your fair Secretary' is probably Wedgwood's daughter Susannah (see letter 84D).

(3) 'Mr French' is Richard French (1738/9–1801), one of Darwin's friends in Derby and a helpful member of the Derby Philosophical Society which Darwin founded. French had been a Captain in the Royal Horse Guards Blue, and he had married Millicent Mundy of Markeaton, sister of Francis Mundy whose poem 'Needwood Forest' had inspired Darwin's verses on 'The Swilcar Oak' (see *Doctor of Revolution*, p. 109). Horace Walpole was well acquainted with French, whose father had

been chaplain to Sir Robert Walpole. French's son Forester (see letters 87I and 94U) helped Darwin with some of his experiments. For a pedigree of the French family, see Glover, *Derby* ii 591.

86F. To Josiah Wedgwood, 12 April 1786

Apr. 12 –86

Dear Sir,

Mr Nicholson is an ingenious and accurate man. The transparency of the flame I had attended to and accounted for the great light from snuffing a candle on that idea and had thought of making a pyramidal lamp. For some optical experiments perhaps his concentric lamp might be of use, for common purposes it would not be so agreable as a pyramidal lamp; but neither of them so useful as the present ones – 1st as their intricacy and [some words omitted?] brilliancy would not be less convenient than two on the common Argands plan.

I have thought of Springs, and made some attempts and hope they will succeed well, but if the reservoir be only 2½ inches diameter it must be 8 or 10 inches high, or 12 to have the spring act with accuracy. But as great exactness in the height of the column of oil is not wanted (I believe) I can make springs for broader reservoirs, as suppose 4 or 6 inches diameter, which will not take up more than 3 inches of space at the bottom of the reservoirs, and then the whole reservoir need not be more than 6 or 8 inches deep. I will immediately execute one on each of these plans and send you them, as they will be much preferable to weights though not quite so accurate.

The Conical bag has an inconvenience viz. which ever end you place upwards, or next to the wick, the narrow end will be compressed first and thence the resistance to its compressure will constantly increase instead of decreasing – this however I have invented a method to counteract, but I believe, so little accuracy is necessary that I can do it by Springs alone.

The consumption of oil is less than I expected – I don't think increasing the thickness or conicallity of the wick will succeed, as we saw by the thick wick I show'd you here.

The pyramidal lamp would be more pleasing to the eye than the concen-

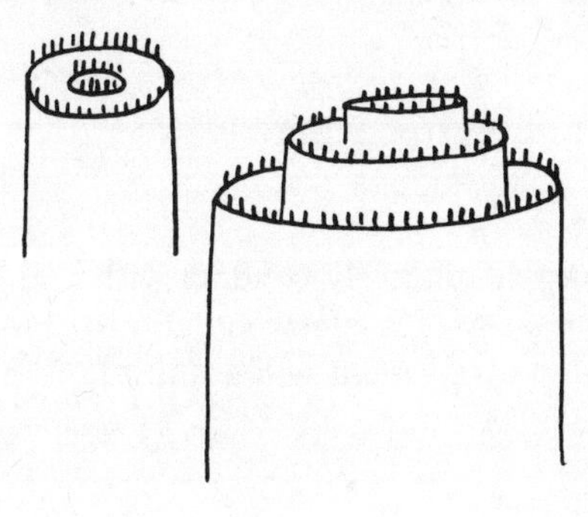

tric one of Mr Nicholson, and is not at all on the same principle as the light of the interior one does not pass through the flame of the exterior one. As he has published it in the Magazine, you have certainly a right to make use of it, as he has given it to the Public.

The defect of the Argan[d] lamp of Mr French is the holes admitting the air, exterior to the flame, being too near the flame; and hence it has not so strong a current, because the heated chimney is so much shorter. But if no smoke is formed I think the lamp preferable, because the consumption is less, as well as the quantity of light, and I should prefer two lamps less luminous, if equally clear, to one of double the luminosity, for common use. But perhaps the contrary may be more agreable for ornamental illumination, for which the pyramidal flame would be more agreable than the concentric, tho' the latter will have a more brilliant eye.

Mrs Darwin begs her compl. to Miss Wedgewood and hopes much to see her at Derby at her return, and will then answer her agreable letter by word of mouth adieu

from your sincere and affect. friend
E Darwin

Mr French said the Cup and saucer you gave Miss Violetta is the most beautiful thing he ever saw in the World. He admires the figures, and the absence of glaze etc, etc.

Original Not traced. There is a MS copy at University College London, Galton Papers, 14.
Printed Unpublished.
Text From MS copy. (*Diagrams* From MS copy.)
Notes
(1) 'Mr Nicholson' is William Nicholson (1753–1815), a talented and wide-ranging scientist and inventor, who devised electrical and printing machines, as well as oil lamps, wrote many scientific books and later (1796–1815) published *Nicholson's Journal*, which brought science to a wider public. For more details of Nicholson, see *DNB*.
(2) For Argand, see letter 86C; for Mr French, see letter 86E.
(3) The cup and saucer were in jasper, and probably a present for Violetta on her third birthday (23 April 1786).

86G. To Josiah Wedgwood, 9 May 1786

Derby May 9 –86

Dear Sir,

Mrs Darwin and myself sincerely condole with you on your late loss; which I hope nevertheless Mrs Wedgewood and Miss Wedgwood bear as well as such a degree of attachment will admit of.

The spring lamps I believe will answer well, I have one now burning by me as a candle, not having added the circular wick. The body is $2\frac{1}{2}$ inches diameter and $10\frac{1}{2}$ high containing the spring and the oil, from thence rises a thin candle 15 inches high, the whole being 26 inches high – grand to behold! like a star among the dusky air! it has burnt 6 hours. I have investigated the law of adapting springs of all strengths, and will begin tomorrow to make one with a circular wick and hope to send it in a week or 10 days.

The bladders I have also I believe found out how to improve and adapt.

The conical bag is not manageable in practice I believe, as the thin stalk of our lamp is its beauty, and when known to be a metallic stem will not appear weak. I think of making it thus and having no wheel and socket but a sliding brass knob.

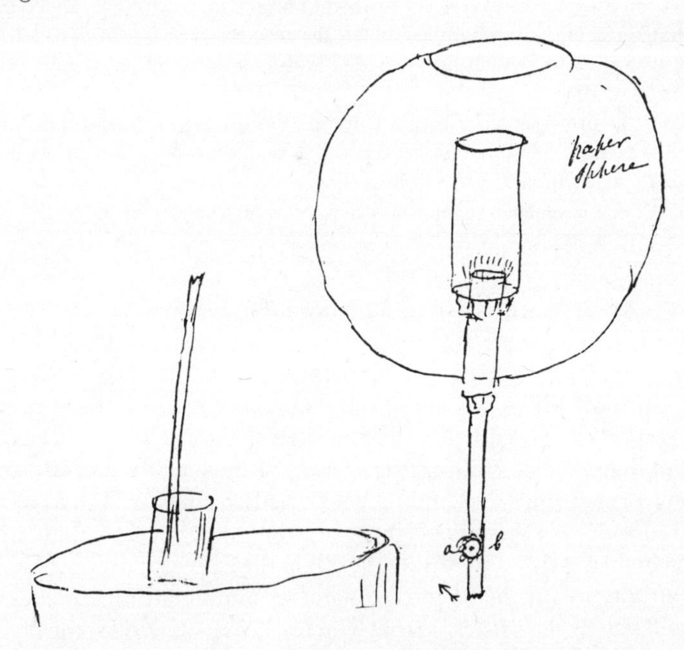

The body parts will be deeper in the spring lamp, as a spring takes up more room than a weight, but by this the lamp becomes as portable as a candlestick.

Mr 's Ideas about lighting rooms are most likely the consequence of his reasoning about the transparency of flame, that is its permeability to light. I don't think they can be any of our ideas. [At] the same time the less he thinks about lamps the less likely he will be to fall upon our ideas, and therefore perhaps it will be better not to hear his ideas at present, till you determine about the Patent.

My new spring-lamp is easily filled and the whole very simple.

The circular wick may either be applied as we did the other, or on a thin stem as above. The latter I shall first attempt.

Adieu, from, Dear W
Your affect. friend
and obed. serv. E Darwin

Original Not traced. There is a MS copy at University College London, Galton Papers, 14.
Printed Unpublished.
Text From MS copy. (*Diagram* From MS copy.)
Notes
(1) The death of Wedgwood's daughter Mary Anne, on 21 April, had obviously been expected by Darwin and Wedgwood. Though nothing could alleviate the immediate grief of the family, they all felt she could never have enjoyed a happy life, because she was 'apparently retarded both mentally and physically' (*Wedgwood Circle*, p. 79).
(2) This and the previous letter reveal Darwin's continuing enthusiasm for invention: he was now 54, but still retained the zeal that kept him inventing so assiduously at Lichfield in the 1770s.
(3) 'Dear W' at the end is unusual and may be a miscopying of 'Dear Sr'.

86H. To Josiah Wedgwood, 12 May 1786

Dear Sir,

Worm-springs, which are the simplest and cheapest, will take up more room in respect to perpendicular height, than springs of a dearer construction (as perhaps 4 or 5 shillings difference). I have made several experiments on worm-springs and find they will easily be managed so as to keep the oil accurately at the same height (excepting some accidental difference of friction in their rising and raising the oil) in the following proportion. The tin-box at bottom will be 6¼ inches diameter and about nine inches tall. The oil will rise eleven inches, so that 3 inches being allowed for the bladder and 6 for the apparatus. The base of the machine will be, as above, nine inches tall and the shaft nine (2 inches being common to both). This I think too low, as a mould-candle and candlestick stands about 2 ft high.

I am about to try springs of another form and also other kinds of worm-springs, but wish to have a line from you to know what size of the base, and what height will be most agreable to your designs, as I shall attempt to come as near that as I can.

In that which I made, the oil rises about 16 inches, but then it is ill-calculated (being the first) and in burning six hours the surface of the oil in the shaft sinks 2 inches, which is not sufficiently accurate because the friction of the bladder and spring can not be calculated, and therefore whatever inaccuracy the machine will allow of must be used for counteracting the inequality of friction.

But by what I have done, I see how to come to perfect accuracy, except friction; but wish to know the height of the base and its diameter; and the height of the shaft which will best suit your designs.

By the additional expense of complicated springs (perhaps 4 or 5 shillings) I expect to make the base with its bladder and springs to rise but $3\frac{1}{2}$ or 4 inches high: and to be about 4 or 6 inches diameter, and the shaft of what height you require, but I shall be necessitated to make 2 or 3 experiments first on the strength and action of these new forms of springs.

Mrs Darwin joins in all good wishes to you and yours with your affect. friend

and obed. serv.

E Darwin

Derby

May 12 –86

Original Not traced. There is a MS copy at University College, London, Galton Papers, 14.

Printed Unpublished.

Text From MS copy.

Notes

Here Darwin is trying to persuade Wedgwood to commit himself to particular sizes suitable for ornamental lamps. But Wedgwood seems to have been reluctant to take the plunge, as he had been with another of Darwin's projects in August 1782: 'I sigh that I am becoming an old man, that age and infirmities overtake and more than whisper in my ear that it is time to diminish rather than increase the objects of my attention' (Meteyard, *Wedgwood* ii 558). In 1786, with the Portland Vase to occupy him, his reluctance was greater, though he still encouraged Darwin to perfect the design of the lamp.

861. To Josiah Wedgwood, 17 September 1786

Dear Sir,

Are you at Matlock? If you are I will come and see you, but should more wish you would come and pass a few days here. Mrs Darwin is [returned] from a week's visit to Lichfield and joins in our wish to seeing you <u>all</u> if you are <u>all</u> at Matlock. Pray say, if Mr and Mrs Clive be at Matlock, and fix what day you will come to Derby.

Adieu

E Darwin

My other engagements here prevented me from consuming oil lately, but I think it in a good train.

Sept. 17 –86

I shall be from here all Monday and half Tuesday.

Original Not traced. There is a MS copy at University College London, Galton Papers, 14.
Printed Unpublished.
Text From MS copy.
Notes
(1) The Wedgwood family often visited Matlock (or Buxton) in September.
(2) Robert Clive (1723–1792), archdeacon of Salop from 1769 to 1792, was a friend of Wedgwood and a patient of Darwin. For Clive, see Scott, *Admissions* Part III, p. 532.
(3) September 17 was a Sunday.

86J. To Jonas Dryander, 5 October 1786

Derby Oct. 5 –86

Sir,

I am indeed much obliged to you for the favor of your letter, and thank you sincerely for the trouble you have taken in accomodating me with the plants from Thunbergh, and particularly for your observations, and other intelligence.

Our translation is nearly half printed, some of these plants from Thunbergh we must therefore add as an appendix. The colour, and real size, shall be omitted in all those, which are not already printed from the Flor. Japonica.

Our work is too far advanced for us to insert the Palms in their proper places, we must therefore retain them in the appendix, but are equally obliged to you for the kind offer of your assistance. Which we shall be much obliged to you for at your leizure; which account of their proper places, we should wish to print after the appendix, that the purchaser of the work may place them there, if He wishes it.

If L'Heritier's Stirpes novae be to be had in London at any shop you know, I shall take it as a favor, if you would direct it to be sent to me – and I will order a friend to call to pay for it, as I shall be glad to insert them on your authority.

And am, Sir, with true regards
your obed. serv.
E Darwin

If you happen to see Dr Blagden I beg to trouble you to present my Compliments and thanks to him. And please to tell him, that I had written to Davis and Elmsly before I troubled him about the copies of Dr Robt

Darwin's paper on Ocular Spectra: but as they never gave me any answer I doubted whether I had acted with propriety.

Original Botany Library, British Museum (Natural History), Dryander Correspondence, No. 41.
Printed Unpublished.
Text From original MS.
Notes
(1) Jonas Dryander FRS (1748–1810) was a Swedish botanist who succeeded Solander in 1782 as librarian to Sir Joseph Banks. Dryander wrote various botanical works, including a five-volume catalogue of Banks's Natural History Library. See *DNB*.
(2) 'Thunbergh' was the Swedish naturalist Carl Per Thunberg (1743–1828), a disciple of Linnaeus, who travelled widely in search of new species. He succeeded Carl Linnaeus the younger as Professor of Botany at Uppsala in 1783. His *Flora Japonica* was published at Leipzig in 1784. See *Biographical Dictionary of Scientists* (Black, 1969).
(3) 'Our translation' is the second translation by the Lichfield Botanical Society, the *Families of Plants*, which was published in 1787 (see letter 87A).
(4) Charles Louis L'Héritier (1746–1800) was a well-known French botanical writer. His *Stirpes novae* was published in six volumes between 1785 and 1791.
(5) Lockyer Davis (1719–1791) and Peter Elmsly (1736–1802) were leading London booksellers, and Davis was official printer to the Royal Society. See *DNB* for biographies of both.
(6) Dr Robert Darwin's 'paper on Ocular Spectra' was published in *Phil. Trans.*, **76**, 313–48 (1786). See *Doctor of Revolution*, pp. 176–7, and letter 87B.

86K. To Josiah Wedgwood, 12 December 1786

Dear Sir,

I have begun again to attend to the lamp, have got over the difficulties, which checked my progress – am about now to try experiments on the bladders, and expect to succeed – the lamp has perform'd well one hour and I am obliged to go from home – the wick is elevated without a wheel, hence it is more simple – the drawing I shall subjoin. We hope to see you all in your road to London or in your return.

Adieu
E Darwin

Decr. 12 –86
Having got free from the rachet wheel, there is nothing to hinder a silver case from being slip'd on over the copper pillar, that the oil which steels [steals] over the edges of the lamp may slide down between them. The lamp is also easily fill'd. Send me word what height of the column, and of the oil

box, and what diameter of the oil box you should like best that I may try the next in that form.

I will write again in a few days.

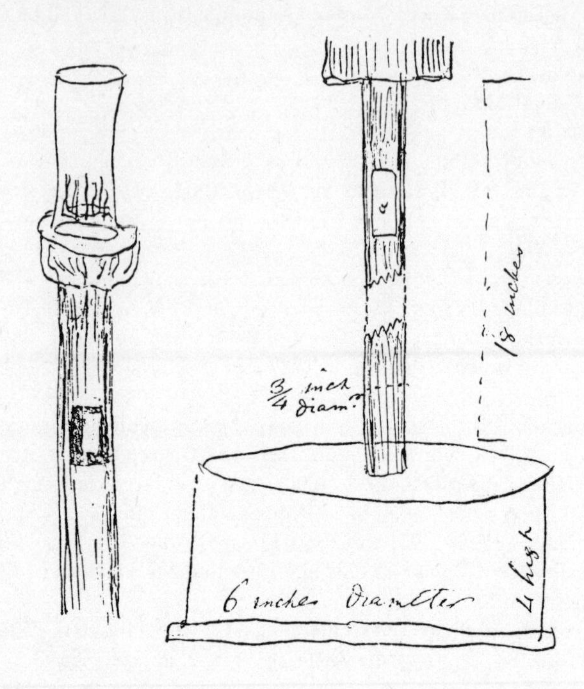

a the air hole drawn too long and not so wide as in the original lamp.

Original not traced. There is a MS copy at University College London, Galton Papers, 14.

Printed Unpublished.

Text From MS copy (*Diagrams* From MS copy.)

Notes

Wedgwood replied as follows on 27 December 1786: 'I am glad you have taken up the lamp again. As for the rest, that difficulties are overcome and fly before your perseverance and happy invention is a thing of course; I expect it, and shall not be disappointed. If I may hazard a conjecture where your prowess will meet with the most obstinate resistance, I would say that you will find the Goliah of the host in the bladder.

Make the height and diameter of the oil box the most convenient to yourself, we will find means to make it ornamental of any dimensions. The height of the pedestal and pillar together may be about 18 inches and higher if convenient.' (From MS copy at the John Rylands University Library of Manchester, English MS 1110.)

87A. To JONAS DRYANDER, 25 March 1787

Dear Sir,

I have the pleasure to acquaint you that our translation of the Genera Plant. is nearly finish'd printing, and in another month or two, I hope will be published. And will be as correct I hope, as can easily be expected, both in the accuracy of expressing the exact {me}aning of Linneus, and in the printing.

We have inserted the plants from Thunberg's Nova genera, and his flor. Japon. and L'Heritier by your kind offices. But Murray in his last edn. has some plants from Jaquin, whose works I believe are very expensive. The plants are but few. I can only now find

Elaodendron p. 241 of Murray System. ed. 14.
Lachenalia p. 314.
Sclerocarpus p. 783.

If you have the works of Jaquin I should be much obliged to you either to take the trouble to transcribe the Natural Character of the few plants Murray has introduced; or to acquaint me where I can see the originals. Perhaps there may be a few more in Murray from Jaquin which I have over-look'd.

I am, Sir, with true esteem,
Your already much obliged
and obed. serv.
E Darwin

Derby
Mar. 25 –87

Your remarks on Vahlia and Russelia and on Myristica, and the Palms, as well as the omission of the colours and real magnitudes of the plants of Thunberg, I intend to take the liberty to mention in our preface, which I hope you will not disaprove.

I shall be much obliged to you to represent my best respects to Sir J. Banks.

Addressed Jon. Dryander Esqr, at Sir J. Banks Bart, Soho Square, London. *Postmark* Derby, Mar 26.

Original Botany Library, British Museum (Natural History), Dryander Correspondence, No. 42.
Printed Unpublished.
Text From original MS.
Notes
(1) Dryander had kindly supplied the information from the books of Thunberg and L'Héritier which Darwin had requested in his letter of the previous October (86J). So Darwin was able to complete the translation of the *Genera Plantarum*.
(2) Nicolaus Joseph von Jacquin (1727–1817) was Professor of Botany at Vienna,

and a prolific author of books on natural history. Darwin was probably referring to Jacquin's *Icones Plantarum Rariorum* (1781–1793). For details of Elaeodendron, Lachenalia and Vahlia, see J. Hutchinson, *The Families of Flowering Plants* (Clarendon Press, Oxford, 1973), pp. 400, 756 and 569.

87B. To Josiah Wedgwood, 8 May 1787

* * *

When I want anything to be done (sais an old tutor of mine), I look out for a man who has the most business of his own; for if I can prevail on him to undertake it, it is sure to be done soon and well! Hence I apply to you, who have more to do of your own than any other man I have the pleasure to be acquainted with. My son, Dr Robert Waring Darwin, of Shrewsbury, I wish to be a member of the Royal Society, as it would be a feather in his cap, and might encourage him in philosophical pursuits; and I flatter myself He will make an useful member of that ingenious society. Now, I am unacquainted with the necessary form, and <u>if it be not either too troublesome or Disagreeable to you</u>, I should be much obliged to you to hang up his name, and take what steps are necessary. I have written to Mr Charles Greville, who has spoken to Sir J. Banks on the subject. Mr Greville desires me to write to my friends, and sais <u>He will receive their certificates</u>. I don't well know what this means; he writes in a great hurry, and sais I may expect success. I intend to write to a few acquaintance I have in town, as Sir G. Baker, Rev. Dr Kaye, Mr Whitehurst – Professor Martyn, I fear, is not a member; Dr Blagden I am not personally acquainted with – Sir Richard Jeb.

* * *

Original Not traced. In the possession of Charles Darwin, 1865.
Printed E. Meteyard, *A Group of Englishmen* (Longmans, 1871), pp. 253–4.
Text From above source.
Notes
(1) Charles Darwin wrote (*Life of Erasmus Darwin*, p. 84): 'Robert Waring Darwin (my father, born 1766), did not inherit any aptitude for poetry or mechanics, nor did he possess, as I think, a scientific mind. He published, in Vol. lxxvi of the "Philosophical Transactions" a paper on Ocular Spectra . . . but I believe that he was largely aided in writing it by his father. He was elected a Fellow of the Royal Society in 1788'.
(2) The paper on Ocular Spectra was published in 1786, when Robert was 20, and by the standards of those days, was enough to justify Robert's election as a Fellow of the Royal Society – or would have been if he had written it himself. Erasmus recognized Robert's lack of enthusiasm for science, and hoped to remedy it by having him elected FRS. Unfortunately his election did not produce the required enthusiasm, but it did begin an overlapping succession of Fellows which lasted for 201 years through five generations of Darwins and is unique in the annals of the

Royal Society. Erasmus was a Fellow from 1761 to 1802, Robert from 1788 to 1848, Charles from 1839 to 1881, Charles's son Sir George Howard Darwin from 1879 to 1912, George's brothers Francis from 1882 to 1925 and Horace from 1903 to 1928; and lastly, George's son Sir Charles Galton Darwin from 1922 to 1962.

(3) Robert's certificate for the Royal Society was read on 15 November 1787 and he was elected on 21 February 1788, at the age of 21. There are eleven signatures on the certificate, the first five being: Charles Greville, Josiah Wedgwood, James Watt, Matthew Boulton, George Baker.

(4) For notes on Charles Greville, see letter 78J. Sir George Baker (1722–1809) was an eminent royal physician whose medical lectures Darwin had attended in 1752 (his lecture notes are in the Wellcome Institute Library). 'Professor Martyn' was probably Thomas Martyn (1735–1825), who was Professor of Botany at Cambridge from 1762 to 1825. However, he was elected FRS in 1786, so Darwin may have been referring to Thomas Martyn (died 1816) who wrote books on natural history and established a private academy in London. For notes on Dr Blagden, see letter 85F. Sir Richard Jebb (1729–1787) was another eminent royal physician. 'Rev. Dr Kaye' was Richard Kaye (1736–1809), chaplain to the King and Prebendary of York, who succeeded to a baronetcy in 1787. See *DNB* for Baker, Martyn and Jebb.

87C. To Jonas Dryander, 21 May 1787

Derby May. 21 –87

Dear Sir,

I take the liberty to return you the Nov. Gen. of Thunberg with many thanks for the use of it. Our translation is nearly printed off, and I hope is much more accurately printed than the other. I will take the liberty to send you a copy, as soon as it is out of the Press, of which I must beg your acceptance.

I must now beg another favor of you, which is to give your suffrage to my Son, Dr Rob[t]. W. Darwin, whom I wish to be a member of the Royal Society, which might add new Spirit to his industry in philosophical pursuits; and should be much obliged to you to present my best compliments to Dr Blagdon, when you see him, and ask for my Son the same favor from him. Mr Ch. Greville has spoken to Sir Joseph Banks in his favor, whom I should otherwise have troubled with a letter on this account.

Doctor Rob[t]. Darwin is settled in Shrewsbury, where he meets with very great encouragement, having been concern'd for near fifty patients in the first six months: and I dare say will make a very useful member of the Society.

I am, Sir your much obliged
and obed. serv.
E Darwin

Original Botany Library, British Museum (Natural History), Dryander Correspondence, No. 43.

Printed Excerpt in *Nature*, **189**, 424 (1961), and D. King-Hele, *Erasmus Darwin* (1963) p. 38.
Text From original MS.
Notes
(1) Darwin is now at the end of his six years of labour on the Linnaean translations. He kept his enthusiasm to the last, as is shown by his borrowing of Thunberg's book to ensure that new species were included.
(2) The remarks about Robert Darwin are similar to those in the letter to Wedgwood (87B), though expressed more politely. Dryander did not sign Robert's certificate, however.

87D. To Benjamin Franklin, 29 May 1787

Derby May 29 –87

Dear Sir,

Whilst I am writing to a Philosopher and a Friend, I can scarcely forget that I am also writing to the greatest Statesman of the present, or perhaps of any century, who spread the happy contagion of Liberty among his countrymen; and like the greatest Man of all antiquity, the Leader of the Jews, deliver'd them from the house of bondage, and the scourge of oppression.

I can with difficulty descend to plain prose, after these sublime ideas, to thank you for your kindness to my son Rob[t]. Darwin in France, and to converse with you about what may arise in philosophy, which I know will make the most agreable part of my letter to you.

1. Mr Bennet, a Curate in my neighbourhood, has found out a method of doubling the smallest concievable quantity of either plus, or minus electricity, till it becomes perceptible to a common electrometer, or increases to a spark. It is thus: He has three brass plates about 4 or 6 inches diameter,

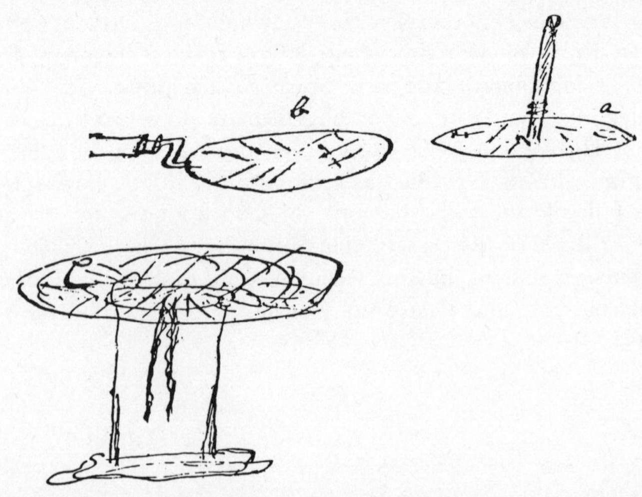

C is a brass plate varnished on its upper surface, lying on an electrometer like Cavallo's except that it has two slips of gold leaf hanging down instead

of threads and balls. Mr Bennet, to try the electricity of the atmosphere, takes a coated Phial in one hand into the open air; and holding up a blazing candle above his head, touches the lower part of the flame with the knob, holding the coating in his other hand. 2. then He comes into the house, and with the knob touches the plate C (which is insulated by being laid on the glass electrometer of Cavallo) and then lays the plate b on C and touches it with his finger. (NB. the handle of this plate, and of a are insulating handles.) Then the electricity recieved by the coated phial is divided between that phial, and the two plates C and b, which have a varnish between them. 3. Then He separates the plate b from C and puts on b the plate a (which is varnish'd on the underside) and touches that plate a. Then He separates a from b [:] now as b was contrary to C – a is the same as C. He then touches C with the edge of the plate a, and at the same time applies the plate b over C and repeats the process 5 – 10 – or 50 times. And thus the electricity is continually doubled, or very nearly so, because one inch of the surface of a coated electric will absorb or recieve perhaps 100 times as much as will continue on such a surface of a conductor in contact with it.

This is likely to be of service in detecting the small quantities of electricity given out in chemical processes, or atmospheric changes. De Luc's idea of two electric fluids, like water and heat, is I think very ingenious, and perhaps true.

In respect to other philosophical news, I have just heard from Mr Wedgewood that Mr Herschel has discover'd three Volcanoes in the Moon now burning.

Since I had the pleasure of seeing you, I have removed from Lichfield to Derby, and have superintended a publication of a translation of the botanical works of Linneus, viz. The System of Vegetables in two volumes 8vo. and the Genera or Families of Plants in 2 vol 8vo. also. I did this with design to propagate the knowlege of Botany. They are sold to the booksellers at 14/ the System of Vegetables – the Genera will be finished in a month, and will be sold to the booksellers at 12/ I believe – but as we are to pay for advertizing and carriage, I expect we shall not clear more than 10/ on each set. If the work had been finished I should have sent you it by the favor of Mr Nicklin, who is so kind as to take the care of this letter. If I thought 20 sets of each were likely to be sold I would send them at 10/ a set of each, that is 20/ for the four volumes. And indeed would now have sent them by Mr Nicklin, had the whole been ready. As I think they would not be worth reprinting in America, and perhaps 20 sets would be as many as would find purchasers.

A Line from you at your leizure, only to acquaint me that you continue to possess a tollerable share of health would be very acceptable to, dear Sir with true esteem

Your most obedient servant
E Darwin.

Addressed Doctor Franklin, America.

Original American Philosophical Society, Philadelphia, Pa., Franklin Papers XXXV 70.
Printed Excerpts in *Science,* **43** (N.S.), 773 (1916) and *Doctor of Revolution*, p. 179.
Text From original MS. (*Diagrams* From original.)
Notes

(1) The opening paragraph of this letter is splendid, and all the more so because it is not mere flattery but Darwin's genuine opinion, and difficult to refute even today: Franklin was foremost among the founding fathers of what is today the world's most powerful state, and what other statesman can claim so great an achievement? Franklin was now eighty-one and was attending the Philadelphia Convention which was to produce the Constitution of the United States. The Convention ran throughout the summer and Franklin was instrumental in framing, and persuading the delegates to accept, the Constitution which has now lasted triumphantly for nearly 200 years.

(2) Abraham Bennet (1750–1799) was curate of Wirksworth near Derby, a keen experimenter in electricity, and the inventor of the gold-leaf electroscope, which Darwin mentions here. In his book *New Experiments in Electricity* (1789) Bennet tells us that he began his experiments in atmospheric electricity at Darwin's suggestion.

(3) For Cavallo, see letter 87H.

(4) Jean André De Luc (1727–1817) was a well-known physicist, born in Geneva, but living for much of his life in England. His chief contributions were in meteorology: see A. Wolf, *A History of Science, Technology and Philosophy in the 18th Century* (Allen and Unwin, 1952) i 288–302. De Luc was a friend of Watt, Boulton and others in the Lunar group: see *Lunar Society*, pp. 240–1.

(5) The observations of bright spots on the lunar surface by William Herschel (1738–1822) were reported in *Phil. Trans.*, **77**, 229–31 (1787). Herschel called them volcanoes, and his observations were the source of Coleridge's 'bright star within the nether tip' of the Moon in the *Ancient Mariner*, and of the modern study of transient lunar phenomena.

(6) For the translations from Linnaeus, the *System of Vegetables* (1783–5) and the *Families of Plants* (1787), see letter 81D and later letters to Banks and Dryander.

(7) I have not identified Mr Nicklin. There are two contenders. The first is 'Edward Nicklin, Gent.', author of the poem *Pride and Ignorance*, printed in Birmingham in 1770: the theme of the poem is that ambition is the source of tyranny and of the horrors of war, and Nicklin (*c.*1740–1805) presents his thesis in vigorous couplets which would have appealed to Darwin. The second contender is Rev. Joseph Nicklin (?1750–*c.*1800) Vicar of Pattingham near Wolverhampton from 1780–96, author of several published Sermons, and a graduate of St John's College, Cambridge (see Scott, *Admissions*, Part III, p. 717).

(8) The gentle courtesy of Darwin's last paragraph matches the splendid fervour of the first. It was his last letter to Franklin, who died in 1790.

87E. To Josiah Wedgwood, 29 July 1787

Dear Sir,

I understand that the Patent lamp is not Mr Keir's of Winson Green. He told Mr Day he knew of no such thing. Pray inquire about it, and send me either an Account of it, or a lamp. Adieu – pray make Derby your road if you can, as nothing gives me so much pleasure, and information I must add, as seeing you. Miss Susan Sneyde is come to our assize ball, but no Miss Wedgewood.

from your affect. friend
E Darwin

Derby
July 29 –87

Original Not traced. There is a MS copy at University College London, Galton Papers, 14.
Printed Unpublished.
Text From MS copy.
Notes

(1) Darwin's old friend James Keir was living at Winson Green in the Smethwick district of Birmingham. Presumably Darwin himself was shy of writing to Keir in case he found his friend had become a rival lamp-designer. So he probably asked his son Erasmus to enquire of Thomas Day, who was probably a regular correspondent of Keir.

(2) The 'Patent lamp' Darwin mentions is that of Peter Kier (*c.*1760–*c.*1820), who took out patent No. 1585, dated 29 January 1787, concerned with a new method of raising the oil. Kier's design is discussed in a letter from Robison to Watt (11 August 1785), in *Partners in Science*, ed. E. Robinson and D. McKie (Constable, 1970), pp. 148–9. See also A. Rees, *Cyclopaedia* (1819), Vol. 20, under 'Lamp'. Kier had also designed a new steam engine of the Savery type, which enjoyed some success, though much less efficient than Watt's. See A. Rees, *Cyclopaedia* (1819), Vol. 34, under 'Steam Engine'. Kier later manufactured engines and coach axles in College Street, Camden Town: see *Post Office London Directory* (1816).

(3) Miss Susan Sneyd (later Mrs Broughton) was the daughter of John Sneyd of Belmont (1734–1809). Susan Sneyd was the subject of one of the best of Darwin's occasional poems, 'On a Target at Drakelow' (1791), quoted in Seward, *Darwin*, pp. 395–6. He also wrote a poem in praise of John Sneyd in 1779: see Seward, *Darwin*, pp. 102–3.

87F. To Josiah Wedgwood, [8–12 September 1787]

Dear Sir,

I cannot conveniently come to Matlock tomorrow, but shall be very happy to see yourself and party, or any of you here, but cannot supply with beds at present.

I have made a lamp, which I think worth a patent; it is pressed with weights as I can not yet succeed with springs, tho' I do not quite despair of them. This new lamp is I suppose 6 or 7 pounds weight. It has not a Bladder nor a conical box or bag, but acts quite well – it requires being carried about with care and to be set down without much jolt, as all other fountain-lamps must. It stands up like a Candlestick, the box at the bottom is 7 inches high, and about 3½ wide, may be made about 5 high and 4 wide conveniently.

I intend to make one more trial with springs, and bladders, neither of which I yet quite despair of. If I do not see you here, you shall hear again from me in a week, but Derby is your last road every way, and our new bridges are the eight[h] wonder of the Peak.

The Madona and Bambino are well. Compliments from all here attend you all. I think of taking the first step towards a patent immediately and will include the springs and bladders also in it, unless in my next trial I shall despair of them.

I burnt my new lamp three hours last night, at the end of which a misfortune happened to it, which will be repaired to day adieu.

Your affect. friend
E Darwin

If I do not see you I will send you a drawing and description – and you will pray send me either Keir's lamp, or an account of its performance and principle.

Original Not traced. There is a MS copy at University College London, Galton Papers, 14.
Printed Unpublished.
Text From MS copy.
Notes
(1) Darwin's wife Elizabeth gave birth to a son, John, on 5 September. This is why Darwin could not travel to Matlock, or provide beds for Wedgwood's party. 'The Madona and Bambino are well', shows that Wedgwood already knew of the birth, so the letter is unlikely to have been written before 8 September, nor after 12 September, since the next letter is dated 14 September.
(2) The letter shows that Darwin is still continuing his experiments on the lamp with undiminished enthusiasm. For Kier's lamp, see previous letter.

87G. To Josiah Wedgwood, 14 September 1787

Derby Sept. 14 –87

Dear Sir,

I send this least you should think of making Derby in your road – I am much press'd by a letter from Mr Bent to go immediately to Mr Moreton, his neighbour. How unfortunate it is that we can't meet! – 1. My new lamp is easily made, fill'd and clean'd – 2. it requires two pounds of mercury – 3. it takes 3 pounds and ¼ of lead – 4. the lamp itself weighs 3¾ pounds – 5. the oil I suppose $\frac{1}{2}$ a pound – nine pounds and a half. If springs can be used, $3\frac{3}{4}$ pounds will be saved and if bladders 2 pounds more, above half the weight.

You shall hear from me in a week or as soon as I can make my ultimate trial with springs and bladders.

Pray don't mention the lamp, as I suspect Professor Robison had heard I was doing something on this subject.

Adieu E Darwin

Dimension of the new Lamp.

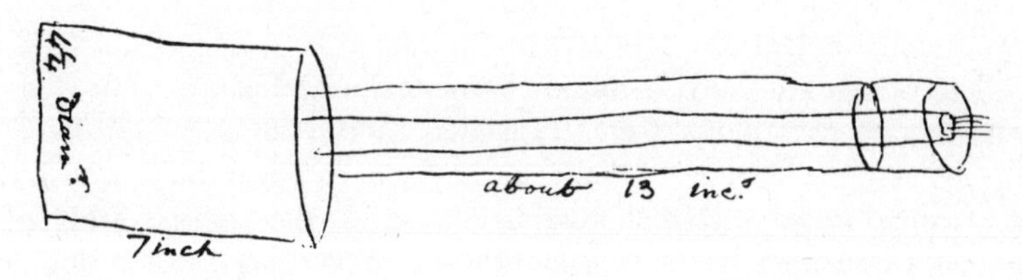

Original Not traced. There is a MS copy at University College London, Galton Papers, 14.
Printed Unpublished.
Text From MS copy. (*Diagram* From MS copy.)
Notes

(1) This is the last of the letters about oil lamps. It seems that Darwin's lamp was never put into production by Wedgwood. Darwin's design was obviously practicable and quite thoroughly tested, however, and probably better than existing designs. To assess Darwin's place in the early history of oil lamps is beyond the scope of these notes.

(2) Mr Bent is probably James Bent (1739–1812), the Newcastle surgeon who amputated Wedgwood's leg in 1768 and served him well for 30 years: see E. Posner, *Pharm. Hist.*, **3** (1) 6–8 (1973).

(3) Rev. Ralph Moreton (*c.* 1720–1803) lived at Wolstanton Hall, two miles north of Newcastle. But Darwin's 'Mr Moreton' is probably his son Ralph Moreton (1753–1787). See P. Adams, *Wolstanton* (1908), p. 40; J. Sleigh, *History of Leek*, 2nd edn (Bemrose, 1883), p. 95; *Alum. Oxon.* iii 978; and *Staffordshire Advertiser*, 16 April 1803.

(4) John Robison (1739–1805), one of Watt's closest and earliest friends, became Professor of Natural Philosophy at Edinburgh in 1773. See *DNB, Doctor of Revolution,* p. 70 and notes to letter 87E. Robison is spelt 'Robinson' in the MS copy.

87H. To John Coakley Lettsom, 8 October 1787

Derby Oct. 8 –87

Dear Sir

I am much obliged to the Gentlemen of your medical society for the honor they have done me in electing me one of their body, and am sorry I have not at present any thing worth communicating to them. I have indeed been lately much engaged in revising the translation of the Families of

Plants of Linneus by a botanical society at Lichfield, but as this work is now out I shall hope to be able to attend to some other investigations at leizure hours.

The public are much obliged to you for the account of the root of scarcity and for your public spirit in distributing the seeds of these [;] if you have still any to spare, I shall be much obliged to you for 2 or 3 or a few plants to try them in my botanical garden.

I am, Dear Sir, with great esteem
Your obed. serv.
E Darwin

P.S. I have just got Curtis's botanical magazine and can not but wish the characters of the plants were translated in the concise english of the translation of the families of Plants and System of Vegetables publish'd by our Lichfield society: – which language is so near the original, and had Dr Johnson's assistance in the settling of it – I am certain it would double or quadruple the sale of Mr Curtis's magazine. I have already recommended it to three ladies, who lament the want of latin to understand it.

I have lately cured two old men, {abou}t seventy each, of the ascites join'd with anasarca, by half an ounce of that decoction of Digitalis mention'd in my paper on that subject in the transactions of the college, given with an ounce of peper-mint water twice a day for 4 or 6 days – and a third of about the same age, with more repeated doses for one day.

If your society proposes questions, I should wish to offer for one. Whether the nervous fever of Huxham (or Fever with debility without petechiae or sore throat, or flush'd countenance, or pungent heat) be the same as the petechiae fever, or jail-fever? The former of these, viz the nervous fever of Huxham prevails much over all this country at this time, but is fatal to I believe hardly any, as yet.

I thought proper to bleed two of these patients in the beginning from violent cough, and pain of the region of the stomach. The blood was very sizey, and cup'd – moderate stimulating diet and medecine at first gradually increased in respect to the quantity of wine as the disease advances, seems to support and save them all. adieu

Addressed Dr J C Lettsom, London.

Original The Houghton Library, Harvard University, Cambridge, Massachusetts.
Printed T. J. Pettigrew, *Memoirs of the Life and Writings of the late John Coakley Lettsom* (Nichols, London, 1817), Vol. 3, pages 117–19.
Text From original MS.
Notes
(1) John Coakley Lettsom, FRS (1744–1815) was an energetic and successful Quaker physician who took up many worthy public causes and was founder of several societies, particularly the Medical Society of London, to which he presented a house and a library. He was also a prolific writer of letters, pamphlets

and articles in journals, particularly the *Gentleman's Magazine*. He helped to promote the use of the mangel-wurzel, or 'root of scarcity' as it was called, and in 1787 translated a French pamphlet on the subject, which Darwin commends. There is a 2½-page biography of Lettsom in the *DNB*; see also T. J. Pettigrew, *Memoirs of J. C. Lettsom*, 3 vols. (1817).

(2) William Curtis (1746–1799) was a botanist and entomologist, best known for his *Flora Londinensis*, begun in 1777. Darwin refers to the *Botanical Magazine*, which he founded in 1781 and maintained with success until his death. For a biography of Curtis, see W. H. Curtis, *William Curtis* (Warren, Winchester, 1941).

(3) Dr Johnson's assistance in the Lichfield Society's translations is mentioned in letter 81F.

(4) Ascites is dropsy of the abdomen, while anasarca refers to dropsy in the limbs. 'My paper . . . in the transactions of the college' is Darwin's paper 'An account of the successful use of Foxglove in some dropsies and in the pulmonary consumption', *Medical Transactions*, **3**, 255–86 (1785). See *Doctor of Revolution*, pp. 169–70.

(5) John Huxham, FRS (1692–1768) was a well-known physician, author of *An Essay on Fevers and their various kinds* (1755), which was regarded as an authoritative classification of the multifarious diseases grouped under the umbrella of 'fever'. Petechiae are red or purple spots on the skin. 'Sizey' [sizy] is thick and viscous. Being cup'd [cupped] was one method of blood-letting: the skin was scarified and a suction cup applied.

(6) Lettsom wrote to Darwin again on 22 November 1787, informing him that on 15 November he was elected to the Medical Society of London (MS letter at Darwin Museum, Down House).

87I. To Thomas Beddoes, [October? 1787]

Dear Sir,

I have put some fossils into a box with the assistance of Mr F. French, but intend to get some more for you in a little time. I write to know whether you would have them sent now to Shifnal or Oxford, or whether you may not probably visit Derbyshire next summer, when I could supply you more amply, and more to your intention, and where I should be happy to see you, and glad, if this motive may attract you into our hemisphere. I shall be likely, perhaps, to send you fossils which you do not care for, and not send you such as you want; I should therefore advise you to come over next summer, and I will load you home again; otherwise I will send you about one hundred weight when you please.

I think Cavallo in his book on magnetism, says that inflammable air from iron filings and vitriolic acid affected the magnetic needle. Can this be true? Has any other person tried the experiment?

Have you any idea, how coals happen to lie stratified, as it were, with clay or sand-stone? Could these alternate strata be made by depositions in the bottom of the sea? or has the phlogistic part of the coal been sublimed from the clay, and condensed in a higher stratum? or lastly, does not the goodness of the different strata of coal arise from their bitumen having been more or less evaporated by fire?

You see, I should endeavour to get much information from you, if I could attract you to Derby next year.

Adieu, from dear Sir,
Your obedient servant,
E Darwin.

Original Not traced.
Printed J. E. Stock, *Life of Beddoes*, Appendix 6, p. xxxvi.
Text From above source.
Notes
(1) Dr Thomas Beddoes (1760–1808) graduated in medicine in 1784 and became a lecturer in chemistry at Oxford University in 1787. Darwin probably first met Beddoes at his home town, Shifnal, Shropshire, shortly before this letter was written. He soon became Darwin's most frequent correspondent.
(2) Since 'next summer' is 'next year', the season must be autumn; and the date cannot be earlier than October 1787 because Beddoes received the letter when he was in the midst of a course of lectures, and Term did not begin until 10 October.
(3) 'Mr F. French' was Forester French, the 16-year-old son of Darwin's friend Richard French (see letters 86E and 94U).
(4) Tiberius Cavallo, FRS (1749–1809) was well known for his researches on magnetism and electricity. Darwin is referring to his *Treatise on Magnetism in Theory and Practice* (1787). Darwin had this book on loan from the Derby Philosophical Society from 15 to 30 August 1787 (Catalogue and charging register, Derby Central Library, MS 9230).
(5) There may be omissions in this letter, as in all those given by Stock.
(6) Beddoes's reply to this letter, also printed by Stock, begins, 'I just receive your letter when I am going through a course of lectures with, I believe, the largest class that ever was assembled in Oxford, at least since the discovery of Justinian's code drew together thirty thousand students'. Beddoes then gives his views on coal, and asks Darwin to send the fossils. The next letter is Darwin's reply.

87J. To Thomas Beddoes, [late 1787?]

Dear Sir,

About ten days ago I sent you a box of fossils. At the top I put some shewy specimens, which I purchased for you of a Mr Brown, parish-clerk and fossil philosopher of this place; for which you are in my debt 1£. 1s. 6d. and for which I shall trust you, till I have the pleasure of seeing you next summer; when I shall beg leave to bring you acquainted with the said Parish Clerk.

In respect to coal I wrote a paper about it and the tar-spring at Colebrook-Dale, about a year or two ago, which I intend for the Royal Society; and to add to it, the borings for coals from six or eight collieries, which I have got in colliery-language; and I have also the specimens of the borings (materials bored) from two mines, wrapped in papers, with their names in the above antient language.

If you analysed the tar (petroleum) at Colebrook-Dale, and are not unwilling to have it sent to the Royal Society, I should be glad to add it to my paper; where it would hang like a diamond in an Ethiop's ear.

All I believe about coal is, that it was not formed beneath the sea, because there are no marine shells in it, in general, or in the strata above it. I have some specimens of coal with impressions of shells; do you know where that coal comes from? if you do, pray inform me at your leisure. The design of my paper is to persuade others to examine the subject, which is yet so imperfectly understood.

1. Sulphur is found in various combinations, was not this once vitriolic acid? and has not the vital air been stolen from it to constitute other acids by uniting with other bases, whence it became solid sulphur? Has not coal been a fluid acid, and the vital air taken from it to compose the acids of flint, and shale?

2. Have not the vegetable recrements, which composed morasses, fermented, and taken fire, and the phlogistic part rose a few feet by a certain degree of heat and left the clay of the vegetables beneath; and hence the coal-strata took their alternations and thickness, hence the tar-spring was sublimed from the coal-bed below?

If you have time and inclination to write to me on this subject, I shall be obliged to you; as I intend, in about three or four weeks, to re-copy, and send the conjectural paper above described, and should be glad to increase its value by your name, in respect to the analysis of the Colebrook bitumen.

If the two boxes of coal-borings would be interesting to you to see, I will either send them to you or shew you them in the summer; they are not above half a hundred weight.

You say a man can only expose himself by asking or answering questions. I think he may shew his ignorance by doing neither, when he corresponds with ingenious men.

I am, my dear Sir,
Your obedient servant,
E Darwin.

Original Not traced.
Printed J. E. Stock, *Life of Beddoes*, Appendix 6, pp. xxxvii–xxxviii.
Text From above source.
Notes
(1) The date of this letter seems likely to be about a month after the previous one. The petroleum at Coalbrookdale was found in October 1786 (*Economy of Vegetation*, Note XXIII), so if Darwin wrote a paper on the subject 'about a year or two ago', this letter cannot be earlier than October 1787 and could belong to 1788.
(2) 'Mr Brown, parish-clerk' is Richard Brown the younger (1736–1816), son of the

founder of the firm of monumental masons, Richard Brown and Sons. See R. Gunnis, *Dictionary of British Sculptors* (1968).

(3) Coalbrookdale, near Ironbridge Gorge in Shropshire, was of course the birthplace of the Industrial Revolution, where iron was first smelted with coke by the Darbys. See B. Trinder, *The Industrial Revolution in Shropshire* (Phillimore, 1973). Darwin's paper on the tar-spring at Coalbrookdale was not sent to the Royal Society but was incorporated in Note XXIII of *The Economy of Vegetation*, which also includes a description of the coal borings in the 'antient language'. Later editions of *The Economy of Vegetation* have a plate with a section of the strata down to 1000 feet at St Antony's Colliery, near Newcastle-upon-Tyne. Darwin's Note suggests that his information about the tar-spring came from William Reynolds (1758–1803), the ironmaster and grandson of Abraham Darby II. Reynolds controlled the Darby iron works and mines at this time. See B. Trinder, *The Darbys of Coalbrookdale* (Phillimore, 1974).

(4) The letter shows Darwin's keen interest in geology: he collected not only fossils but also borings from mines. His speculations about coal (see also letter 91E) are generally most perceptive.

(5) Beddoes's reply, printed by Stock, shows that he disagreed with Darwin about the origin of coal: Beddoes believed it was laid down beneath the sea.

88A. To JAMES PILKINGTON, 5 February 1788

Derby, Feb 5 –88

Dear Sir,

In compliance with your request I send you on paper, what has occur'd to me about the natural history of the Buxton and Matlock waters.

[There follows a 20-page scientific paper, which it would not be appropriate to reproduce here. Darwin contends that the heat of the springs is caused not by chemical reactions, as had been widely suggested, but by 'subterranean fires deep in the Earth'. He ends his 'letter' as follows:]

If you think any part of this long letter worthy a place in your expected history of Derbyshire, I desire you will do with it whatever is agreable to you.

I am, dear sir,
With great respect, etc
E Darwin

Original Not traced.
Printed J. Pilkington, *A View of the present state of Derbyshire* (Drewry, Derby, 1789), Vol. 1, pp. 256–75.
Text From above source.
Notes

(1) James Pilkington (1752?–1804) was educated at Warrington Academy, and in the 1770s became Minister at the Unitarian Chapel at Friar Gate, Derby (an engraving of the Chapel is given in Glover, *Derby* ii 498). He also conducted a

school and published some elementary books for Sunday schools. In addition to the useful *View of Derbyshire*, he wrote *Doctrine of Equality of Rank Examined* (1795). For information on Pilkington, see W. Turner, *The Warrington Academy* (Warrington, 1957), p. 65, and R. Simpson, *History of Derby* (1826) i 439.

(2) Pilkington introduces Darwin's paper as follows: 'I shall now subjoin a letter . . . with which Dr Darwin has been so obliging, as to favour me. The respectable name of the writer, I am persuaded, will be sufficient to excite the curiosity and attention of the public, and from the novelty and ingenuity, which appear in some of the remarks, it may be presumed, that they will not be disappointed in their expectations.' (Pilkington, *Derbyshire* i 255)

(3) Darwin's 'letter' is well-argued and correct. He gives numerous reasons for believing that the springs are warmed by subterranean heat rather than chemical action: (a) the heat of the springs has been invariable for many centuries; (b) they do not decrease during droughts; (c) the limestone rocks abound in clefts which might connect to great depths; (d) subterranean heat is well attested from volcanoes and in mines; (e) there are hot springs in volcanic areas; (f) water heated by chemical action would taste of the chemicals; etc. He also discusses the artesian well in his garden. See *Doctor of Revolution*, pp. 157–9, 198.

88B. To Richard Lovell Edgeworth, 20 February 1788

Derby Feb. 20 –88

Dear Edgeworth

You write short scrawling letters full of questions, which take up one line, and expect me to send you dissertations in return, on academies, hot-houses, philosophers, attorneys, etc. etc. – and then, without being at the trouble of acquainting me with your previous knowlege of all these intricate subjects, you tell me, after I have been laborious in inquiry, that you knew all this before.

I have seen a Nottingham Quaker, who speaks much in favor of the Nottingham Master, and his academy, and have heard of several parents, who have sons there, and approve of their progress in their respective lines of study – and that there are 80 scholars more or less, and that there are but two beds in a room, and but two boys in a bed. I don't know if He has vacancies, this you must write to him about, if you approve.

Of Hothouses I am ready to answer any questions you will ask, written fairly on paper, and number'd, so as to be easily dispatched by a yes or a no – but can not write a volume, for you to light tapers with – because you will say in answer "my dear friend, I knew all this before".

1. My Hothouse has a fire 4 months in a year only. 2. consumes about six tons of coals by conjecture. 3. it is about 82 ft. long and about 9 ft. wide. 4. it has only oblique sashes, no perpendicular ones. 5. it is divided into two, that one half may be a month forwarder than the other. 6. it produces abundance of Kidney-beans, cucumbers, Melons, and Grapes, not pines.

Plate I

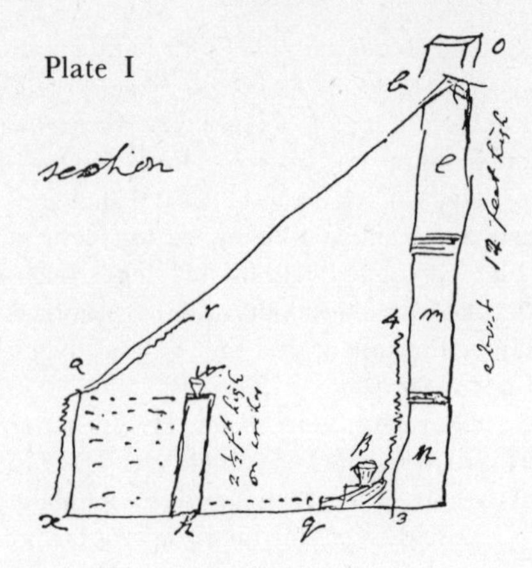

ab glass panes 8 inches square each at 3d½ a foot square, ax brick-wall, wk a flue, 1mn flues, o a chimney, kq a wooden walk made with old barrel staves to prevent the vine roots from injury, 34 a vine, xar another vine. {aw xk bark-bed for melons 3ft½ wide. The flue is 7 inches wide and about [2ft 6] [?] inches deep built on arches. B pots of flowers between the vines, the same on w.

Plate II

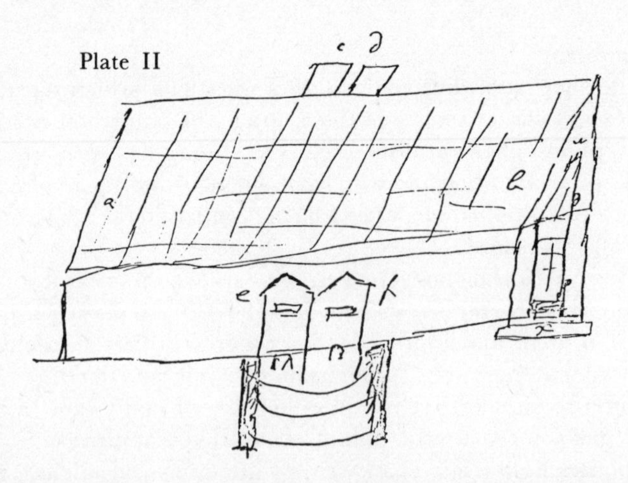

ab glass roof, cd two chimneys, ef two fire-places with piger[?] Henricus or athanor in each, wx one end door. NB. the fire enters the flues from ef two feet below the surface of the earth, and passes to the ends of the house, one eastward, the other westward; along the flue wk (plate I) then the flue dives under the step at the door at the end of x (Plate II) and then rises again into the flue n (Pl.I) then it goes to the midle of the building and returns through m and again return to the chimney through 1.

My fires are in front, as I have no property {behind the} back wall – hence the smoke {*gap of about 5 words*} under it very {*gap of 3 lines, cut by scissors from the manuscript*} kidney-bean plants now 8 inches {*gap of about 5 words*} and melons one inch high, and vine-shoots 5 inches long with fruit appearing upon them. I put fire to the other division this day Feb 20 –88.

Best aspect is about 20 degrees from the S. towards the E.

Philosophers.

Mr Keir amuses his vacant hours by mixing oil and alcaline salts together, to preserve his Majesty's subjects clean and sweet – and pays 1000 Guineas every six weeks to an animal call'd an Exciseman; who would otherwise trouble this diversion of his. And He is now printing a new improved edition of his chemical dictionary.

Attorneys

Mr Erasmus keeps free from the jaundice, but is obliged to use equitation now and then to preserve his health.

I drink water only, and am always well, we have five young creatures, 3 male, and two female, all tall (and hansome as Mrs Darwin thinks) the boys strong, the girls less so.

In your first letter I believe you never mentioned Nottingham academy – your vines should be raised from cutting this spring, even now, and planted in your house in June or July from the pots they are raised in. Plant them two feet asunder in your house and one against every rafter on the outside. My house cost 100£ and I don't repent the expense. I had glass 8 inches square from Stourbridge. Square panes are preferable, because you can chuse which side fits best to that below it. But all these you are already possessed of I dare say – and my labour in writing is vain!

Answers to other questions. Mr Pole is at Cambrige. Miss Pole, at Derby, goes to London with Mrs and Miss Wedgewood for some weeks. Miss M. Pole at Cambden-House. Erasmus is indifferent well, and writes law-parchments. Keir makes soap and dictionaries, and such things, and thrives I dare say. He has one daughter only at once – when this dies he begats another. You put fewer words in a line than a lawyer, and fewer lines in a page. God mend you.

{*Ending and signature cut out with scissors*}

Addressed R. L. Edgeworth Esq, Edgeworth-town, Ireland.

Original Cambridge University Library, Dar. 218.
Printed Excerpts in *Doctor of Revolution*, pp. 186–7.
Text From original MS. (*Diagrams* From original.)
Notes

(1) The manuscript is complete apart from a slice one inch deep cut out with scissors containing the signature and three lines on the reverse. In transcribing the key to the hot-house, some commas have been inserted for clarity. In the manuscript the paragraphs beginning 'Philosophers' and ending 'less so' come after '(Plate II) and then rises'. See facsimile between pages 192 and 193.

(2) Apart from its contents, this letter is interesting in showing how much trouble

Darwin would take to help his friends. The playful complaint in the first paragraph also reveals how secure was his friendship with Edgeworth, who had been away in Ireland for nearly seven years.

(3) The Nottingham Academy was founded in 1777 by Charles Wilkinson, who conducted the school for 6 years. His successor was Rev. John Blanchard, who was Master from 1783 to 1814. See F. A. Wadsworth, *Trans. of the Thoroton Society*, **45**, 57–75 (1941); *Gent. Mag.*, **56**, 1002 (1786) & **97**, 376 (1827).

(4) Darwin's description of the hot-house is so detailed that a replica could be constructed. (In Plate I, the height is, I believe, 12 feet.)

(5) The letter is notable for its biographical sketches. There is a glimpse of James Keir and his thriving alkali factory at Tipton, where he made alkali, soap, white and red lead, and other products. The factory was one of the show-places of the Midlands and the excise duties on soap reached £100 000 a year: see B. M. D. Smith & J. L. Moilliet, *Notes and Records of the Royal Society*, **22**, 144–54 (1967).

(6) 'Mr Erasmus' is Darwin's eldest surviving son, now 28 and doing well as a solicitor in Derby. 'Mr Pole' is Sacheverel Pole (1769–1813), Elizabeth Darwin's son by her previous marriage, who was now 18 and had entered St John's College, Cambridge in 1787. 'Miss Pole', her eldest daughter Elizabeth (1770–1821), was now 17, and Milly ('Miss M. Pole', 1774–1857) was 13.

88C. To Robert W. Darwin, 21 February 1788

* * *

I am printing the 'Loves of the Plants', which I shall not put my name to, tho' it will be known to many. But the addition of my name would seem as if I thought it a work of consequence.

* * *

Original Not traced. In the possession of Charles Darwin, 1879.
Printed C. Darwin, *Life of E. Darwin*, p. 92.
Text From above source.
Notes

(1) Dr Robert Waring Darwin (1766–1848), the youngest of Darwin's three sons by his first marriage, had begun practice as a physician at Shrewsbury in 1786, and was elected a Fellow of the Royal Society (see letters 87B and 87C) on the day this letter was written; so it may be a letter of congratulation. Robert continued his medical practice for 60 years and was extremely successful: he was said to have had the highest income of any provincial doctor.

(2) This is the first of many letters to Robert from which excerpts are quoted by Charles Darwin in his biography of Erasmus.

(3) Erasmus was very shy about publishing his poem *The Loves of the Plants*. It was frivolous, and he feared, not without reason, that it might damage his reputation as a serious author.

88D. To James Watt, 18 November 1788

Dear Sir,

Mr Wedgewood sent me the inclosed, and desired me to sign it, and to forward it to you – who he hopes will get it signed by the Birmingham F.R.S.s and that you will then send it to Dr Darwin of Shrewsbury, and desire him to remit to Mr Wedgewoods in London Greek Street Soho.

I have some design, if not prevented, of {com}ing to see the next Lunar Meeting – pray acquaint me, whether yourself and Mr Boulton and Dr Priestly are likely to attend it and likewise, what day it will be celebrated upon.

Pray take care of your health, and use the warm bath, and try Balsam of Canada for your stomach complaints, about 20 grains or drops mix'd with honey, and then with water according to art, twice a day.

And remember the philosophical experiment Monsr. Rabelais determined to try upon himself for the good of the Public – which was "to try how long an ingenious and agreable man might last, if taken good care of".

Where did you say manganese was got, and whither sent to, and at what price?

Pray read the 40 first pages of the Introduction to Fourcroy's Elements, and tell me, if the facts are in general true – if they be, the theory holds them nicely together.

When steam is passed through red hot iron scrapings, if the water be not decomposed, whence comes the vital air, which unites to the iron? – does the water vanish, or is it annihilated? pray explain this experiment.

In Dr Priestly's Liquor of conviction[?] there is much nitrous acid, some marine acid, and tenfold the quantity of water. What follows? – did the water exist dissolved in the two airs? or did the nitrous acid exist in either of the airs? or the marine acid in either of them? or are they all produced in the process? If the inflammable air was not pure, if it contain'd mephitic (phlogistic air) air the water and nitrous acid might both be form'd.

I shall wait with Patience to see this great dispute decided, which involves so great a part of the theory of chemistry – and thank the Lord, that chemical Faith is not propagated by fire and sword, – at present I am inclined to the heterodox side of {the} question. The Lord take you into his holy keeping, adieu

from your affect. friend

E Darwin

Mrs D. begs to be remember'd to Mrs W.

Mr Bennet is going on with his new discoveries of adhesive electricity.

Addressed James Watt Esq, Birmingham. (*Endorsed* Doctor Darwin Nov 18 1788.)

Original Birmingham Public Libraries, Boulton and Watt Collection.

Printed Excerpts in *Doctor of Revolution*, pp. 189, 199.
Text From original MS.
Notes

(1) 'The inclosed' was probably a petition about the slave trade: Wedgwood and Darwin were strong supporters of the campaign against the trade (see letters 89B and 89D).

(2) The second paragraph shows that, after seven years away from the Birmingham area, Darwin was still trying to attend Lunar Society meetings.

(3) Watt took Darwin's Rabelaisian advice and 'lasted' well into his eighties.

(4) During the 1780s Darwin abandoned the phlogiston theory of combustion, according to which a burning substance gives off 'phlogiston', and adopted the 'French heresy' that burning was combination with oxygen. The Lunar chemists, Priestley, Keir and Watt, continued to believe in phlogiston, however, and this letter shows Darwin tactfully trying to 'soften up' Watt. Darwin's strength of mind, and his overriding devotion to scientific truth, are nowhere more obvious than in his rejection of phlogiston. Two of his closest friends, James Keir, a 'mighty chemist' and pioneer of the chemical industry, and James Watt, whose work with steam led the world, were both strongly committed to phlogiston, and so was Joseph Priestley, the great experimental chemist. They would regard Darwin as a renegade if he abandoned phlogiston: yet he did so, because he was convinced that the phlogiston theory was wrong.

(5) 'Fourcroy's Elements' is the *Elements of Natural History and Chemistry* (4 volumes; Robinson, London, 1788), by A. F. de Fourcroy (1755–1809). The catalogue and charging register of the Derby Philosophical Society (at Derby Central Library) shows that Darwin borrowed Fourcroy's book from the Society's Library on 20 October 1788 and returned it on 25 November. Darwin was not alone in being influenced by Fourcroy, who probably did more than anyone else to publicize the new chemistry. See W. A. Smeaton, *Fourcroy, chemist and revolutionary* (1962).

(6) There are long discussions of 'acidic liquors' by Priestley, and by Keir and Withering, in *Phil. Trans.*, **78**, 147–57 and 313–30 (1788).

(7) For Mrs Ann Watt, see letter 94D.

88E. To Edward Johnstone, 16 December 1788

Dear Sir

You may have heard of a dispute between my Son, Dr Darwin of Shrewsbury and Dr Withering, and have probably recieved a printed paper, which I find has been dispersed about it. My Son acquaints me that Dr W has written him a second letter, and that he thinks the way to prevent future ill-usage is to retaliate the present injury by mentioning more cases of Dr W's mistakes in practise, and thinks of inserting the following, but wished I would learn whether you desired to have any part of it alter'd or omitted. The sentence is this. Was not you consulted along with Dr Johnson in the case of Mr Francis of the Moat Birm? And did not you by the solemn quackery of large serious promisses of a cure get the management of this patient into your own hands, and did not you, Sir on the night before He died congratulate his son on his perfect recovery?

I was at Birmingham yesterday and ment to have waited on you, to return the visit you once favor'd me with at Derby.

I dined at the Hotel with the philosophical (or lunar) society, and was sorry to see no physician there but Dr Withering. – How does this happen? when philosophers are liberal-minded and agreable; and there are so many ingenious of the faculty at Birmingham.

You will oblige me by not mentioning this inquiry of you, about the case of Mr Francis.

I am sir
your obed. serv.
E Darwin

Derby
Dec 16 –88

Addressed To Doctor Johnsone, Birmingham. (*Postmark* Lichfield 16 Dec 1788.)

Original Royal Society of Medicine, London, Withering Letters, pp. 103–4.
Printed Excerpts in *Proc. Roy. Soc. Medicine*, **22**, 1089 (1929); *Lunar Society*, p. 324; and *Doctor of Revolution*, p. 185.
Text From original MS.
Notes

(1) Dr Edward Johnstone (1757–1851) was a prominent Birmingham physician, who was appointed with Dr Ash as physician to the General Hospital opened in 1779. Johnstone lived at 15 Temple Row (see Pye's *Directory of Birmingham* for 1785) and was the son of Dr James Johnstone (1730–1802), who knew Darwin. For the Johnstones, see *DNB* and C. L. Johnstone, *History of the Johnstones, 1191–1909* (Johnston, Edinburgh, 1909).

(2) This quarrel between his son Robert and Dr Withering really upset Erasmus and led to two letters with touches of malice. The reasons were deep-seated, going back ten years, to the death of Darwin's brilliant son Charles at the age of 19 when he was a medical student at Edinburgh. In 1780 Darwin published the pamphlet on 'Pus and Mucus', which included Charles's medical thesis and an account of five medical cases treated by digitalis – the first published account of the use of digitalis in heart disease. Withering was angry because he regarded this as his discovery, and Darwin was incensed at Withering's attack on his dead son. There was another skirmish in 1785, when Withering accused Darwin of stealing his botanical nomenclature. Then in 1787 Withering made an unprovoked attack on Darwin's younger son Dr Robert Darwin (see letter 88C), newly started as a physician at Shrewsbury: Withering deliberately countermanded Robert's treatment of a patient without consulting him. This could have gravely damaged Robert's career – he was only 21 at the time – and his father urged him to fight back. He did so, and began an acrimonious correspondence with Withering, later published as a pamphlet entitled *Appeal to the Faculty concerning the Case of Mrs Houlston* (Shrewsbury, 1789). Apparently Withering had an unlucky habit of

congratulating patients on their recovery on the day before they died, and Mr Francis of the Moat was one such case. Dr Johnstone was a friend of Withering, and he chose not to 'oblige' Darwin: instead he showed Darwin's letter to Withering, thus further fanning the flames of controversy. An excellent account of the quarrel is given by E. Posner, *History of Medicine*, **6** (1), 51–7 (1975).

(3) 'Mr Francis' was John Francis senior of Moseley, who died on 4 October 1788. The obituary in *Aris's Birmingham Gazette* (6 October 1788) says: 'upon his demise his body was opened, and a very large stone was found in his bladder'. Darwin would have known of the post-mortem, which discredits Withering's optimism.

88F. To Josiah Wedgwood, [1788?]

* * *

Mrs Darwin has commissioned me to write to thank you for the very excellent Bath you have been so kind as to send her. But what was the astonishment of the family when on striking it there was heard a rattling of armour within! Some of us began to think it like the Trojan Horse, and fear'd it might contain armed warriors; others that like Pandora's box, it might contain many evils at the top and hope at the bottom. These fears ceased on recollecting who was the kind donor, and that presents from a friend were not to be suspected like those of an enemy. So out came the straw, and with it the bowels of the pestiferous animal, beyond any power of numbering or naming, for each individual of which I am commanded to thank you over and over, and to add all our kindest compliments to you and yours.

* * *

Original Not traced. In the possession of Charles Darwin, 1865.
Printed Meteyard, *Wedgwood* ii 560.
Text From above source.
Notes

(1) The date of this letter is doubtful. Meteyard states that 'In April 1786, Wedgwood sent Darwin's little daughter Violetta a jasper cup and saucer of extreme beauty; and a few years later the Doctor thus writes:' There follows the text given above. So the letter presumably belongs to 1788, 1789 or 1790. The choice is open, but there are already five letters to Wedgwood recorded in 1789, so I have chosen 1788, when there are otherwise none.

(2) The letter shows Darwin in his domesticated role, writing a thank-you letter on behalf of his wife. The 'armed warriors' inside were presumably presents for the children.

89A. To Matthew Boulton, 18 January 1789

Derby Jan. 18 –89

Dear Boulton

When I had the pleasure of seeing you, you mention'd to me that Dr Withering had congratulated Mr Francis of the Moat at Birmingham on the perfect recovery of his father on the day before He died. Dr Robt. at Shrewsbury acquaints me, that He is about to answer Dr W's printed paper, and wishes to know {the} fact about Mr Francis, whether Dr W congratulated him on the recovery of his father the day before his death, or the day but one before his death. And I wish to know whether Mr Francis himself told you so – you may depend on your name not being mention'd, least it should bring a war upon you. I hear the faculty in Birmingham are on Dr Robt's side of the question; and will I trust be more so, the more it is canvass'd.

I often lament being at so great a distance from you – your fire-engine for coining money gave me great pleasure for a week after I saw it – and the centrifugal way of regulating the fire. If you ever come near Derby pray give me a day of your company. I hope your gravelly complaint does not now molest you – pray write me an account how you are in that respect.

Adieu from, dear Boulton,
Your old affectionate friend
E Darwin

Original Birmingham Reference Library, Matthew Boulton Papers, Darwin 34.
Printed Short excerpts in *Doctor of Revolution*, pp. 185, 202.
Text From original MS.
Notes
(1) The first paragraph is a continuation of the dispute between Robert Darwin and Withering (see letter 88E). Probably Boulton did not respond, not wishing to be involved.
(2) Darwin was most impressed by Boulton's coining machine and gives it a prominent place in *The Economy of Vegetation* (see letter 89C). Watt's centrifugal governor for rotative steam engines was introduced in 1788 and was presumably new to Darwin.

89B. To Josiah Wedgwood, 22 February 1789

Dear Sir,

Herewith you will recieve the Botanic Garden, of which I am the supposed, not the avow'd author. After you have read the page on the Slave-trade 117, and the eulogy on Mr Howard's Humanity in visiting prisons p. 80 – I do not insist on your reading any more. What I wish is, if it can be done without much trouble, that you would do me the favor to use your interest with Mr Griffiths to get it review'd early; as that I am told may influence the sale of it. If this can be done, you will please either to send the book I herewith send you, (which I will replace for your library) or tell me how to direct one to be sent from Johnson St Pauls, who is to sell them.

After having asked this favor I intend to scold you for not making Derby in your way to and from London, this I shall however postpone till I have the pleasure of seeing you here, as I shall hope to inlist Mrs and Miss Wedgwood, as well as Mrs Darwin against you.

I saw yesterday a book written by Daniel Defoe author of Robinson Crusoe, it is called The history of the life of Colonel Jaque. There is a long story in it at p. 140 of the generous spirit of black slaves, when kindly used; it is told rather too diffusely, so as to be almost tedious, but would well suit a magazine, or a newspaper, if it was curtail'd, or publish'd in parts. If the Society have not reprinted this story, it is worth their attention: tho' found in a novel.

Mr Boothby has put up a french paper in his double room, which He wishes you to see; and I wish Miss Wedgwood to see it; as it is full of imagination. He has also a french stove, made of delf[t]; I am not certain whether such stoves might not become in use in this country; as they have not the smell (as he assures me) that iron-ones have. There is some ingenuity in the structure of it – more than at first view meets the eye.

All here beg to be kindly remember'd by all at Etruria, with

dear Sir

your affect. friend

E Darwin

Derby
Feb 22 –89.

Original Darwin Museum, Down House, Downe, Kent.
Printed Short excerpts in *Lunar Society*, p. 205 and *Doctor of Revolution*, p. 200.
Text From original MS.
Notes

(1) *The Loves of the Plants*, Part II of *The Botanic Garden*, was published in April 1789, and Darwin is here sending Wedgwood a pre-publication copy. He seems to have thought Wedgwood would be bored by the catalogue of plant life and directed his attention to two digressions on social reform. Page 117 on slavery includes the lines:

E'en now in Afric's groves with hideous yell
Fierce SLAVERY stalks, and slips the dogs of hell;
From vale to vale the gathering cries rebound,
And sable nations tremble at the sound! –
–YE BANDS OF SENATORS! whose suffrage sways
Britannia's realms, whom either Ind obeys;
Who right the injured, and reward the brave,
Stretch your strong arm, for ye have power to save! (III 373–80)

In the passage on John Howard beginning on page 80, Darwin tells us that the 'Spirits of the Good', on seeing Howard,

> Mistook a Mortal for an Angel-Guest,
> And ask'd what Seraph-foot the earth impress. (II 465–6)

(2) Darwin had no inkling of the poem's imminent fame, and was anxious to ensure that it received some reviews and did not fall dead from the press. 'Mr Griffiths' was Ralph Griffiths (1720–1803), editor of the *Monthly Review*, and formerly a close friend and neighbour of Thomas Bentley, Wedgwood's partner (who died in 1780).

(3) Wedgwood replied to Darwin's letter on 7 March (Wedgwood, *Letters*, iii 77), offering 'a thousand thanks for the pleasure and instruction I am receiving in the perusal of these beautiful and charming Cantos'. He also says he has written to Griffiths. On 15 March Wedgwood wrote to Darwin again: 'Mr Griffiths tells me that the Botanic Garden is not yet published and until that is the case it cannot be admitted into the review. Otherwise I believe it would have made its appearance the next month and perhaps if you publish immediately it may do so still'. (From MS copy at the John Rylands University Library of Manchester, English MS 1109, p. 7). The review appeared in April, the month of publication.

(4) Wedgwood and Darwin both keenly supported the Society for the Abolition of the Slave Trade, and Darwin was on the lookout for possible propaganda to pass on to Wedgwood (see letter 89D). The instance he quotes here is from Daniel Defoe's *The History and remarkable life of the truly honourable Col. Jacque, commonly called Col. Jack* (Brotherton, London, 1723; reprinted: Constable, London, 1923). In the story Jack becomes manager of a plantation in Virginia: the slaves respond favourably when he treats them well (pp. 169–88 in the 1923 reprint).

(5) Brooke Boothby's home improvements seem to be mentioned in the hope of luring the Wedgwoods into Derbyshire. Miss Wedgwood is Josiah's eldest daughter Susannah, who often visited the Darwins: see letter 84D. The Delft-ware stoves were less smelly than iron stoves because they were cooler on the outside and did not fry dust, hairs or other debris settling on them: see Wedgwood, *Letters* iii 78.

89C. To Matthew Boulton, 12 March 1789

Dear Boulton

I was much obliged to you for your kind letter from London, which breath'd the true spirit of our long and antient friendship; which I dare say will not cease on either side, till the earthy tenement of our minds becomes decomposed. I fail'd in getting an evidence from Dr Johnson, to whom I wrote; and least your name should have been mentioned, you see by Dr Robts paper the fact was not made use of. My letter to Dr Johnson was simply a letter of inquiry; and Dr W. by shewing it only published his own disgrace. I think Dr Robt Darwin has given him a dressing, He will not soon forget. Whether He designs to write again I have not heard – I should suppose He must already feel his inferiority in this controversy.

How He came to suspect your having told me the affair of Mr Francis of the Moat I can't concieve – but He knew I had called on you that morning.

I hope you have got your contract for coining; or will get it, as soon as the King can do business. Your machinery gave me great pleasure and reminded me much of antient days.

I beg my compliments to your son, and to Miss Boulton both whom I should be glad to see, if any occasion should bring you into this part of the country.

Adieu from dear Boulton
your affect. friend E Darwin

Mar 12 –89.

I hope you are free from your gravelly complaint. The way to keep free is to drink much aqueous potation – to occasionally lie down in the day on a Sopha – to use the warm bath once or twice a week, and to drink as little wine or malt-liquor or spirit, as you can.

adieu

If you have no gravelly symptoms I see no use in your continuing to take aerated water, whether with alcaline salt in it or not.

Original Birmingham Reference Library, Matthew Boulton Papers, Darwin 35.
Printed Excerpt in *Doctor of Revolution*, p. 202.
Text From original MS.
Notes

(1) The first paragraph continues the story of the quarrel between Dr Robert Darwin and Dr Withering. 'My letter to Dr Johnson' is 88E. See also letter 89A.

(2) Boulton's machinery for coining greatly impressed Darwin, and in *The Economy of Vegetation* (I 281–8) he celebrates Boulton's success with these red-hot technological verses:

> With iron lips his rapid rollers seize
> The lengthening bars, in thin expansion squeeze;
> Descending screws with ponderous fly-wheels wound
> The tawny plates, the new medallions round;
> Hard dyes of steel the cupreous circles cramp,
> And with quick fall his massy hammers stamp.
> The Harp, the Lily and the Lion join,
> And GEORGE and BRITAIN guard the sterling coin.

(3) Boulton's son Matthew Robinson Boulton (1770–1842), though lacking his father's genius, had good business sense and began to take over the management of Soho in the 1790s. Boulton's daughter Anne (1768–1829) was beginning to suffer more severely from the congenital malformation that was to cripple her in the 1790s (see letter 69D).

(4) 'Gravelly complaints', passing small particles of stone in the urine, were much more common in the eighteenth century than today. See also letter 94S.

89D. To Josiah Wedgwood, 13 April 1789

* * *

I have just heard that there are muzzles or gags made at Birmingham for the slaves in our islands. If this be true, and such an instrument could be exhibited by a speaker in the House of Commons, it might have a great effect. Could not one of their long whips or wire tails be also procured and exhibited? But an instrument of torture of our own manufacture would have a greater effect, I dare say.

* * *

Original Not traced. In the possession of Charles Darwin, 1879.
Printed C. Darwin, *Life of E. Darwin*, p. 46.
Text From above source.
Notes
Wedgwood and Darwin both strongly supported the Society for the Abolition of the Slave Trade, and Darwin was on the lookout for any useful propaganda. See also letter 89B.

89E. To Robert W. Darwin, 19 April 1789

Apr. 19 –89

Dear Robert,

I am sorry to hear you say you have many enemies, and one enemy often does much harm. The best way, when any little slander is told one, is never to make any piquant or angry answer; as the person who tells you what another says against you, always tells them in return what you say of them. I used to make it a rule always to receive all such information very coolly, and never to say anything biting against them which could go back again; and by these means many who were once adverse to me, in time became friendly. Dr Small always went and drank tea with those who he heard had spoken against him; and it is best to show a little attention at public assemblies to those who dislike one; and it generally conciliates them.

* * *

Original Not traced. In the possession of Charles Darwin, 1879.
Printed C. Darwin, *Life of E. Darwin*, p. 37.
Text From above source.
Notes
Unlike many sons, Robert seems to have valued his father's advice, of which this is a sample.

89F. To Josiah Wedgwood, June 1789

Dear Sir,

I have written the enclosed at your desire and taken as much from Thomson as I could, to guard it from minor Critics. If you wish to have the speech of Hope longer, send me what materials you would have added, as suppose:

Here future Newtons shall explore the skies.
Here future Priestleys, future Wedgewoods rise.

Now in return remember you are to send me a drawing of Cupid warming a butterfly, and also to tell me, what Engraver I should apply to do it not excellently, but tollerably, because the other was done intollerably. I have written to Alken about making Flora and Cupid smile, but have yet had no answer.

Seriously, if you wish to have any other matter added to the speech of Hope, acquaint me and I will endeavour to add it. When I print the first part of the Botanic Garden, which as this book sells, I probably may, about Christmas, I intend to introduce a few lines on the Pottery, and a note on it. Now in these lines or note I wish to mention the principal improvements which you have made. Pray answer me the following questions.

Do you pretend yourself to have improved the antique forms or only pretend to have copied them? What you have <u>really</u> done is no part of my question, for I believe you have very greatly improved them.
2. Was anything of consequence done in the medallion or Cameo kind before you? I am thinking of celebrating China for colours and early Art, Antique Etruria [for] forms and modern Etruria for living medallions, giving about four lines to each. But I shall wait till I hear your opinion. A short historic note on the subject of Potteries I should wish to add of about half a page, particularly of your own Pottery which you must supply me with.

Adieu, with the Ladies' compliments
Your much obliged friend
E Darwin

Addressed Josiah Wedgewood Esqr., Etruria, near Newcastle, Staffordshire.

Original Not traced. There is nineteenth-century MS copy at the John Rylands University Library of Manchester, English MS 1110, pp. 49–50.
Printed Excerpt in *Doctor of Revolution*, p. 201.
Text From MS copy at Rylands Library.
Notes
(1) 'The enclosed' is the 'Visit of Hope to Sidney Cove, near Botany Bay', 26 lines of verse printed at the end of the later editions of *The Economy of Vegetation*, in which

Darwin offers his famous prophetic vision of the future of Sydney Cove, including the city –

> *There* shall broad streets their stately walls extend –

and the Sydney Harbour bridge –

> *There* the proud arch, colossus-like, bestride
> Yon glittering streams, and bound the chafing tide.

The extra couplet about Newtons, Priestleys and Wedgwoods was not included in the published version, probably because Wedgwood would agree only if *Wedgewoods* was changed to *Darwins* – an embarrassing amendment for the author. The verses were intended to accompany Wedgwood's medallion of Hope at Sydney Cove, attended by Peace, Art and Labour. The idea was 'to show the inhabitants what their materials would do, and to encourage their industry', as Darwin put it (*Economy of Vegetation* II 315 note). Wedgwood was very pleased with the verses: 'I thank you for some of the prettiest lines I have read a long time, except in the botanic garden, and perfectly suitable for the occasion', he wrote to Darwin on 28 June (Wedgwood, *Letters* iii 84).

(2) The first edition of *The Loves of the Plants*, published in April 1789, had a rather ugly drawing of Cupid by Edward Stringer, the Lichfield topographical artist. The illustrations of the Linnaean classes of plants were also drawn by Stringer; so probably Darwin had been favourably impressed by Stringer's illustrations for the *Gentleman's Magazine* in 1785 (Vol. 55, pp. 25 and 412). (Most of the Stringer drawings in the Stringer Collection at the Bodleian Library are by his son C. E. Stringer. See *Trans. Lichfield Arch. Hist. Soc.*, **4**, 27–46 (1962).) Wedgwood provided Darwin with a much more elegant Cupid, who adorns the second and subsequent editions of *The Loves of the Plants*.

(3) 'Alken' refers to Samuel Alken (1750–1815), the engraver of the picture of Flora and Cupid which formed the frontispiece of the first edition of *The Loves of the Plants*.

(4) Darwin's hope that *The Economy of Vegetation* would be printed 'about Christmas' was not fulfilled: it was published in June 1792. The verse and notes on pottery are in Canto II, lines 291–348, and the long Additional Note XXII on the Portland Vase (see letter 89H). The following lines (340–6) are typical:

> Whether, O Friend of Art! your gems derive
> Fine forms from Greece, and fabled Gods revive;
> Or bid from modern life the Portrait breathe,
> And bind round Honour's brow the laurel wreath;
> Buoyant shall sail, with Fame's historic page,
> Each fair medallion o'er the wrecks of age.

(5) Like everyone else at the time, Darwin over-rated Etruscan art, and did not know that much of the 'Etruscan' work was really Greek.

(6) The date 'June 1789' is written on the MS copy, and is almost certainly correct, because Wedgwood replied on 28 June (*Letters* iii 84–7), giving a full answer to Darwin's queries about pottery.

89G. To James Keir, 12 September 1789

Bar[r], at Mr Galton's,
Sept. 12 –89.

Dear Keir,

I wrote you a hasty letter yesterday, before I had had time to peruse your very ingenious and excellent preface, which I have not now time to admire, but only to mention what I think an error of the flying pen.

Page 8, at the bottom line, you seem to say that common attraction or gravitation is the property of matter, by which its quantity has been always ascertained. Now I always understood that it was the vis inertiæ that measures the quantity, and not the gravity. Gravity is variable in different parts of the earth, as here or at the line; vis inertiæ is everywhere the same, as the different lengths of pendulums necessary to measure the same portions of time, here and at the line, evince. Is it too late to alter this?

You have successfully combated the new nomenclature, and strangled him in the cradle, before he has learnt to speak.

Adieu,
From your affectionate friend,
E Darwin

Original Not traced. Probably in the possession of the Moilliet family in 1868.
Printed Moilliet, *Keir*, pp. 95–6.
Text From above source.
Notes

(1) James Keir (see letters 67E and 88B) was friendly with Darwin for nearly fifty years, but only two letters from Darwin to Keir have survived, both on rather peripheral topics. Keir's house was burnt down in 1807 and Darwin's letters may have been destroyed then. There are eleven letters from Keir to Darwin, at dates between 1755 and 1798, printed in Moilliet, *Keir*.

(2) The preface Darwin mentions is that of Keir's book *The First Part of a Dictionary of Chemistry* (Elliot & Kay, London, 1789), not to be confused with Keir's admirable translation of Maquer's *Dictionnaire de Chymie* (1771). Keir's preface runs to 20 pages, and he vigorously defends the phlogiston theory of chemistry, which Darwin had already abandoned in favour of the 'new nomenclature' – with 'oxygen' and 'hydrogen' instead of 'dephlogisticated air' and 'inflammable air'. Keir's *Dictionary* is alphabetical, and the First Part only goes as far as 'Acid (vegetable)'. So about twenty Parts were presumably intended, but never published, probably because Keir recognized he was wrong. See B. M. D. Smith & J. L. Moilliet, *Notes and Records of the Royal Society*, **22**, 144–54 (1967).

(3) Darwin was right to object to Keir's confusion of weight with mass (*vis inertiæ*), though Keir contested the correction: see Moilliet, *Keir*, pp. 96–7.

1. Sir Joseph Banks (1743–1820)

2. Thomas Beddoes (1760–1808)

3. Thomas Bentley (1730–1780)

4. Charles Blagden (1748–1820)

5. Matthew Boulton (1728–1809)

6. Thomas Brown (1778–1820)

7. Thomas Byerley (1747–1810)

8. Thomas Cadell (1742–1802)

9. Joseph Cradock (1742–1826)

10. James Currie (1756–1805)

11. Robert W. Darwin (brother) (1724–1816)

12. Robert W. Darwin (son) (1766–1848)

13. Thomas Day (1748–1789)

14. Georgiana, Duchess of Devonshire (1757–1806)

15. Fifth Duke of Devonshire (1748–1811)

16. Jonas Dryander (1748–1810)

17. Richard Lovell Edgeworth (1744–1817)

18. Benjamin Franklin (1706–1790)

19. Lucy Galton (1757–1817)

20. Samuel Tertius Galton (1783–1844)

21. Thomas Garnett (1766–1802)

22. Charles F. Greville (1749–1809)

23. Lady Harrowby (*c.* 1735–1804)

24. Lord Harrowby (1735–1803)

25. William Hayley (1745–1820)

26. Phoebe Horton (1755/6–1814)

27. James Hutton (1726–1797)

28. Edward Jenner (1749–1823)

29. Edward Jerningham (1727–1812)

30. Joseph Johnson (1738–1809)

31. James Keir (1735–1820)

32. John Coakley Lettsom (1744–1815)

33. Samuel More (1725–1799)

34. Elizabeth Pole (1747–1832)

35. Richard Polwhele (1760–1838)

36. Joseph Priestley (1733–1804)

37. Richard Pulteney (1730–1801)

38. Dudley Ryder (1762–1847)

39. Anna Seward (1742–1809)

40. Sir John Sinclair (1754–1835)

41. Peter Templeman (1711–1769)

42. Gilbert Wakefield (1756–1801)

43. John Walcott (1755?–1831)

44. William Watson (1715–1787)

45. James Watt (1736–1819)

46. Josiah Wedgwood (1730–1795)

47. Thomas Wedgwood (1771–1805)

48. William Withering (1741–1799)

49. Letter from Darwin to Richard Lovell Edgeworth dated 20 February 1788 (Letter 88B)

In your first letter I believe you never mentioned Nottingham academy — your vines should be raised from cuttings this spring, even now, & planted in your back [illegible] in July [illegible] the pots they are raised in. plant them the first summer in your house, & one against every rafter on the outside. my house cost 100£ & I don't repent the expense. I had glass & [illegible] from Stourbridge. Square panes are preferable, because you can close which side [illegible] best & fix the [illegible] below it. But all these you are already [illegible] if I dare say, — so my labour in writing is vain!

Derby Feb. 20–88

Dear Edgeworth

You write short scrawling letters full of questions, which take up one line, & expect me to send you dissertations in return, on academies, hot-houses, philosophers, attornies, &c. &c. — & then, without being at the trouble of acquainting me with your previous knowledge of all these intricate subjects, you tell me, after I have been laborious in inquiry, that you knew all this before.

I have seen a Nottingham Quaker, who speaks much in favor of the Nottingham master, & his academy, & have heard of several parents, who have sons there, & approve of their progress in their respective lines of study. — & that there are 80 scholars more or less, & that there are but two beds in a room, & but two boys in a bed. I don't know if the boys undress; this you must write to him about, if you approve.

Of Hothouses I am ready to answer any questions you will ask, written fairly on paper, & numbered, so as to be easily dispatched by a yes or a no. — but not write a volume, for you to light tapers with. — because you will say in answer "my dear friend, I knew all this before".

1. My Hothouse has a fire 4 months in a year only. 2 consumes about six tons of coals by conjecture. 3 it is about 32 ft. long & about 9 ft. wide 4. it has only oblique sashes, no perpendicular ones. 5. it is divided into two, that one half may be a month forwarder than the other. 6. it produces abundance of kidney beans, cucumbers, melons, & grapes. not pines

R. L. Edgeworth Esqr.
Edgeworthstown
Ireland

Feb 20 1788 Dr Darwin
playful answer
Hothouses

5.18 9
7.5 4

section

a b glass panes 8 inches square each at 3½ a foot sq.

a x brick-wall with a flue

b c chimney k q a wooden wall made with old

barrel staves to prevent the vine roots from injury

3 4 a vine x a r another vine.

{a m x k} bark-bed for melons 3 ft. ½ wide. the flue is 7 inc. wide & about 16 inches deep built on arches.

13 pots of flowers between the vines. the same on W.

c d

a b glass roof

c d two chimneys

e f two fire-places with pigeon-heaters or Athanor in each.

w x one end door

NB. the fire enters the flues from e f two feet below the surface of the earth, & passes to the ends of the house, one eastward the other westward; along the flue x k (plate I.) then the flue dives under the step at the door at the end at x (Plate II) & then rises

philosophers.

Mr. Keir amuses his vacant hours by mixing oil & alcaline salts together, to cleanse his majesty's subjects clean, & sweet & pays 1000 Guineas every six weeks to an animal called an Exciseman; who would otherwise trouble this diversion of his. & he is now printing a new improved edition of his chemical dictionary.

Attorneys,

Mr. Erasmus keeps free from the jaundice, but is obliged to use equitation now & then to preserve his health.

I drink water only, & am always well, we have five young creatures, 3 male, & two female, all tall, (& handsome as Mr. Darwin [illegible]) the boys strong, the girls [illegible]

again into the flue n (Pl. I) then it goes to the midle of the building & returns through m. & again return to the chimney through l.

My fires are in front, as I have no property [illegible] back wall. — hence the smoke [illegible] under it any

[illegible]

kidney-bean plants are 8 inches high, & vine-shoots 5 inches long, & melons one inch high, with fruit appearing upon them, I put fire to the other division this day. Feb. 20 — 88

this house is about 20 degrees from the S. towards the E.

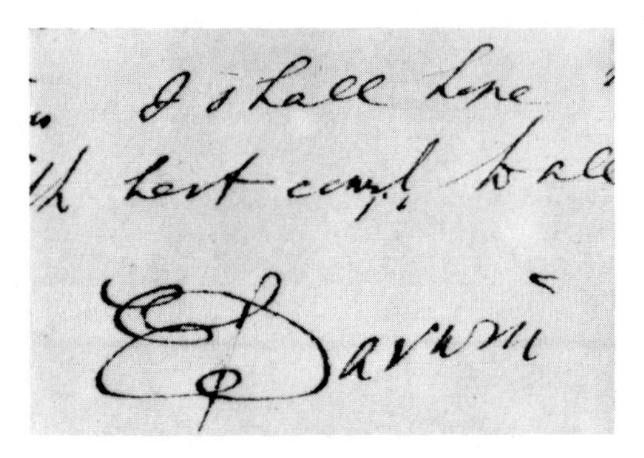
I shall hope
best compl. to all
E Darwin

50. Signature of Erasmus Darwin, taken from his letter to Richard Lovell Edgeworth dated 24 April 1790 (Letter 90B)

89H. To Josiah Wedgwood, October 1789

* * *

I have disobeyed you and shown your vase to two or three, but they were philosophers, not cogniscenti. How can I possess a jewel, and not communicate the pleasure to a few Derby philosophers?

* * *

Original Not traced. In the possession of Charles Darwin, 1865.
Printed Meteyard, *Wedgwood*, ii 581.
Text From above source.
Notes
(1) For the history of the Portland Vase, see letter 86A.
(2) After three years of work and numerous defective copies, Wedgwood at last made his first perfect copy of the Portland Vase in October 1789. He at once sent it to Derby for Darwin to inspect, and possibly to keep (to judge from the wording). Certainly, Darwin was given one of the earliest copies, now in the Fitzwilliam Museum, Cambridge.
(3) Enclosed with this letter Darwin probably sent his verses on the Portland Vase (*Economy of Vegetation* II 319–40) and his explanation of the ritual depicted on it (*Economy of Vegetation*, Additional Note XXII). Darwin believed the scenes on the Vase represented the Eleusinian mysteries, an explanation which is still in favour today. Wedgwood was pleased with the verses and notes, and wrote (October 1789): 'I thank you for a page of charming poetry, and many pages of very ingenious and learned notes, which I have read over and over with great pleasure, and could not do so without much improvement . . . With respect to the *Eleusinian lore*, I do not know much of those sacred mysteries, and therefore cannot judge so well how far you are founded in your references to them'. (Wedgwood, *Letters* iii 101–2.) On 17 November Wedgwood wrote again, giving Darwin permission to include engravings of the Vase in *The Economy of Vegetation*. (Wedgwood, *Letters* iii 108.)

89I. To Matthew Boulton, 22 October 1789

Derby Oct. 22 –89

Dear Boulton

I write this to introduce to your acquaintance my friend Mr Archdale, a gentleman [of] classical abilities, and will therefore not steal any of your inventions, and who is going to make speeches in the Irish parlament – and wishes to see all things visible on earth.

I had design'd to have written to you on another subject, if this opportunity had not offer'd; and that is, that as an idle book of mine, call'd the Loves of the Plants, sells so well, as to pay me; I think of publishing another part of the same work in the spring, and shall have occasion to speak of fire-engines. Now what I wish is that you would tell me as much about the superiority of your fire-engines, as may fill half a page or more by way of note.

Likewise I want much to know, if you have the Coinage, and something about your coining fire-engine for a note also. Viz, a short account of what number of halfpence He makes in a minute, and at how many operations, etc. The Lord be with you, adieu, I hope your Son and Daughter are well

your affect. friend E Darwin

Original Birmingham Reference Library, Matthew Boulton Papers, Darwin 36.
Printed Unpublished.
Text From original MS.
Notes
(1) Industrial spies were a menace to Boulton and Watt, so Darwin assumed Boulton would be suspicious of any new acquaintance. Darwin's friend was Richard Archdall (1746–1824) of Spondon near Derby, a member of the Derby Philosophical Society and later MP for Kilkenny (1801) and Dundalk (1802) in the first United Kingdom Parliaments.
(2) To his Lunar friends, Darwin pretended to be rather ashamed of *The Loves of the Plants*: while they were forging the sinews of industry, he was writing frivolous verses. He hoped to mollify them by mentioning that he was making money – which at least showed there was a demand for his product.
(3) Boulton had not replied to this letter by mid November, so Darwin wrote to Watt: see letter 89L. He wrote to Boulton again on 31 December and received a splendid reply from Boulton: see notes to letter 89N.

89J. To THOMAS BYERLEY, 31 October 1789

Dear Sir

I am desired by Mr Wedgewood to write a line to you about the note to be inserted at the bottom of the first page of the verses on Hope for his medallion.

He has sent me two specimens of this {no}te, but desires me to write a third, if I do not approve of either of them. I think the following one is best, to be inserted at the bottom of the page.

The above lines were written by the author of the Botanic Garden to accompany the medallions of Hope, which Mr Wedgewood designs to distribute at Sydney Cove. Which were made from clay brought from that place, and presented to him by Sir Joseph Banks Bart.

I shall be glad to recieve a proof sheet by the post before it is printed off, that I may correct any press-errors, or the punctuation.

I am Sir your obed. serv.
E Darwin

Derby
Oct. 31 –89

Addressed Mr Byerley, at Mr Wedgewoods, Greek Street, Soho, London.

Original Keele University Library, Wedgwood Archive, 2–226.
Printed Unpublished.
Text From original MS.
Notes
(1) Tom Byerley (1747–1810) was Josiah Wedgwood's nephew, and had led rather an unsettled life in his twenties. He calmed down later, however, and after Bentley's death in 1780 took over the management of Wedgwood's business in London. In 1790 Josiah took his three sons and Byerley into partnership, and the firm was re-named Josiah Wedgwood, Sons and Byerley. For Byerley, see *Wedgwood Circle, passim.*
(2) This letter refers to the 26 lines of verse about the future of Sydney that Darwin wrote to go with Wedgwood's medallion of Hope at Sydney Cove: see letters 89F and 89M. The verses were much admired and are printed in the later editions of *The Botanic Garden*, at the end of Part I (after the Additional Notes). The proof of the verses, mentioned at the end of the letter, is also preserved in the Wedgwood Archive at Keele University Library, but differs only marginally from the published version.

89K. To Robert W. Darwin, [October? 1789]

* * *

I much lament the death of Mr Day. The loss of one's friends is one great evil of growing old. He was dear to me by many names (multis mihi nominibus charus[carus]), as friend, philosopher, scholar, and honest man.

* * *

Original Not traced. In the possession of Charles Darwin, 1879.
Printed C. Darwin, *Life of E. Darwin*, p. 30.
Text From above source.
Notes
(1) Darwin's friend Thomas Day was killed in Berkshire on 28 September 1789 on being thrown from a horse that was only half broken-in, which he was trying to tame by kindness, in accordance with the precepts in his famous book *Sandford and Merton*. At the time of his death Day was only 41, and he never gained distinction

in politics, as his friends had expected because of his great eloquence and powers of argument.

(2) It seems likely that the letter would have been written within a month of Day's death.

89L. To James Watt, 20 November 1789

Derby, Nov. 20 –89

Dear Sir,

As the Loves of the Plants pays me well, and as I write for pay, not for fame, I intend to publish the Economy of Vegetation in the Spring: – now in this work I shall in a note mention something about steam-engines, which may occupy 2 or 3 pages. The historical part, as far as relates to the Marquis of Worcester, Capt. Savery, Messrs. Newcomen and Cawley, I think to abstract from Harris's Lexicon and Chambars's Dictionary. But what must I add about Messrs. Watt and Boulton? This is the question. Now, if you will at a leisure hour tell me what the world may know about your improvements of the steam-engine, or anything about your experiments, or calculated facts about the power of your engines, or any other ingenious stuff for a note, I shall with pleasure insert it, either with or without your name, as you please.

If you do not take this trouble, I must make worse work of it myself; and celebrate your engines as well as I can. I wish the whole not to exceed 2 or 3 quarto pages, and to consist of such facts, or things, as may be rather agreable; I mean, gentlemanlike facts, not abstruse calculations, only fit for philosophers.

I hope you use the warm bath, and enjoy better health, and ride some furious hobby-horse. I am sorry I cannot see you oftener; it is a great loss to my understanding, and to my happiness. The Lord keep you! Adieu!

From yours affectionately,
E Darwin

Original Not traced.
Printed Muirhead, *James Watt* ii 230–1.
Text From above source.
Notes

(1) Boulton had not yet replied to letter 89I, so Darwin wrote to Watt instead. Apparently Watt did not reply immediately and Darwin wrote to Boulton again on 31 December: see letter 89N.

(2) Again Darwin pretends to be writing for money, but Watt would not have been deceived. Publishing *The Economy of Vegetation* 'in the spring' was no more than an intention, probably inserted to give Watt a sense of urgency. The poem eventually appeared in June 1792.

(3) The Marquis of Worcester (1601–1667), Thomas Savery (1650–1717), Thomas

Newcomen (1663–1729) and John Cawley (*c.*1660–*c.*1720) were all early pioneers of the steam engine. See H. W. Dickinson, *A Short History of the Steam Engine* (CUP, 1939).

(4) 'Harris's Lexicon' is the *Lexicon Technicum: or, an Universal English Dictionary of Arts and Sciences* compiled by John Harris, DD, FRS (*c.*1666–1719). It was first published in 1704, with a second edition in two huge volumes in 1708–10. It was the first example of what was later called an encyclopaedia, and contains a long account of a steam engine (under 'Engine'). For Harris, see *DNB*.

(5) 'Chambars's Dictionary' is the *Cyclopaedia: or an Universal Dictionary of Arts and Sciences* compiled by Ephraim Chambers, FRS (*c.*1680–1740) and issued in two huge volumes in 1728. The *Cyclopaedia*, inspired by Harris's *Lexicon*, went through five editions by 1746, and the French translation of it gave rise to the *Encyclopédie* of Diderot and D'Alembert. For Chambers, see *DNB*.

89M. To Josiah Wedgwood, 1 December 1789

* * *

I have received great pleasure from your excellent medallion of Hope. The figures are all finely beautiful, and speak for themselves.

* * *

Original Not traced. In the possession of Charles Darwin, 1865.
Printed Meteyard, *Wedgwood* ii 570.
Text From above source.
Notes

The first copies of Wedgwood's medallion of Hope addressing Peace, Art and Labour at Sydney Cove were produced in November 1789, and Darwin is here expressing his appreciation on receiving one. He had written the lines about Sydney Cove (letters 89F and 89J) to go with an engraving of the medallion in a book about Botany Bay.

89N. To Matthew Boulton, 31 December 1789

Derby Dec 31 –89

Dear Boulton

I intend in a few months to print another poem, as the first pays me so well – now I shall make a note on the steam-engine, part of which Mr Watt will supply me with, but I wish to say something <u>popularly</u> about your application of it to Coining and shall be glad if you will send me some account of it, that may fill about a page, viz. the processes the copper goes through by your steam-engine, and the number it can make in a minute or hour, etc.

It would give me great pleasure to hear of your own health, and the welfare of your Son, and daughter, to whom I beg my Compliments if they are with you, and am, dear Boulton

your affect. friend
E Darwin

Addressed Matth. Boulton Esqr., Soho, near Birmingham.

Original Birmingham Reference Library, Matthew Boulton Papers, Darwin 37.
Printed Unpublished.
Text From original MS.
Notes

(1) Here Darwin repeats the request he made in his letter of 22 October (89I). His persistence was rewarded, for Boulton replied immediately, on 4 January 1790, with a long and masterly letter of great historical interest, which begins as follows: 'Dear Doctor, I should have answered your first letter before this time had I not been obliged to go to London, Bath etc. Your last of the 31 December I will now answer, but I have not yet considered of any arangement or what part of the subject to begin it. I will therefore give you only a Catalogue of Facts and leave you to make any use of them you please.

Steam Engins of B & W construction differ from Newcomens in 6 principles:

1. The acting power is Steam on the piston, and Not the Atmosphere as in the Common Engine
2. The Steam is condensed in a separate Vessell, and Not in the Cylinder as in D^o^
3. The Cylinder is kept as hot as Steam, and Not alternately heated and coold as in D^o^
4. The air is extracted by an air pump at each Stroke, and not blown out at Snifting Cla[c]k
5. The packing of the piston is preserved air and Steam tight by oyls Snift[?] and not by Water
6. The Vacuum is alternately made above and below the piston and thereby obtain double power.

Convenience and properties of B & W improved Engines.

a. It is the most *powerful* Machine in the World
b. It is the most *tractable*
c. It is the most *regular* and hath the property in itself of conforming to the least work
d. It shews by a Mercurial Vacuum the degree of Vacuum in the Cylinder
e. It shews the Strength of the Steam by another Mercurial Gage
f. It registers (sans error) the number of Strokes it makes in 100 Years
g. It is applicable to every purpose that requires either Rotative or Reciprocating motion.'

(From facsimile copy on Watt's copying-machine, Birmingham Reference Library, Matthew Boulton Papers, Darwin 38.) Boulton then goes on to justify each of the statements (a)–(g) with detailed calculations. His breath-taking confidence was well-founded: some of the engines *were* in use for more than 100 years.

(2) Darwin's own sustained enthusiasm for Watt's engine is expressed in *The Economy of Vegetation* (Note XI and Canto I, lines 253–96 and notes). He foresees steam power being used to pump piped water and drive corn-mills (lines 271–8):

Here high in air the rising stream He pours
To clay-built cisterns, or to lead-lined towers;
Fresh through a thousand pipes the wave distils,
And thirsty cities drink the exuberant rills. —
There the vast mill-stone with inebriate whirl
On trembling floors his forceful fingers twirl,
Whose flinty teeth the golden harvests grind,
Feast without blood! and nourish human-kind.

The celebration of technology continues with the lines about Boulton's coining machine, quoted in the notes to letter 89C; and Darwin concludes with his prophecy of powered flight, already quoted in the Introduction.

(3) For Boulton's son and daughter, see letter 89C.

89P. To [Anna Seward?], [March 1789?]

* * *

celebrity.

I am happy to hear, we are likely to see Mr Hayley shortly in this part of the country.

Adieu

from your affect. friend
E Darwin

Original In the possession of the editor.
Printed Unpublished.
Text From original MS fragment.
Notes

The recipient and date of this fragment are uncertain. The poet William Hayley (1745–1820) stayed a fortnight at Lichfield in December 1781, and Anna Seward visited him in Sussex for six weeks in the summer of 1782. For several years they indulged in strong mutual admiration of each other's poems. Hayley found his wife Eliza very difficult to live with, and he tried unsuccessfully to persuade his old friend Dr John Beridge of Derby (1743–1788) to look after her for him. Beridge, who was a member of the Derby Philosophical Society, died in October 1788, and his widow agreed to let Mrs Hayley live with her. So Hayley brought his wife to Derby on 27 April 1789, and left her with Mrs Beridge (he never saw her again). See M. Bishop, *Blake's Hayley* (Gollancz, London, 1951). My dating assumes (perhaps wrongly) that Darwin is referring to this visit. If so, Anna Seward seems the most likely recipient, (a) because she seems to be the only friend of Darwin interested in Hayley and (b) because Darwin wrote her at least two letters in the spring of 1789: the first accompanied a copy of *The Loves of the Plants*, which she received with rhapsodic admiration ('I admire, beyond expression, the skill and happiness . . .' and so on for eight pages); the second letter from Darwin acknowledged her fervent reply. (See A. Seward, *Letters*, Vol. 2, pp. 263–70 and 272–8.) However, it is questionable whether Darwin would, in writing to Anna, have signed off as 'your affect. friend', which was usually reserved for Boulton, Watt, Wedgwood, Keir, Edgeworth and occasionally one or two others.

90A. To James Watt, 19 January 1790

Derby Jan. 19 –90

Dear Watt,

I am much obliged to you for your letter, and am impatient for the further paper you promise me. And if you will give me leave to put your name to the paper you send me (by saying that it is part of a letter you wrote me at my request) it would be more agreable to me – but this shall be as you please.

For my part I court not fame, I write for money; I am offer'd 600£ for this work, but have not sold it. I have some medico-philosophical works in MS. which I think to print sometime, but fear they may engage me in controversy (which I should not much mind) and that they will not pay so well (which I mind much more).

Why will not you live at Derby? I want learning from you of various kinds, and would give you in exchange chearfulness, which by some parts of your letter, you seem to want – and of which I have generally a pretty steady supply.

Why the d——l do you talk of your mental faculties decaying, have not you more mechanical invention, accuracy, and execution than any other person alive? – besides an inexhaustible fund of wit, when you please to call for it? So Misers talk of their poverty that their companions may contradict them.

Your headachs and asthma would recieve permanent relief from warm bathing I dare say – but perhaps you are too indolent to try its use? – or have some theory against it?

What I said about your steam engine, I believed; I said it was the most ingenious of human inventions, can you tell me of one more ingenious? I can think of none unless you will except the Jew's harp, which is a combination of wind and wire instruments – or the partridge-call, which is a combination of the drum and fiddle.

Seriously, I do think the inventor of a wheel for a carriage had wonderful luck, or wonderful genius. The bow {and} arrow is also a curious invention, which the people of New South Wales, a continent of 2000 miles square, had not discover'd.

Do you not congratulate your grand-children on the dawn of universal liberty? I feel myself becoming all french both in chemistry and politics. Adieu, the Lord keep you from Megrim and asthma and believe me, your affect. friend

E Darwin

Addressed James Watt Esq, Engineer, Birmingham.

Original Birmingham Reference Library, Boulton and Watt Collection.
Printed Excerpts in *Doctor of Revolution*, pp. 204–5 and Plate 11, which gives a facsimile of the second page of the MS.
Text From original MS.
Notes
(1) This is one of the finest of Darwin's letters, with just the right mixture of sympathy, cheerfulness and wit to combat Watt's tendency towards depression and hypochondria.
(2) The letter from Watt, which Darwin mentions in the first line, was to give Darwin details of the history of the steam engine, in reply to his request in letter 89L.
(3) Darwin often said he wrote for money, but Edgeworth insisted that this was just a pose: it seems that Darwin was ashamed of being an author, at least among his non-literary friends, and tried to divert the suspicion that he was enjoying it by pretending to write for money.
(4) The final paragraph shows Darwin's strong commitment to the French Revolution. 'Megrim' is migraine.

90B. To Richard Lovell Edgeworth, 24 April 1790

Derby Apr 24 –90

Dear Edgeworth

I much condole with you on your late loss, I know how to feel for your misfortune! The paper you sent me is a prodigy, written by so young a person with such elegance of imagination – nil admirari may be a means to escape misery, but not to procure happyness – there is not much to be had in this world, we expect too much. – I have had my loss also! – – the letter of Sulpitius to Cicero, Melmoth's transl. V.III. p. 6, is fine eloquence, but comes not to the heart, it tugs but does not draw the arrow – pains and diseases of the mind are only cured by Forgetfulness; – Reason but skins the wound, which is perpetually liable to fester again.

I am much obliged to you for your second Hygrometer, which is very ingenious, yet I prefer the former as being simpler, easier to make, and easier to compare with each other, or with some third criterion, as by soaking the wood first, and then baking it, and thus making two points of expansion, and is more certain than where the friction of pullies are concern'd.

Your three wards ['cousins' *crossed out*] were a week with us at Easter. I mention it to add, that I think them all fine sensible good-humour'd boys; and they seem'd so grateful for the little notice taken of them by Mrs Darwin, that she quite admires and pities them. Their parents, she sais, must have iron hearts neither to send for them, nor to visit them annually at least! I believe they will think the time long before they hope to come again to Derby.

I will sometime make a shelf for your animal to walk upon, but not till I repair my study, which wants a new roof. And He shall have a walk for a year or two at 2 inches a month – but I prefer long-back for a race; and would wager him against brazen-wheels for a cool hundred.

Poor Sneyd I really am [*six lines crossed out and largely illegible*].

I should be very happy to see you this summer, if any thing here can attract such a comet from its orb, and any of your family.

I now come to your second letter, which is very flattering to me indeed – it was your early approbation that contributed to encourage me to go on with the poem, which I have now sold to Johnson as thus[?]. I got 100£ by the first edition of this 2d part, and have sold the copy right of it for 300£ more, and the copy right of the first part, which will be published in 2 or 3 months for 400£ – in all 800£. The first part is longer than the second, and two of my critical friends think it will be more popular. I wish it were possible for me to shew you the MS. for your criticisms, but I dare not trust it – especially as I have sold it. I am obliged by your criticisms on the Loves of the Plants. I alluded to Moore of Moore hall on purpose to give a little air of burlesque, to shelter myself under. For the same purpose the proem was written, to the best part of which, "the festoon of ribbon" I think you helped me to. Your designs for new poetic pictures are very good ones. Franklin I have celebrated in the Economy of Vegetation, and also the Irish Voluntiers as defensores libertatis. Chaos and gasses are also in the Economy of Vegetation. Landau, axle, vegetable lamb were all felt to be objectionable, but not easily mended. – I am glad of any criticisms, and if I can eas{ily} mend them, I do. I intend to write no more v{erse} and to try a medico-philosophical work next, called Zoonomia.

Why do not you publish something wonderful, who have so much invention? a century of new machines with plates, or a decade, would be an agreable work and well recieved I dare say. If you were to propose some as practicable, others as tried, etc etc, I should send you any, which have occur'd to me. I think such a book would sell, and procure fame, and would be new in its way. Adieu I shall hope to see you this summer, and am with best compl. to all yours

Your affect. friend and serv.

E Darwin

Pray think of a decade of mechanic inventions with neat drawings, by R. L. Edgeworth Esq, F.R.S., F. R. S. Hibern etc etc to the end of the alphabet.

Addressed R. L. Edgeworth Esq, Edgeworth's town, Ireland.

Original Cambridge University Library, Dar. 218.

Printed Excerpts in Edgeworth, *Memoirs* ii 129–30, 133–4 and in *Doctor of Revolution*, pp. 205–6.

Text From original MS.

Notes

(1) 'Your late loss' refers to the death of Edgeworth's daughter Honora (1774–1790), who had died of consumption at the age of fifteen. 'My loss' refers to the death of Darwin's son Charles twelve years earlier. 'Nil admirari' is, literally, 'nothing to be admired'; but Darwin was using it on the assumption that Edgeworth knew the complete couplet from Horace, which may be freely translated: 'Never to involve your feelings too deeply is, Numicius, almost the one and only thing to make and keep happiness'. The letter of Sulpicius to Cicero (see also letter 80C) is, as Darwin says, in Volume III of W. Melmoth's *The Letters of Marcus Tullius Cicero* (2nd edn, Dodsley, London, 1772), pages 6–13.

(2) Edgeworth's first hygrometer was an ingenious wooden 'crocodile' with sharp backward-pointing claws on its feet. In moist weather the wood lengthened and the front feet slid forward. In dry weather the back feet were drawn forward. So, with every change in humidity, the animal crawled forwards. (See *Loves of the Plants*, p. 99.) The second hygrometer was evidently a metal device with pulleys.

(3) I have not identified 'your three wards'. Sneyd Edgeworth (1786–1864) was the second son of Edgeworth and his third wife Elizabeth (née Sneyd). Maria Edgeworth was probably responsible for crossing out the six lines about Sneyd.

(4) Edgeworth's second letter, printed in his *Memoirs* (ii 131–2), begins with praise of *The Loves of the Plants*: 'I have felt such continued, such increasing admiration in reading the "Loves of the Plants", that I dare not express my enthusiasm, lest you should suspect me of that tendency to exaggeration, with which you used to charge me. I may, however, without wounding your delicacy, say that it has silenced for ever the complaints of poets who lament that Homer, Milton, Shakespeare, and a few classics, had left nothing new to describe, and that elegant imitation of imitations was all that could be expected in modern poetry.' After more comments in the same vein, Edgeworth offers some verbal criticisms, saying that the *quantity* of Landau (I 344) is unusual [the line reads 'Glides the gilt Landau o'er the velvet lawn']; and that the axle of a weathercock (I 159) is usually fixed [whereas Darwin had written, 'the polished axle turns']. Edgeworth did not like Moore of Moore Hall (I 165) because 'it is associated with burlesque'. [Darwin had written, 'So arm'd, immortal Moore uncharm'd the spell/And slew the wily dragon of the well'.] Also Edgeworth objected to 'vegetable lamb' (I 290) ending the line. [The first edition of the poem does not have 'axle' and the line-numbers given by Edgeworth refer to the second edition (they are the same in the third too). Evidently the first edition was sold out before Edgeworth could obtain a copy in Ireland. That is why he was so late in reading the poem.] Edgeworth ends his letter by suggesting subjects for new poetic pictures, including 'the creation of climate in hot-houses and ice-houses', the thermometer and barometer, Franklin, 'Chaos and the Fata Morgana', etc. Darwin replies to each of the comments and several of the suggestions. The items he mentions appear as follows in *The Economy of Vegetation*: Franklin, I 383–98 and II 361–70; Irish Volunteers, II 371–6; Chaos, I 103–6; gases, IV 25–46. In the proem to which Darwin refers, he asks the reader to look on his poem as 'diverse little pictures suspended over the chimney of a Lady's dressing-room, *connected only by a slight festoon of ribbons*'. It was two years, and not '2 or 3 months' as Darwin hoped, before *The Economy of Vegetation* was published.

90C. To Dudley Ryder, 27 May 1790

Dear Sir,

When I had the pleasure of seeing you at Lichfield races, you was so kind as to say that you had read my poem on the Loves of the Plants; I have now finished the other part, and have herewith sent you the last Canto of it (which I have just now got copy'd fair over), and shall be much obliged to you, if you are at leizure to give yourself the trouble of reading it, for any remarks which m[a]y occur to you; and which you will please to note with a pencil on the blank pages – some passages will want explanatory notes, which refer to natural philosophy; which I would also have troubled you with, had they been written. I have sold the copy-right of this and the former part to Johnson of St Pauls for 800£ and wish therefore to make it as accurate as I can.

This however is not the principal design of my troubling you with this letter – which is to sollicit your opinion whether I might have any chance for the vacant Laureatship, if I was to apply; and also by what channels I could apply, if you should think I may stand any chance of succeeding. The Fame of it is no object to me, but as I have a large and increasing family, and from my time of life am not likely to live long enough to provide well for them, I should be glad of a 100£ a year addition to my income, which I understand the Laureatship is worth. I should hope Lord or Lady Stafford would give me their patronage, if you think that might avail. At all events I hope you will excuse my giving you this trouble, and present my most respectful compliments to Lord and Lady Harrowby, and all your good family, and believe me with true respect your

much obliged and most obed. serv.
E Darwin

Derby
May 27 –90

About line 540, you will see I have endeavour'd to pay a compliment to the royal family; but am in doubt whether to insert the lines on the left hand page at 580, as perhaps the best compliment on account of the King's health <u>now</u>, is to forget that he was ever ill.

Addressed The Rt Hon. Dudley Ryder.

Original Sandon Hall, Stafford, Harrowby MSS, Vol. VIII, ff. 71–2. (Privately owned)
Printed Unpublished.
Text From original MS.
Notes
(1) The Hon. Dudley Ryder (1762–1847), the eldest son of the first Baron Harrowby, had been MP for Tiverton since 1784, and early in 1790 became Comptroller

of the Household and a Privy Councillor. He was a close friend of the Prime Minister William Pitt, and acted as Pitt's second in his duel with Tierney. Ryder was Foreign Secretary in 1804, was created Earl of Harrowby in 1809, and was President of the Council from 1812 to 1827. He was asked to be Prime Minister in 1827, but declined. For details, see *DNB*.

(2) Ryder made a lengthy and most courteous reply to Darwin on 31 May (Harrowby MSS, Vol. VIII, ff. 201–2). He begins with nearly a page in praise of Darwin's poem, and then turns to the Laureateship. Ryder consulted Lord Salisbury, the Lord Chamberlain, 'in whose patronage it lies' and found that the appointment was already settled. 'I trust you will believe', he wrote, 'how heartily I wish, both for the credit of the place, and for your own satisfaction, that I had been able to give a more encouraging answer to your question; in the mean time, I must rejoice with you, that Johnson is a more efficient patron than even my Lord Salisbury . . .' The Laureateship had become vacant on the death of Thomas Warton (1728–1790) on 21 May. His successor was Henry James Pye (1745–1813), whose poems were to suffer much ridicule.

(3) Though many of Darwin's letters were written hurriedly, few are so inconsistent as this one. Perhaps he was suffering some financial stringency at this time, just before the birth of Harriot, the seventh and last child of Darwin and Elizabeth; and perhaps he had doubts about his health. But if Darwin wanted to convince Ryder that he was in need of cash, it was rather naive to mention that he had just received £800 from Johnson: as Ryder noticed, this was equal to eight years' salary as Laureate. Darwin, with his flair for painting pageantry in verse, would have been a better Poet Laureate than Pye; but his support for the French Revolution would have caused problems.

(4) Lord Stafford (1721–1803) was previously Earl Gower, and his active support had been crucial in the promotion of the Trent-and-Mersey canal in the 1760s. He was created the first Marquess of Stafford in 1786. Lady Stafford (*c.* 1735–1805) was his third wife Susannah, daughter of the sixth Earl of Galloway. See Burke's *Peerage*, under 'Sutherland'. Their daughter Susan married Dudley Ryder in 1795.

(5) King George III suffered his first serious 'derangement' (or attack of porphyria, as is now thought) in November 1788, but recovered in 1789. Darwin did remove the reference to the King's health in Canto IV of *The Economy of Vegetation* and contented himself with a picture of the King and Queen walking through the gardens at Kew, in the lines beginning

> So sits enthron'd in vegetable pride
> Imperial KEW by Thames's glittering side . . .

90D. To Dudley Ryder, 12 June 1790

Derby Jun. 12 –90

Dear Sir

I am much obliged to you for your criticisms on the part of the poem I troubled you with; and hope you will allow me at your leizure to send you a part of the remainder, when you come into this country. Permit me also to thank you in the best manner I can for your kindness in endeavouring to procure for me the Laureatship. Mr Hayley writes in a letter to Mrs Hayley, who lives in this town (there being a friendly seperation existing between them), that the Laurel was offer'd him, that he has had a conference with Mr Pitt on this subject; and that he has refused it. What reasons could induce him to refuse it, who I understand is not in very opulent circumstances, I don't concieve: as I am conscious I should not have done so in similar circumstances.

I beg my most respectful compliments to Lord and Lady Harrowby, and to all your family, and am Sir with true esteem, your much obliged friend and servant,

E Darwin

Original Sandon Hall, Stafford, Harrowby MSS, Vol. VIII, f. 73. (Privately owned.)
Printed Unpublished.
Text From original MS.
Notes
(1) Dudley Ryder had returned Canto IV of *The Economy of Vegetation* with some marginal notes, 'all confined to mere verbal criticism' (letter of 31 May 1790, Harrowby MSS, Vol. VIII, ff. 201–2).
(2) For William Hayley, see letter 89P. Hayley's refusal of the Laureateship was surprising, and the reasons remain conjectural. See M. Bishop, *Blake's Hayley* (1951), p. 120.
(3) William Pitt (1759–1806) had been Prime Minister since 1783, but this is his first mention in Darwin's letters.

90E. To Robert W. Darwin, 17 December 1790

Derby, Dec. 17 –90

Dear Robert,

I cannot give any letters of recommendation to Lichfield, as I am and have been from their infancy acquainted with all the apothecaries there; and as such letters must be directed to some of their patients, they would both feel and resent it. When Mr Mellor went to settle there from Derby I took no part about him. As to the prospect of success there, if the young man who is now at Edinburgh should take a degree (which I suppose is probable), he had better not settle in Lichfield.

I should advise your friend to use at first all means to get acquainted with the people of all ranks. At first a parcel of blue and red glasses at the windows might gain part of the retail business on market days, and thus get acquaintance with that class of people. I remember Mr Green, of Lichfield, who is now growing very old, once told me his retail business, by means of his show-shop and many-coloured window, produced him £100 a year. Secondly, I remember a very foolish, garrulous apothecary at Cannock, who had great business without any knowlege or even art, except that he persuaded people he kept good drugs; and this he accomplished by only one stratagem, and that was by boring every person who was so unfortunate as to step into his shop with the goodness of his drugs. "Here's a fine piece of assafoetida, smell of this valerian, taste this album graecum. Dr Fungus says he never saw such a fine piece in his life." Thirdly, dining every market day at a farmers' ordinary would bring him some acquaintance, and I don't think a little impediment in his speech would at all injure him, but rather the contrary by attracting notice. Fourthly, card assemblies – I think at Lichfield surgeons are not admitted as they are here; – but they are to dancing assemblies; these therefore he should attend. Thus have I emptied my quiver of the arts of the Pharmacopol. Dr K——d, I think, supported his business by perpetual boasting, like a Charlatan; this does for a blackguard character, but ill suits a more polished or modest man.

If the young man has any friends at Shrewsbury who could give him letters of introduction to the proctors, this would forward his getting acquaintance. For all the above purposes some money must at first be necessary, as he should appear well; which money cannot be better laid out, as it will pay the greatest of all interest by settling him well for life. Journeymen Apothecaries have not greater wages than many servants; and in this state they not only lose time, but are in a manner lowered in the estimation of the world, and less likely to succeed afterwards. I will certainly send to him, when first I go to Lichfield. I do not think his impediment of speech will injure him; I did not find it so in respect to myself. If he is not in such narrow circumstances but that he can appear well, and has the knowlege and sense you believe him to have, I dare say he will succeed anywhere. A letter of introduction from you to Miss Seward, mentioning his education, may be of service to him, and another from Mr Howard. Adieu, from, dear Robert,

Yours most affectionately,
E Darwin

Original Not traced. In the possession of Charles Darwin, 1879.
Printed C. Darwin, *Life of E. Darwin*, pp. 37–40.
Text From above source.
Notes
(1) 'Mr Mellor' appears as 'Samuel Mellor, surgeon' among the Lichfield residents recorded in the *Universal British Directory* (1793).
(2) 'Mr Green' is Richard Greene (1716–1793), the well-known surgeon-apothecary of Lichfield, who also created a museum, much admired by Dr Johnson. For Greene and his museum, see *DNB* and H. S. Torrens, *Newsletter of Geological Curators Group*, No. 1, pp. 5–10 (1974).
(3) 'Dr Fungus' was a 'fun-name' for a doctor, rather like 'Dr Strabismus'.
(4) I have not identified Dr K——d.
(5) For Miss Seward, see letter 80B. 'Mr Howard' was the brother of Darwin's first wife Mary (Robert's mother): see letter 92A.
(6) This letter provides a good example of Darwin's 'benevolence'. After saying he cannot help, he proceeds to give the most detailed and helpful advice to Robert's friend.

91A. To Phoebe Horton, 16 February 1791

Madam

I have found a case of the measles appearing in the course of inoculation in a work of credit of Mr John Hunters. He inoculated a child on March 16th [,] the arm show'd the appearance of infection on the 19th. He became feverish on the 20th and 21st but the inflamation in the arm made no advance since the 19th. The measles appear'd on the 22d and the sores in the arm appear'd to go back being less inflamed. Was full of the measles on the 23rd, punctures in the arm made no advance. 25th the measles began to disappear. 26 and 27th punctures began to look a little red. 29th inflammation of the punctures increased, and a little matter form'd. 30th a fever commenced, the smallpox appear'd at the regular time, and He recover'd.

This case from Hunter (on vener. disease p. 3 of the introduction) is so similar to that of Miss Louisa that I thought proper to send you an account of it – as I suspected that Miss Louisa might have the smallpox when the measles were over; as the puncture on her arm I thought was not quite heal'd.

And it is probable Miss Horton's puncture has had its inflammation delay'd from the infection of the measles, tho no measle-eruption has appear'd.

A journal should be made out and kept of the time they were both inoculated, and the appearances of the measles, and of the inflammation of the punctures of their arms, like the above, as it may give satisfaction at some future time.

I shall hope to be favor'd with an account of their progress, as I can not be without some anxiety about them till I hear further.

And am, Madam, with true esteem, and most respectful compl. to Mr Horton, your much obliged and obed. serv.

E Darwin

Derby
Feb. 16 –91

Addressed Mrs Horton, Catton, near Burton (to be sent from Burton or from Wichnor Bridge).

Original Derby Central Library, Catton Collection (owner: Mr D. W. H. Neilson, Catton Hall, Barton-under-Needwood, Derbyshire).
Printed Unpublished.
Text From original MS.
Notes
(1) Mrs Phoebe Horton (née Davenport) (1755/6–1814) was the wife of Eusebius Horton (1747–1823), who inherited Catton Hall from his father Christopher Horton (1701–1764), via his elder brother Christopher, who died in 1769. (I am indebted to Mr D. W. H. Neilson for some of this information on the Hortons.)

(2) John Hunter FRS (1728–1796), the founder of scientific surgery and a pioneer in many branches of medicine, was certainly an author 'of credit', as Darwin remarks. Hunter's *Treatise on the Venereal Disease* was published in 1786, and the case of measles appearing in the course of smallpox inoculation, which Darwin summarizes accurately, is on pages 3–4, as Darwin indicates.

(3) 'Miss Louisa' was Mrs Horton's younger daughter Frances Louisa (later Mrs Carleton): see Glover, *Derby* ii 204. 'Miss Horton' was Mrs Horton's elder daughter Anne Beatrix (1787–1871), who married R. J. Wilmot in 1806. On 12 June 1814, when Anne was dressed in black after her mother's death, she met Byron and inspired his famous poem, 'She walks in Beauty, like the Night . . .'

(4) This letter might be framed with the title, 'How to give patients confidence in their doctor'. Confronted with a puzzling case, Darwin seeks a parallel, finds one, and immediately gives full details to the parent of the children involved, taking a hopeful view and at the same time expressing caution and concern about the children. No wonder Darwin was so highly regarded by his patients.

91B. To Richard Dixon, 30 March 1791

Derby Mar 30 –91

Dear Sir,

I have often experienced the readiness with which you serve your friends: as to these girls they ought to esteem you as a father. Now my conscience will not permit me to trouble your son Robert with them, unless he will permit me to pay for their board, a guinea a week a piece while they stay with him – this I hope he will not take amiss that I offer; because I have no other way of repaying him by any kind of obligation at this distance from him. On these conditions I will send them up for 3 or 4 weeks, which will be an improvement to them, whether they get proper situations or not.

Your goodness to them requires that I should tell you my whole design about them. I think to leave them when I die (NB, that is not till the next century) the value of 500£ a piece, part in money and part in annuities – which last I design to prevent their coming to absolute poverty in case of unhappy marriage. If they marry with my approbation I shall give them 2 or 300£ a piece at the marriage and an annuity of the value of the remainder at my death. By this sum and some employment as Lady's Maid or teacher of work they may be happier than my other girls who will have not much more than double or treble that sum, and brought up in more genteel life, for I think happiness consists much in being <u>well</u> in one's situation in life – and not in that situation being higher or lower. As soon as I have your answer to this letter I shall give further directions to the girls at Birmingham. Pray send me a particular direction to your Son Robert that I may write a line to him with them, to whom and his family I beg my best respects and am dear old friend

Yours affectionately
E Darwin

Addressed Mr Richard Dixon, Broomfield Mill, near Chelmsford, Essex.

Original Not traced.
Printed Two excerpts in *Doctor of Revolution*, p. 210.
Text From MS copy in Cambridge University Library (Dar. 218) sent by R. W. Dixon to Charles Darwin in 1879.
Notes
(1) 'These girls' are Darwin's daughters Susan and Mary Parker, who apparently often stayed with Dixon. Susan was now 18 and Mary 16, and, as the letter shows, they were at Birmingham, presumably staying with their mother Mrs Day (née Parker). Previously they had lived most of their lives in Darwin's house and he was now trying to let them 'see the world'.
(2) Darwin's plans for their future, as revealed here, are kind and sensible; but in fact he did far more for the girls, buying a house for them at Ashbourne and setting them up in a school there, which they ran very successfully for more than twenty years.

91C. To Edward Jerningham, 31 March 1791

Derby Mar 31 –91

Dear Sir,

I am much obliged to you for the pleasure of perusing your poem; which I should sooner have acknowledged, but could not till today find your direction.

In my next poem, which is now printing, and I suppose will be out in 6 or 8 weeks, I shall borrow something from your Il Latte, but not without acknowledgements. I hope you will continue to write. I think I have exhausted my stock.

I am Sir with great respect
Your obed. serv.
E Darwin

Mr French begs his compliments and admires your poem.

Original St John's College Library, Cambridge.
Printed Excerpt in *Doctor of Revolution*, p. 209.
Text From original MS.
Notes
(1) Edward Jerningham (1727–1812) was a poet, dramatist and man of fashion, and a close friend of Horace Walpole. See *Edward Jerningham and his Friends*, ed. L. Bettany, Chatto & Windus, London, 1919. Among his poems was *Il Latte* (1767), in which he exhorts ladies to suckle their own children. Darwin duly quoted six lines from the poem in his *Economy of Vegetation* (Canto III, lines 367–72), with a footnote to acknowledge the borrowing.
(2) Darwin was quite wrong in thinking that his poem would be published in '6 or 8 weeks': it did not come out until 14 months later, chiefly because of delays over the engravings of the Portland Vase (see letter 91F). Darwin's exhaustion is scarcely

surprising: *The Economy of Vegetation* has notes running to 100 000 words, as well as the 2500 lines of verse.
(3) 'Mr French' was Richard French (see letter 86E).

91D. To Matthew Boulton, 9 May 1791

May 9 –91

Dear Sir,

I saw Mr Gregory of Spondon and Mr Haden, an ingenious Surgeon of Derby, who had attended Mr Laws I believe by his own desire, or that of Mr Gregory, instead of Mr Fowler, who saw him at first – they both assured me yesterday that he went on well.

This morning, Monday, I wrote a line to Mr Haden, and desired he would go over to Spondon this morning. He has sent me word, that he has been, and that all goes on quite well. So I see no occasion for my going over, should otherwise immediately have done so.

You are so concern'd for Mr Laws, that you have said nothing about yourself – it would have been very agreable to me to have heard that you are well in your own health; and that your young scions are in vigour like olive branches round your table.

The conflagration of the Albion mills grieved me sincerely, both as it was a grand and successful effort of human art; and also because I fear you was a considerable sufferer by it. I well remember poor old Mr Seward comparing the immortality of the Soul (in a devout sermon) to a fire-engine. He might now have made it a type of the mortality of this world, and the conflagration of all things.

If any thing ever brings you into this part of matter, I hope you will pass a day or two amongst the animated beings of this country. I dare say neither yourself nor myself have changed many of our opinions, since we formerly used to compare them.

Adieu
from dear Boulton
your affect. friend
E Darwin

I will again inquire about your serv[ant?] in 2 or 3 days, and if I should hear any thing amiss, will see him, and acquaint you.

Addressed Mat. Boulton Esqr., Soho, near Birmingham.

Original Birmingham Reference Library, Matthew Boulton Papers, Darwin 39.
Printed Unpublished.
Text From original MS.
Notes
(1) Boulton seems to have been much concerned about the health of Mr Laws, apparently a servant of his.

(2) I have no information on Mr Gregory.

(3) Mr Haden is the surgeon Thomas Haden (1761–1840), whose house was in St Michael's Churchyard, Derby. He was Mayor of Derby in 1811 and 1819. Two of his five sons became surgeons. See Glover, *Derby* ii 601.

(4) Mr Fowler is the surgeon William Tancred Fowler (1764–1821). He attended Derby School, and his memorial in St Werburgh's Church, Derby, tells us that 'meekness, humility, resignation, integrity and holiness characterized the life of this consistent Christian'.

(5) Albion Mill was Boulton and Watt's showpiece in London, a huge flour mill worked by a steam engine, which was in operation from 1786 until it was destroyed by fire in March 1791. Darwin was right in thinking that Boulton lost heavily on it; Schofield (*Lunar Society*, p. 335) says he lost £6000. 'Poor old Mr Seward' was Anna's father, Rev. Thomas Seward (see letter 80B).

91E. To George Gray, 9 May 1791

Dear Sir

I am glad to hear you are going upon a journey which may be both agreable, safe, and profitable. Mr French is in London.

Articles of my faith about coal. 1. that coal is never to be look'd for beneath limestone (except perhaps a thin stratum a few inches thick of liver stone made from fresh-water muscle shells etc.) 2. nor beneath Granite, moorstone, porphery, basalt, and other great original, submarine volcanic productions; nor beneath a silicious sand-stone, which rests upon these. 3. I believe that all coal has been made from morasses; and argilaceous grit-stone and clay, and iron, are stratify'd with it – the coal has often been sublimed by heat, into a stratum above the grit-stone but beneath the clay – and that the 5 or 7 strata of these materials are owing to the different periods of the world in which the successive woods and morasses grew and perish'd.

The sublimation of coal oil accounts for jet, amber, cannel-coal, naphtha, Petroleum – and for the upper part of some coal-beds being better, if they lie under clay, which prevented the sublimation of the oil; and worse, if they lie under grit-stone, which favour'd its evaporation.

The heat, which coal-beds have experienced, has been owing to the fermentation of the morass: if it has been very great, the whole of the coaly matter burnt away. This is the case in those countries where beds of red clayey marle (or marly clay) abound. Where Sagar-clay (grey-clay) abounds it is I think a sign of coal, and also where shale (shistus) abounds, as these have not undergone so much heat, as to sublime the coal, and dissipate it into air. These criterions are not certain; the appearance of one edge of a coal bed coming up to day, by its inclination is I suppose the surest criterion.

In this country there are 5 or 6 strata of coal, with clay, iron-stones with fern in them, argilaceous grit, which burns red, alternately between them.

This is all I know or conjecture upon the subject, and shall be glad to hear from you at your return. Fosile tar or bitumen is an indication of coal beneath it, as exists in coal-brook dale.

From your sincere Friend
E Darwin

May 9 –91

If your friend Mr Macnab should ever come into Derbyshire I shall be glad of the pleasure of his company – and if I exactly knew his design, and what he had done, or meant to do on the subject, would give him any intelligence on the subject I am acquainted with. Ammon, Wallis, Dr Franklin, and Haller have all written something on articulation. I believe myself that there are not more than 17 simple sounds in all the European languages, including the Welch – but this subject is too long for a letter. I shall read with pleasure, and give my opinion of any treatise your friend may favour me with.

Adieu

But if He comes this way at any time, I would shew him what I have some years ago done on articulation, as I once made a wooden head, which spoke distinctly p.a. and m. so that it said pam, map, papa, mamma; as plain as most human heads.

Adieu

Addressed Mr G. Gray, To the Care of Mr H. Macnab, Byker Point Pleasant, near Newcastle upon Tyne, Northumberland.

Original Yale University, New Haven, Conn.: Beinecke Rare Book and Manuscript Library, Osborn Collection.
Printed Unpublished.
Text From original MS.
Notes
(1) George Gray (1758–1819), of Newcastle-upon-Tyne, worked for some years as a painter, of portraits, fruit and signs. He was an ingenious man, also skilled in botany, geology and chemistry, and in 1787 he made a botanical journey to North America. In 1791 he was commissioned to report on the geology of Poland, with Macnab as his companion. This is the 'agreable' journey mentioned by Darwin. It turned out badly: Gray was disgusted by Macnab's luxurious mode of travel and abandoned the expedition. In later years Gray became rather eccentric, but was respected for his abilities, according to the short biography in E. Mackenzie, *A Descriptive and Historical Account of Newcastle upon Tyne* (Newcastle, 1827) ii 577–8. See also *DNB*.
(2) For Mr French, see letter 86E.
(3) Darwin's views on coal are mostly correct, and his comments on petroleum etc are prescient. See also letter 87J. For cannel-coal, see letter 78H.
(4) 'Mr H. Macnab' is probably Henry Gray MacNab MD (1761–1823), educationist and miscellaneous writer. *The Gentleman's Magazine*, in its obituary of

MacNab (i 378, 1823), states that he 'was at an early age Professor of Elocution in the University of Glasgow': this is consistent with Darwin's belief that 'Mr H. Macnab' was writing a treatise on speech. Though there was no official 'Professor of Elocution' at Glasgow, MacNab possibly lectured unofficially, since he was a friend and disciple of Thomas Reid (1710–1796), the Professor of Moral Philosophy at Glasgow. MacNab was author of *A Plan of Reform in English Schools* (1787), *Analysis and Analogy in Education* (Paris, 1818), etc, and is also described as Physician to the Duke of Kent. See *DNB*.

(5) 'Ammon' is John Conrad Amman, a Swiss physician born in 1669, who published two much admired short tracts on speech: *Surdus Loquens* (Harlem, 1692), translated into English as *Talking Deaf Man* (London, 1694); and *De Loquela* (Amsterdam, 1700). See R. Watt, *Bibliotheca Britannica* (Constable, Edinburgh, 1824).

(6) 'Wallis' is John Wallis (1616–1703), the eminent mathematician, who wrote *Grammatica linguae anglicanae* with a *Praxis grammatica*, and also a treatise on the production of the sounds of speech, *De Loquela* (1652, sixth edition 1765).

(7) Franklin's interest in sounds arose chiefly from the development of his Armonica, a musical instrument made of 37 ringing 'wine glasses' – actually graded nesting hemispherical glasses. The instrument was popular: Mozart played one and wrote music for it. See *The Papers of Benjamin Franklin*, Vol. 10 (Yale University Press, 1966) pp. 116–30.

(8) The literary and scientific work of the great physiologist Albrecht von Haller (1708–1777) is almost unparalleled in range and volume. His influential *Elementa Physiologiae* (1759–1766) covered all aspects of physiology, including the mechanics of breathing and the production of sounds.

(9) Byker is the name of a district in the eastern part of Newcastle.

91F. To Josiah Wedgwood, 9 July 1791

Dear Sir,

Mr Johnson's engraver now wishes much to see Bartolozzi's plates of the vase, and will engrave them again if necessary – I told Johnson in my own name, <u>not in yours</u>, that I thought the outlines too hard, and in some places not agreable.

Now if you could be so kind as to send Bartolozzi's prints to Mr Johnson St Paul's Churchyard – or let him know if He can have them at your house in Greek Street – as he said he can not anywhere procure them. He promises to take great care of them. The name of the engraver I don't know, but Johnson said He is capable of doing anything well.

I am sorry to give you so much trouble on this account, but always apply to you in all difficulties in matters of taste.

* * *

Your affect. friend and serv.
E Darwin

Original Not traced.
Printed G. Keynes, *Blake Studies* (2nd ed OUP, 1971), p. 60.
Text From the printed version.
Notes
(1) The engravings of the Portland Vase to be included in *The Economy of Vegetation* delayed the publication of the poem for many months. Darwin had hoped it would be out in May 1791, but it did not appear until June 1792. The sequence of events was probably as follows. Darwin assumed that engravings would be available. Johnson, his publisher, told him that the only ones were those by Bartolozzi. Darwin mentioned this to Wedgwood, who condemned the Bartolozzi engravings in no uncertain terms. (Wedgwood to Darwin, 27 June 1791; Rylands Library, Manchester, English MSS 1109.) Darwin told Johnson of this condemnation, in his own name, as this letter says; at the same time Darwin tried to speed up the snail's-pace progress by asking Wedgwood to let Johnson have prints of Bartolozzi's engravings. However, Johnson wrote to Darwin on 23 July to say that Bartolozzi's prints could 'not be copied without Hamilton's consent, being protected by act of parliament' (Keynes, *Blake Studies*, p. 60). Back to square one, where fortunately Mr Johnson's engraver, who was 'capable of doing anything well', was waiting. This omnicapable engraver, whose name Darwin did not know, was of course William Blake, and Johnson said 'Blake is certainly capable of making an exact copy of the vase . . . if the vase were lent to him for that purpose'. So Blake obtained access either to the Vase itself or to a Wedgwood copy, and made the superb engravings which appear in *The Economy of Vegetation*.
(2) Francesco Bartolozzi (1727–1815) was a well-known engraver, who came to England in 1764 and was an original member of the Royal Academy. For details see *DNB*.

91G. To Joseph Priestley, 3 September 1791

An Address to Dr Priestly
Agreed upon at a Meeting of the Philosophical Society at Derby
Sept. 3, 1791

Sir,

We condole with yourself and with the scientific world on the loss of your valuable library, your experimental apparatus, and your more valuable manuscripts: at the same time we beg leave to congratulate you on your personal safety in having escaped the sacrilegious hands of the savages at Birmingham.

Almost all great minds in all ages of the world, who have endeavoured to benefit mankind, have been persecuted by them; Galileo for his philosophical discoveries was imprisoned by the inquisition; and Socrates found a cup of hemlock his reward for teaching 'there is one God'. Your enemies, unable to conquer your arguments by reason, have had recourse to violence; they have halloo'd upon you the dogs of unfeeling ignorance, and of frantic fanaticism; they have kindled fires, like those of the inquisition, not to illuminate the truth, but, like the dark lantern of the assassin, to light the murderer to his prey. Your philosophical friends therefore hope, that you will not again risk your person amongst a people, whose bigotry renders them incapable of instruction: they hope you will leave the unfruitful fields of polemical theology, and cultivate that philosophy, of which you may be called the father; and which, by inducing the world to think and reason, will silently marshall mankind against delusion, and with greater certainty overturn the empire of superstition.

In spite of the persecution you have sustained we trust that you will persevere in the exertions of virtue, and the improvements of science. Your fame, already conspicuous to every civilized nation of the world, shall rise like a Phenix from the flames of your elaboratory with renovated vigour, and shine with brighter coruscation.

R. Roe, Secretary

Original Not traced.
Printed The Derby Mercury, 29 September 1791.
Text From above source.
Notes

(1) The inclusion of this letter may seem odd, because Darwin did not sign it personally. However, he almost certainly wrote most of it, because the second paragraph is very much in his style, and because, as President of the Society, he was by far the most likely author.

(2) Joseph Priestley (1733–1804), one of the greatest of all experimental chemists and a Unitarian Minister, had come to Birmingham in 1780, and during the 1780s the meetings of the Lunar Society were largely occupied in discussing his researches (see letter 81A). Priestley greatly valued the Lunar meetings: 'I consider my settlement at Birmingham as the happiest event in my life . . .' (*Memoirs* i 97; see also *A Scientific Autobiography of Joseph Priestley*, ed. R. E. Schofield, MIT Press, 1966, pp. 264, 266, 287). But after ten fruitful years there, Priestley's work came to an abrupt end when his house and all his belongings were burnt in the savage Birmingham riots on 14 July 1791. The riots showed that the tide had turned against the French Revolution and all forms of dissent. Shouts of 'No philosophers – Church and King for ever' were heard during the riots, and from now onwards Darwin had to be careful not to express radical opinions too publicly.

(2) The only clergyman in the Derby Philosophical Society, the Rev. Charles Hope (1733–1798), complained to the newspapers that not enough members were present when the message was agreed. Darwin had him thrown out: the Society, having heard Mr Hope's explanation, resolved that 'he be desired to withdraw his name from the list of the Society' (from Broadsheet dated 10 October 1791 at Derby Central Library). Trying to 'overturn the empire of superstition' was from now onwards interpreted as an attack on the Church.

(3) Richard Roe (*c*.1750–1813?), the Secretary of the Derby Philosophical Society, whose name appears at the end of the letter, is described as a schoolmaster in the *Universal British Directory* of 1791. He was paid twelve guineas a year for his work as Secretary.

(4) Priestley's reply to the letter, dated 21 September 1791, was also printed in the *Derby Mercury* of 29 September. He expressed his thanks, said he was not discouraged and was 'setting about the re-establishment of my philosophical apparatus'. He concluded, characteristically: 'Excuse me, however, if I still join theological to philosophical studies, and if I consider the former as greatly superior in importance to mankind to the latter'.

91H. To Anna Seward, 2 October 1791

Derby Oct 2 –91

Dear Miss Seward

I send for your criticism an epitaph, which I have written for Mr French to put over the tomb of Mrs French. You know how difficult it is to write an epitaph, as everything has been written on the deceased, that can be written, from Petrach's lamentations to Miss Seward's "on a coffin lid".

I never attempted an epitaph before, except on Dr Small, which Mr Keir has printed in the life of Mr Day. Pray send me your criticisms or emendations, as they will come [soon?] enough for the marble, tho' I fear not soon enough for the proofs, as I have directed it to be printed at the end of the Economy of Vegetation. I believe the ingenious Mrs Damer is to give Mr French a design for the tomb.

Adieu, go on to write good verses, and believe me with true esteem your obed. serv.

E Darwin

[There follow 16 lines of verse, a draft for the lines 'Sexton! oh, lay beneath . . . angel-tongue for Me!' printed on page 212 of *The Economy of Vegetation* (first edition) and incorporated in Canto III (lines 325–40) in later editions.]

Addressed Miss Seward, at the Palace, Lichfield.

Original Library of Fitzwilliam Museum, Cambridge, S. G. Perceval Bequest, General Series.
Printed Unpublished.
Text From original MS.
Notes
(1) Mrs Millicent French (née Mundy) (1745/6–1789) was the wife of Darwin's friend Richard French (see letter 86E). According to Darwin's note in later editions of *The Economy of Vegetation* (III 308), Millicent French was 'a lady who to many other elegant accomplishments added a proficiency in botany and natural history'. She died in August 1789, and there is a memorial tablet in Mackworth Church, 2 miles west of Derby, but apparently no tomb.
(2) Anna Seward's 'To a Coffin Lid', written in March 1790, is her 97th sonnet: *Poetical Works* (ed. W. Scott, Ballantyne, Edinburgh, 1810) Vol. III, p. 218.
(3) James Keir's *Life* of Thomas Day (Stockdale, London, 1791) was a circumspect short biography. Darwin's epitaph on Small is on page 112.

(4) 'The ingenious Mrs Damer' was the sculptress Anne Seymour Damer (1749–1828). See *DNB*. In the *Economy of Vegetation* (II 113), Darwin writes:

> Long with soft touch shall Damer's chisel charm,
> With grace delight us and with beauty warm.

(5) Darwin still hoped *The Economy of Vegetation* would be published before the end of 1791, but the engravings of the Portland Vase delayed it until June 1792.

92A. To Robert W. Darwin, 5 January 1792

Derby Jan. 5 –92

Dear Rob[t].

I do not remember your having before asked me the questions about Mr Howard and your mother; which I am sure I should openly without any scruple have answer'd. The late Mr Howard was never to my knowlege in the least insane, he was a drunkard both in public and private – and when he went to London he became connected with a woman and lived a deba[u]ched life in respect to drink, hence he had always the Gout of which he died but without any the least symptom of either insanity or epilepsy, but from debility of digestion and gout as other drunkards die.

In respect to your mother, the following is the true history, which I shall neither aggravate nor diminish any thing. Her mind was truly amiable and her person hansome, which you may perhaps in some measure remember.

She was siezed with pain on the right side about the lower edge of the liver, this pain was follow'd in about an hour by violent convulsions, and these sometimes relieved by great doses of opium, and some wine, which induced intoxication. At other times a tempor[ar]y dilirium, or what by some might be term'd insanity, came on for half an hour, and then she became herself again and the paroxism was terminated. This disease is called hysteric by some people. I think it allied to Epilepsy.

This kind of disease had several returns in the course of 4 or 6 years and she then took to drinking spirit and water to relieve the pain, and I found (when it was too late) that she had done this in great quantity, the liver became swelled and she gradually sunk; a few days before her death, she bled at the mouth, and whenever she had a scratch, as some hepatic patients do.

All the drunken diseases are hereditary in some degree, and I believe epilepsy and insanity are produced originally by drinking. I have seen epilepsy produced so very often – one sober generation cures these dr[unkards] frequently, which one drunken one has created.

I now know many families, who had an insanity on one side, and the children now old people have had no symptom of it. <u>If it was otherwise, there would not be a family in the kingdom without epileptic gouty or insane people in it!</u>

I well remember when your mother fainted away in these hysteric fits (which she often did) that she told me, you, who was not then 2 or 2½ years old, run into the kitchen to call the maid-servant to her assistance.

I have told every thing just as I recollect it, as I think it a matter of no consequence to yourself or your brother, who both live temperate lives, keeping betwixt all extreams.

I have lately taken to drink two glasses of made-wine with water at my dinner, instead of water alone, as I found myself growing weak about 2 months ago; but am recover'd and only now feel the approaches of old age.

I shall not mention your letter to Erasmus, you may always depend on secrecy when you require it.

My next book will come out in May. Adieu

From your truly affect. father
E Darwin

Original Not traced. There is a typed transcript in the collection of Sir Geoffrey Keynes.

Printed The Autobiography of Charles Darwin, edited by Nora Barlow. Collins, London, 1958, pp. 223–5.

Text From the typed transcript.

Notes

(1) Robert was worried about the possibility of his inheriting, or passing on to children of his own, the debilities of the Howards, or rather the Foleys (see letter 57B): Robert's mother, née Mary Howard, died at the age of thirty, as a result of the illnesses described in this letter. Erasmus replies with complete frankness, as he promises, and convincingly dispels any fears of hereditary drunkenness. But he does not mention the other debilities. In fact Robert's two sons, and some of his daughters, suffered from physical or psychosomatic illness which seems to have been at least partially hereditary. For a full discussion, see R. Colp, *To be an Invalid: the Illness of Charles Darwin* (University of Chicago Press, 1977).

(2) This letter was first printed in Nora Barlow's edition of Charles Darwin's *Autobiography*, and Lady Barlow identified 'Mr Howard' as Robert's grandfather Charles Howard (1706–1771), an identification which I accepted in *Doctor of Revolution*. But I now think that 'Mr Howard' is Robert's uncle Charles Howard (1742–1791). There are four reasons for preferring this identification. (a) Charles Howard junior had died only 12 months before, so 'the late Mr Howard' would fit him much better than his father, who died 20 years earlier. (b) Darwin refers to Charles Howard junior as 'Mr Howard' in letter 90C. (c) The dubious hereditary qualities came not from Charles Howard senior, but from his wife. (d) Charles Howard junior *was* a drunkard – he was found lying dead drunk in a ditch on the evening of Edgeworth's first visit in 1766 (see Edgeworth, *Memoirs* i 163). His father, on the other hand, seems to have been sober and respectable – 'a cool and wise man', as Dr Johnson called him. Also he apparently lived in Lichfield all his life. (My apologies to Charles Howard senior for this mistake.)

(3) The text of this letter is nearly the same as that given by Nora Barlow, but I have based my text on Sir Geoffrey Keynes's typescript, which is nearer Darwin's

usual style. The two greatest differences are in the first line of the third paragraph, where Lady Barlow's text has 'left side' instead of 'right side', and in the last paragraph, where she has 'home-made wine' instead of 'made-wine'.

(4) Medically, the letter is puzzling. Mary seems to have died from liver disease, which was aggravated by alcohol. But the pre-existing disease which caused the paroxysms is obscure. Hydratid disease, caught from sheep, offers one possibility, and gall-stones another; but hereditary disease is perhaps more probable, in view of the bad health of the Foleys. Acute intermittent porphyria has been suggested by T. K. With. See S. Thiele and T. K. With, *Scandinavian Journal of Clinical and Laboratory Investigation*, **16**, 465–9 (1964).

92B. To Robert W. Darwin, 7 February 1792

* * *

I always console myself by saying 'a man must do something, otherwise he becomes moped and low spirited', and therefore he may as well do something for the advantage of himself and family, as labour in fox-hunting or card-playing. So life rubs on!

As to fees, if your business pays you well on the whole, I would not be uneasy about making absolutely the most of it. To live comfortably all one's life, is better than to make a very large fortune towards the end of it.

* * *

I am studying my 'Zoonomia', which I <u>think</u> I shall publish, in hopes of selling it, as I am now too old and hardened to fear a little abuse. Every John Hunter must expect a Jessy Foot to pursue him, as a fly bites a horse.

* * *

Original Not traced. In the possession of Charles Darwin, 1879.

Printed First paragraph: unpublished, but printed in first proofs of C. Darwin, *Life of E. Darwin*, now in Cambridge University Library. Second and third paragraphs: C. Darwin, *Life of E. Darwin*, pp. 65, 102.

Text From sources specified above.

Notes

(1) The first paragraph is probably a response to a suggestion by Robert that his father might retire. Erasmus made a very similar reply later (see letter 93G).

(2) As the second paragraph suggests, Robert's medical practice at Shrewsbury was flourishing, and he was becoming quite affluent, though still only 25.

(3) Erasmus had been very reluctant to publish *The Botanic Garden*, but the acclaim that greeted it had softened his antipathy to authorship. Here we see him plucking up courage to publish *Zoonomia*, which he had been working on for 20 years.

(4) Jesse Foot (1744–1826) was a surgeon who greatly disliked John Hunter and continually attacked him in pamphlets (and a biography). See *DNB*.

(5) The second extract is from a letter written in February 1792, but it may not be the letter of 7 February.

92C. To Robert W. Darwin, 13 April 1792

Derby, Apr. 13 –92

Dear Robert,

I think you and I should sometimes exchange a long medical letter, especially when any uncommon diseases occur; both as it improves one in writing clear intelligible English, and preserves instructive cases. Sir Joshua Reynolds, in one of his lectures on pictorial taste, advised painters, even to extreme old age, to study the works of all other artists, both ancient and modern; which he says will improve their invention, as they will catch collateral ideas (as it were) from the pictures of others, which is a different thing from imitation; and adds, that if they do not copy others, they will be liable to copy themselves, and introduce into their work the same faces, and the same attitudes again and again. Now in medicine I am sure unless one reads the work of others, one is liable perpetually to copy one's own prescriptions, and methods of treatment; till one's whole practice is but an imitation of one's self; and half a score medicines make up one's whole materia medica; and the apothecaries say the doctor has but 4 or 6 prescriptions to cure all diseases.

Reasoning thus, I am determined to read all the new medical journals which come out, and other medical publications, which are not too voluminous; by which one knows what others are doing in the medical world, and can astonish apothecaries and surgeons with the new and wonderful discoveries of the times. All this harangue lately occurred to me on reading the trials made by Dr Crawford.

* * *

Original Not traced. In the possession of Charles Darwin, 1879.
Printed C. Darwin, *Life of E. Darwin*, pp. 50–51.
Text From above source.
Notes

(1) The advice of Reynolds is in his famous sixth Discourse, on imitation: 'Study, therefore, the great works of the great masters, for ever . . .' (J. Reynolds, *Discourses* (Seeley, 1905) pp. 141–176). Darwin's own emphasis on the importance of imitation, the theme of Canto III of *The Temple of Nature*, probably derives from Reynolds.

(2) Dr Adair Crawford (1748–1795) was a well-known physician and chemist, whose *Experiments and Observations on Animal Heat* (1779) had aroused great interest. Darwin is probably referring to his treatise *On the Matter of Cancer and on the Aerial Fluids* (1790).

(3) As Charles Darwin remarks, this letter shows Erasmus's 'continued zeal in his profession' at the age of sixty.

92D. To William Hayley, 15 July 1792

My Dear Sir,

Nothing is so flattering to a cultivator of any branch of art as the praise of those, who best understand the value of his exertions. I am therefore particularly happy in the approbation of yourself and your valuable friend Mr Cowper; both whose works I have studied with great delight and improvement.

It would give me the greatest pleasure to meet Mr Cowper, whom you soon expect at Eartham! – this can not be, but if the wilds or the waters of Derbyshire should ever attract you both, or either of you, into this part of the country, I can not easily express how happy a visit from you would make me here.

I believe it is [not?] common for Booksellers to print more editions than one at once, but if my book undergoes another really-new edition, I shall take the liberty of prefixing the two elegant little poems, you have favour'd me with. Go on, both of you! to instruct and amuse the world in variety of composition both of prose and verse, and believe me, with true regard and affection amongst your numerous admirers.

E Darwin

Derby July 15 –92

Addressed W. Hayley Esqr., Eartham, near Chichester.

Original Library of Fitzwilliam Museum, Cambridge, S. G. Perceval Bequest, General Series.
Printed Unpublished.
Text From original MS.
Notes

(1) The publication of Darwin's *Botanic Garden* was completed when Part I, *The Economy of Vegetation*, appeared in June 1792. One of the reviewers was the poet William Cowper (1731–1800), who wrote a long and laudatory notice in the *Analytical Review*: 'The verse has afforded delight to all who delight in verse', Cowper remarked. As if this was not enough, Cowper and his friend and fellow-poet William Hayley (see letter 89P) wrote their 'Lines Addressed to Dr Darwin', consisting of six stanzas of almost fulsome praise, beginning

Two poets (poets by report
Not oft so well agree)
Sweet Harmonist of Flora's court!
Conspire to honour thee. . . .

This poem, over Cowper's name, appears at the beginning of later editions of *The Botanic Garden*, as does a poem by Hayley, again of six stanzas, beginning

As Nature lovely Science led
Through all her flowery maze
The volume she before her spread
Of Darwin's radiant lays . . .

These were the two poems Hayley sent to Darwin, who replied with this extremely courteous letter.

(2) Darwin knew very well that Mrs Hayley was living in Derby with Mrs Beridge, and sometimes he met them (see Seward, *Letters* iii 66). So he was being either over-tactful or perhaps a little sarcastic in implying that Hayley was unlikely to be attracted to Derbyshire.

92E. To Richard Polwhele, 17 July 1792

Derby, July 17 –92

Dear Sir,

I have waited some weeks in vain for a frank to thank you for an elegant collection of poems by bards of Devonshire and Cornwall, and for a more elegant complimental Idyllium, written in the first page, on my Botanic Garden. Approbation from poet to poet is more grateful, as they who have laboured in the art can best judge of its excellencies and difficulties; and praise therefore comes with a better grace from one who has himself written so well. I shall take the liberty of prefixing your elegant lines to the next edition of my work, along with others I have received from Mr Hayley and Mr Cowper.

My son Erasmus, who remembers passing some days on a journey with you, speaks of you with great pleasure. He is settled as an attorney in Derby, and is in great business, built on the most lasting foundation of ingenuity and integrity.

If any thing should ever call you into this country, your company would give great pleasure to us both. I am, Sir, your obedient servant,

E Darwin

Original Not traced.

Printed R. Polwhele, *Traditions and Recollections*. 2 vols. Nichols, London, 1826. Vol. I, p. 299.

Text From above source.

Notes

(1) Richard Polwhele (1760–1838) was a prolific writer, quite well known as a poet in the 1790s. He lived in Devon and Cornwall, and wrote histories of both counties, as well as compiling the collections of verse by their poets, which Darwin mentions. The later editions of *The Botanic Garden* include the complimentary sonnet by Polwhele, as well as the poems of Cowper and Hayley (see letter 92D). Polwhele was influenced by Darwin, but was never a disciple. By 1798 he had begun to turn against Darwin, and his poem *The Unsex'd Females*, though primarily aimed at Mary Wollstonecraft and the women's rights movement, has some side-swipes at Darwin, e.g.

> Thrilled with fine ardours Collinsonias glow,
> And, bending, breathe their loose desires below.

The *DNB* has a three-page biography of Polwhele.

(2) Darwin's son Erasmus, who was now 32, was doing well as a solicitor in Derby.

Polwhele was the same age as Erasmus junior, and they had met in 1780 when they travelled together from Truro to Bristol after Erasmus had been on a visit to Boulton in Cornwall. Polwhele called Erasmus 'an intelligent and enlightened companion. We parted at Bristol with tears reciprocally shed! So rapid in their growth are the friendships of the young!' (Polwhele, *Traditions and Recollections* i 87.) They remained friendly until Erasmus's death in 1799.

92F. To James Edward Smith, 12 September 1792

Dear Sir

I should sooner have written to thank you for the books you were so kind as to send me, but you mention'd your intended absence for 6 weeks from town, which I now {supp}ose are about expired. I mention'd {your} society to Sir Brooke Boothby of Ashbourn, and to Dr Johnson of this town, two ingenious men, who wish to become fellows of the Linnean Society. Sir B. Boothby has a great collection of plants at his house at Ashbourn, which probably you saw, when you was in this country.

You will please to acquaint me, when you admit me, and Dr Johnson, if it [be] agreable to your society that we shall be members, and we can write in sending a small bill of the admission fee. Sir B. Boothby is now at Ashbourn, [so] that I can acquaint him also if it be agreable for you to admit him of your society.

I recieved much pleasure from your thesis on generation, tho' our theories will not agree.

With true esteem,
I am Sir your obed. serv.
E Darwin

Derby
Sept 12 –92

Original Linnean Society, London, Smith MSS 4, Scientific Correspondence of Sir J. E. Smith, Vol. IV, f.8.

Printed Edited version in *Memoir and Correspondence of the late Sir James Edward Smith*, ed. Lady Smith (Longman, 1832) Vol I, p. 401.

Text From original MS.

Notes

(1) James Edward Smith (1759–1828) was a wealthy botanist who in 1784 bought the collections of Linnaeus for 1000 guineas. In 1788 Smith founded the Linnean Society, and was re-elected President annually until his death, after which the Society purchased his collections from his widow. He wrote extensively on botanical subjects, his *English Botany* in 36 volumes (1790–1814) being his longest work. He was knighted in 1814 and after the death of Banks was the leading figure among British botanists.

(2) Darwin became a Fellow of the Linnean Society earlier in 1792, and it was natural that he should propose his old friend and botanical co-author, Sir Brooke Boothby, as a Fellow.

(3) 'Dr Johnson of this town' is William Brookes Johnson MB (1763–1830), a member of the Derby Philosophical Society, who collected botanical specimens during an expedition to America and wrote about plants in Pilkington's *View of Derbyshire*. Johnson was author of *History of the Progress and Present State of Animal Chemistry*, 3 vols. (London, 1803), and lived at Coxbench Hall, 4 miles north of Derby. See J. Stokes, *Botanical Commentaries* (Simpkin and Marshall, 1830), p. cxvii; *Alum. Cantab.* ii 585; and R. P. Sturges, *Midland History*, **4**, 212–29 (1978).

(4) Smith obtained a medical degree at Leyden in 1786 with a thesis entitled *De Generatione*, of which he had apparently sent a copy to Darwin.

92G. To Richard Dixon, 25 October 1792

Derby Oct 25th –92

My dear old friend,

I should have written to you much sooner but waited for a frank, as Erasmus promises at the same time to send you a scrawl – I wish him to send you a subpoena to Derby, where you well know I shall at any time be very glad to see you or your's. I hope you will come next summer and see your Cousin Sumner; you who are now a gentleman, at large, and not confined at home all the year, as I am, should not neglect your old Elston friends; – if I were Cousin Sumner, I believe I should strike you out of my will for not coming down this last summer! – she expects an annual kiss from you at least; – whether your mouth is drawn on one side or not – you must mind on which side you approach lest you should kiss her ear. I am glad to find your spirits are so good, as to joke upon your infirmity of having your mouth drawn a little to one side. My next door neighbour, a young man to you, I suppose not 50, has had just such a seizure, which he is slowly recovering from. You say the only inconvenience you find, is that you cannot now whistle to amuse yourself, and he says all the inconvenience he finds, is, that he cannot blow out his candle, when he goes into bed. If your Doctor says you are well, how dare you think to the contrary? The success of the French against a confederacy of kings gives me great pleasure, and I hope they will preserve their liberty, and spread the holy flame of freedom over Europe. For my part I go on as usual to practice physic, and to write books – I sold a work called "The Botanic Garden" for 900£ to Johnson the bookseller near St Paul's, it is a poem; perhaps you may borrow it from some circulating library; it is in two parts and sold for 1–13–0. I intend to publish another work next in prose which will be chiefly on physic, I fear it will not sell so well as the last. The worst thing I find now is this d——n'd old age, which creeps slily upon one, like moss upon a tree, and wrinkles one all over like a baked pear. But I see by your letter that your juvenility will never fail you; you'l laugh on to the last, like Pope Alexander, who died laughing, on seeing his tame monkey steal to bed side, and put on the holy Tiara, the triple crown, which denotes him king of kings. Now Mr Pain says that he thinks a monkey or a bear, or a goose may govern a kingdom as well, and at a much less expense than any being in Christendom, whether idiot or madman or in his royal senses; adieu dear Citizen from thy affectionate equal

E Darwin

Mrs Darwin and all here beg to be remembered to you and send compliments to Mrs Dixon. Brother John is returned to Carlton, and preaches furiously, he prays as usual, and advises his parish, and makes up differences and advises the poor as he used to do – He'll hold the Devil a good tugg, I hope yet, for there are few such clergy to be found. I don't believe amongst the 8000 French parsons, whom you are now feeding in London, and whom France has spewed out of her mouth, that you can find one equal to your old playfellow at Carlton Scroop. Pray give my compliments to all your sons. Mary[?] Day is a teacher at Mrs Ton's boarding school at Chesterfield, and Susannah is going to be governess in a family near Chesterfield. And seem both very happy in their situations.

Addressed To Richard Dixon Citizen. Hartfordend, Felsted, Essex.

Original Not traced.
Printed Excerpts in *Doctor of Revolution*, p. 229.
Text From MS copy in Cambridge University Library (Dar. 218), sent by R. W. Dixon to Charles Darwin in 1879.
Notes
(1) This is one of the most relaxed and revealing of Erasmus's letters. His old friend Dixon, now 61, had suffered a slight stroke, and the letter is outwardly jokey; but it is also incidentally one of the clearest statements of Darwin's political views, which from this time onwards he had to suppress in his published writings. The sentences about 'the success of the French against a conspiracy of kings' and the views of Tom Paine, combined with the address to 'Richard Dixon Citizen', tell us where Darwin stood. He belonged to the radical Derby Society for Political Information, which advocated votes for all adult males. See E. Fearn, *Derby. Arch. Journ.*, **88**, 47–59 (1968).
(2) The Sumner family frequently appears in the marriage registers of Elston Church and Elston Chapel. 'Cousin Sumner' could have been one of these (e.g. Mary Sumner, née Bell, married 1761), or she may have been unmarried.
(3) *The Botanic Garden* was the most famous poem of the day, and that exacting critic Horace Walpole had written, 'Dr Darwin has destroyed my admiration for any poetry but his own' (H. Walpole, *Letters*, ed. Mrs Paget Toynbee, xv 41 (OUP 1905)). But Darwin was not impressed by the furore and told Dixon, 'I have sold a work called "The Botanic Garden" . . . it is a poem'. For Johnson, see letter 84G.
(4) Darwin's brother John (1730–1805) was rector of Elston and of Carlton Scroop, 12 miles to the east.
(5) 'Mary Day' is Darwin's daughter, usually called Mary Parker. Her mother was of course now Mrs Day (see letter 85C). Susannah is Susan Parker (see letter 75G).

93A. To Matthew Boulton, 5 January 1793

Jan 5 –93 Saturday

Dear Boulton

I am glad to see your handwriting as it assures me you are in the land of the living. I was a day lately at Mr Keir's to visit Miss Keir, who was ill, and wished to have come to Soho for ½ a day, but Time and Fate, two great Despots, withstood my efforts.

Tomorrow (Sunday) I go into Leicestershire, and return on Monday even. and shall after that time be glad to see your friend, whenever it is agreable to him.

Now I think you promised to send me a chemical lamp, which you have not perform'd – don't let it be a magical lamp, with Genii and Daemons subservient to it, like the lamp of Aladin in the arabian tales, least I should not be able to lay the devils I may raise.

The Lord preserve you from every kind of Devil

adieu

E Darwin

Pray come and see us. I am glad to hear your son is well. I think your daughter is married? I beg my compliments to them both. Mrs Darwin joins her best respects, and shall be glad to see your daughter with you.

Addressed Math. Boulton Esqr., Soho, near Birmingham.

Original Birmingham Reference Library, Matthew Boulton Papers, Darwin 41.
Printed Unpublished.
Text From original MS.
Notes

(1) This is a reply to a letter from Boulton earlier in the week, which begins with Lunar banter: 'Dear Doctor, My old Horses (one of whom is now very ill) my old Servants, and my old Friends, I never forsake, because I find the older we all grow, the more my affection grows. Witness our selves at Westminster[?]. And for these reasons I shall send you an old Friend in hopes you will be able to make him an older one'. He goes on to say he has a poor opinion of the young doctors in Birmingham, and 'the Flower of Physick was in such a Withering way as to render a transportation to Lisbon necessary' (i.e. Dr Withering had gone to Lisbon). The old friend who wished to consult Darwin was John Scale, formerly one of Boulton's clerks, and for 22 years his partner in button-making. Boulton asks when Darwin would be free to see him.

(2) Boulton continues: 'My son is a very honest and reasonable Fellow, but is not so greedy of Work as his Father. My daughter is also well. I want to come to see Mrs Darwin and your fine family . . .' He ends: 'God bless and preserve your Body says your old and affect. friend Matthew Boulton'. (From facsimile copy on Watt's copying-machine, Birmingham Reference Library, Matthew Boulton Papers, Darwin 40.)

(3) For Mr Keir and Miss Keir, see next letter.

(4) Boulton's daughter Anne (see letter 89C) was not married, as Darwin thought.

93B. To James Keir, 17 January 1793

Derby,
Jan. 17 –93

Dear Keir,

I have much availed myself of your observations, and have corrected my work accordingly. If you will be at the trouble of reading some more of the work I shall have time to send you a part, as I have not yet absolutely begun to print. Pray give me a line on this head; I mean whether you have leisure to read any more of it at present.

I hope Miss Keir continues well, and am, with Mrs Darwin's united compliments to the ladies,

Yours affectionately,
E Darwin

Original Not traced. Probably in possession of the Moilliet family in 1868.
Printed Moilliet, *Keir*, p. 99. Excerpt in *Doctor of Revolution*, p. 231.
Text From Moilliet, *Keir*, p. 99.
Notes
(1) Keir was not only Darwin's oldest friend but also his most judicious adviser, and he commented on Darwin's books before publication. Darwin usually accepted his advice, as this letter shows. On this occasion Keir was reading *Zoonomia*: Darwin accepted his recommendation to publish it in two volumes. Keir had himself been a medical student at Edinburgh (though he went into the Army before taking his degree), so his comments on *Zoonomia* would have been particularly well-informed.
(2) 'Miss Keir' is Keir's daughter Amelia (1780–1857), who in 1801 married John Lewis Moilliet of Geneva. Their son James married Lucy Galton, granddaughter of both Samuel Galton (see letter 93E) and Erasmus Darwin, thus making the Moilliets a triply Lunar family.

93C. To Thomas Beddoes, 17 January 1793

Derby Jan. 17 –93

Dear Sir,

Your treatise on Consumptions I have read with great pleasure, and am glad to find you are about to combat this giant-malady, which has hitherto baffled the skill, and withstood the prowess of all ages; and which in this country destroys whole families, and, like war, cuts off the young in their prime of life, sparing old age and infirmity.

The few observations I have made on this disease, since you request them, are at your service; as I wish to contribute even a mite to your great design. I hope you will be led to try a variety of experiments with mixtures of airs; your very ingenious reasonings from the scarlet colour of the blood in consumptive patients, and from the inflammatory size or coagulable part of it, and from the delay of the progress of consumption in pregnant women, indicate indeed hyperoxygenation to be the cause of this fatal disease. But by instituting other experiments with different kinds or proportions of airs, you will, at least, if your first experiments should not succeed to your utmost wish, hold out hopes to those unfortunate young men and women; who, if they knew the general fatality of their disease under the present modes of treatment, would despond at the commencement of it, or wish to try some new kind of medicine. For though catarrhs are sometimes mistaken for consumptions by the ignorant, or are designedly called so by the crafty, and are hence supposed occasionally to have been cured; yet it is well known to attentive practitioners, that those patients, who are truly consumptive from pulmonary ulcers, which have arisen spontaneously (and are thus distinguishable from a single ulcer owing to violent peripneumony, or the wound of a sword) – whether they are kept on vegetable or on animal diet; and whether, with the usual quantity of wine and beer, or with water and milk alone – at length submit to fate; and that generally with many of their relations by heriditary predisposition, or with their nearest friend by contagion.

The immediate cause of Pulmonary Consumption consists in ulcers of the lungs; these ulcers, whether they arise spontaneously, or from previous infection, may vary in respect to their seat, as well as to their remote cause; but I suppose the immediate cause of these ulcers to consist in the inirritability of either the lymphatic or of the venous system. First, if the mucaginous fluid, which is poured into the cells of the mucous membrane of the lungs be not perfectly absorbed, it will produce ulcers by its accumulation, or by its chemical change. Secondly, if the blood brought to the internal surface of the bronchia by the bronchial arteries be not perfectly absorbed, or taken up by the correspondent bronchial veins, haemoptoe will be induced, and small ulcers succeed in consequence.

Hence there has always appeared to me to be two kinds of Pulmonary Consumption, one which begins with slight haemoptoe and which is generally seen in dark-eyed people with large pupils; and the other, which commences without haemoptoe, and which is generally seen in light-eyed people with large pupils. The aperture of the iris in both these kinds of Consumption is generally large, which evinces the inirritability of the eye, and thence perhaps in consequence the inirritability of the whole system. The former of these consumptions is generally heridītary without any appearance of scrophula; and the latter with appearance of scrophula in the present, preceding, or third generation upwards. The former commences more certainly between seventeen and seven and twenty; the latter attacks people of all ages.

I believe both these Consumptions to be infectious to those who sleep with such patients in the last stage of the disease; as I have observed a husband in two cases begin to be diseased soon after the death of his wife; and in one a wife, who became consumptive soon after the death of her husband; in all which cases there was no reason to suspect hereditary predisposition.

The haemoptoe generally begins at its first attack during sleep, for as respiration is in part a voluntary action during our waking hours, and as the power of volition is totally suspended in perfect sleep, the blood in the lungs of inirritable or weak people is liable to accumulate for want of the aid of the voluntary power at that time, and hence haemorrhage ensues; and the ulcers are produced in consequence of the rupture of the bronchial veins, and of some part of the effused blood stagnating in the air-vessels, and undergoing a chemical change.

Large abscesses frequently exist many weeks – so long as they are precluded from any access of air without occasioning hectic fever; but on their surfaces being exposed to the contact of the air by opening them, hectic fever is occasioned in a very few hours. From this cause arises the advantage of opening large abscesses by means of a seton passed through them. Hence where a wound is required to heal by the first intention (as it is called) as in compounded fractures, it must carefully be confined from any access of air. And hence lastly the reason, why ulcers of the lungs are so difficult to heal, viz. because they are perpetually exposed to a current of air. What part of atmospheric air is hurtful to ulcers, whether it be the oxygene in too large a quantity, as your ingenious reasonings seem to countenance; or too great a proportion of azotic air, I hope your cautious experiments and your particular attention to consumptive patients will soon decide.

One agreable circumstance attending your application of different kinds of air to consumptive people is, that they may at the same time persist in any plan of medicine or diet, which theirselves or their friends for them are solicitous to pursue, without interfering with your remedy which is immediately applied to the seat of the disease, and the ulcerations of the lungs.

Go on, dear Sir, save the young and the fair of the rising generation from premature death; and rescue the science of medicine from its greatest opprobrium.

[From] your [sincere friend?]
E Darwin

Addressed To Dr Beddoes.

Original Not traced.
Printed T. Beddoes, *A Letter to Erasmus Darwin, M.D., on a new Method of treating Pulmonary Consumption.* Murray, 1793, pp. 61–7.
Text From above source.
Notes
(1) Beddoes was an unrestrained writer: some of his 'pamphlets' have more than 200 pages, and his 'letter' to Darwin (including the letter from Darwin) extends to

72 pages. In the *Letter*, Beddoes enthusiastically propounds his ideas for treating consumptive patients with 'a mixture of airs': a mixture of hydrogen and air seems to be his favourite at the time of writing (June 1793), but he may soon have changed his mind. Beddoes's great achievement was to promote the idea of pneumatic medicine and by his energy and enthusiasm to bring it into operation. It seemed a good idea at the time, especially to those who recognized how futile the existing medicines were; but the results were, in the end, disappointing, and many of the 'cures' claimed must have been spontaneous.

(2) Darwin accepts Beddoes's conclusion that consumption thrives on oxygen and should be treated by reducing the oxygen supply (as in the twentieth-century mountain TB clinics). Darwin also rightly pointed out that consumption was (a) hereditary and (b) also infectious in its last stages. But he failed to incorporate microbial infection into his medical credo, although he often speculates about the possible role of micro-organisms (see *Temple of Nature*, Additional Note I). His idea that people having eyes with large pupils are subject to consumption has not been validated.

(3) Darwin's letter, written at a time when he was hard at work on *Zoonomia*, reveals his continuing enthusiasm for medical advance, and particularly for any promising method of combating the scourge of consumption. Beddoes had a good idea and immense energy, so Darwin took the trouble to write a long letter to encourage him.

(4) Glossary: 'Your treatise on Consumptions' was probably a draft of Beddoes's 278-page *Observations on the Nature and Cure of Calculus, Sea Scurvy, Consumption, Catarrh and Fever* . . . (Murray, 1793), which was apparently not published until later in 1793. The 'size' of blood is the semi-solid coagulate. 'Haemoptoe' refers to spitting of blood. 'Scrophula', also known earlier as 'King's evil', was a disease with chronic enlargement and degeneration of lymph glands. A seton is a thread down through a fold of skin to maintain an opening (for an issue, or discharges of pus).

(5) I have corrected some obvious misprints and replaced 'Yours, etc' by a more likely ending. In paragraph 6, line 4, *irritable* has been changed to *inirritable*, in accordance with Beddoes's own *erratum*, given in his *Letters from Dr Withering* . . . (Bristol [1794]).

93D. To Benjamin S. Barton, 3 July 1793

Derby July 3. 1793

Dear Sir,

I have the favor of your letter dated Apr. 12 last, but have not yet recieved the Diploma with which the American Philosophical Society has honor'd me.

The first part of the Botanic Garden, call'd the Economy of Vegetation, has been out about a year, a copy of which I would do myself the pleasure to send you, if I knew how. I am also beginning to print a prose-work, which will be call'd Zoonomia, or the laws of organic life, in two volumes, the first of which I expect will be out about Christmas.

A few years ago I was concern'd with a society in Lichfield in translating the Genera plantarum, and the Systema vegetabilium of Linneus into english and sent about a dozen copies of each to Philadelphia about two years ago, together with a few copies of the Loves of the Plants, but have not yet heard whether they are sold.

I beg you will present my due thanks for the great honor the society has confer'd upon me, and believe me dear Sir,

Your much obliged
Erasmus Darwin

Addressed For Benj. Smith Barton Esq., Professor of natural philosophy in the University of Pensylvania (to go by the British packet).

Original Historical Society of Pennsylvania, Philadelphia.
Printed Unpublished.
Text From original MS.
Notes
(1) Benjamin Smith Barton (1766–1815), physician and naturalist, was Professor of Natural History and Botany at the University of Pennsylvania from 1789 to 1815, and author of the textbook *Elements of Botany* (1803). See *Dictionary of Scientific Biography* (Scribners, New York, 1970–6) i 484–6.
(2) The American Philosophical Society at Philadelphia, originating from an initiative by Benjamin Franklin in 1743, was the first academy of the western world. Its *Transactions* date from 1771 and Franklin was President until his death in 1790. The President in 1793 was the astronomer David Rittenhouse (1732–1796), and the third President was Thomas Jefferson.

93E. To Robert W. Darwin, 19 September 1793

Sept. 19 –93

Dear Robt

Mrs Galton of Birmingham has written me the inclosed letter desiring me to inquire of you about the Beringtons. I shall send you her letter, tho' it cost you double postage, because you will see both an elegant hand-writing and an elegant style. And you will please in return to send me any account you can collect, which you think may be interesting to her.

You have sent me no remarks on my book. Whether I am to esteem your silence as a mark of approbation or the contrary I can not tell. Adieu from dear Robt yours affectly

E Darwin

Please to return me the letter, as I shall not answer it, till I recieve yours.

Addressed Dr Darwin, Shrewsbury. (*Postmark* Derby.)

Original British Library, Add. MS 30262, ff. 73–4.
Printed Unpublished.

Text From original MS.

Notes

(1) Mrs Galton was the wife of Samuel Galton (1753–1832) the Quaker and armament manufacturer, whose son Samuel Tertius Galton was later to marry Darwin's daughter Violetta (Sir Francis Galton being one of their sons). Mrs Galton was frequently ill, and Darwin rather enjoyed being called to see her because he was so well entertained by the Galtons and could obtain the latest Lunar news from Galton, who was a leading light of the Lunar Society in the 1780s and 1790s. Also Galton was rich, and Darwin felt no qualms in charging high fees for his long journeys to Birmingham. See also letter 98C for Mrs Galton.

(2) 'The Beringtons' probably refers to Joseph Berington, the 'liberal Catholic' priest at Oscott, near Barr, where the Galtons lived. See letter 98C.

(3) 'My book' is probably a set of proofs of *Zoonomia*, Volume I. See also letter 94C.

93F. To Thomas Wedgwood, 15 November 1793

Derby Nov. 15 –93

Dear Sir,

People with black eyes are liable to gutta serena, which consists of a paralysis of the retina, or immediate organ of vision, and not to cataracts. Those, who have light eyes, are subject to cataracts, and not to gutta serena; but cataracts seldom come {so} early in life, except from some external violence; and frequently so imperceptibly, that the patients lose the sight of one eye even without being conscious of it.

On those considerations I do not suppose that the complaint of your eyes is either a beginning gutta serena, or a beginning cataract; – but that the wearyness and uneasy feels of your eyes are symtomatic, that is, owing to the sympathy of these parts with the diseased action of some other part of the system.

One very common sympathy is between the stimulus of worms in the intestines and the nostrils, which occasions itching – often with grat[er]ing the teeth – and I believe sometimes with the production of polypus in the nose in children. Now as you have formerly been troubled with ascarides, and probably may still have them, I should advise a course of Harrowgate water for 3 or 4 weeks; and this should be urged more energetically, if any ascarides appear – which should be attended to.

For this purpose half a pint of Harrowgate water should be drank 3 or 4 times in the morning part of the day, so as to induce daily 5 or 7 stools. If to each of these glasses of water be added a common teaspoonful of a solution of liver of sulphur, and about ten grains of bay salt, and five grains of bitter purging salt (Magnesia vitriolata), both in fine powder, the effect would be increased.

The solution of Hepar sulphuris is made by roasting three ounces of salt of tartar, with one ounce of flowers of sulphur in an iron ladle, till they become liver-colour, and then putting them into half a pint of water and when settled, pour it from the subside.

An issue in either arm may be tried, but I should previously use the Harrowgate water.

The headachs are probably owing to the same cause with the complaint of your eyes; both of which I esteem to be more troublesome than dangerous; and that time will gradually either remove the cause of them, or render the constitution insensible to the effects of the unremoved cause.

Pray let me hear, when you have tried the water about a week, whether any worms can be detected.

Thank [you] for your additional proof of the tremella nostoc being digested frogs, as I mention'd in a note at the end of Canto I of Part II of Botanic Garden.

Mr Darwin is now in London, but returns in 4 or 3 days. I will then acquaint him.

All parts of the body become strengthen'd by moderate or natural use – hence you should use your eyes but not to great fatigue. The way of curing squinting is by covering the good eye, which weakens it; and using the worse eye, which strengthens it. But when smarting or fatigue ensues, the exercise of it should be intermitted. Mrs Darwin thanks your brother for the snow-shoes, and hopes some time to see him and repay him.

All our best compliments to all at Etruria.

Adieu E Darwin

Addressed Thos. Wedgwood Esqr., Etruria, Newcastle, Stafforshire. (*Postmark* Derby.)

Original Keele University Library, Wedgwood Archive, 26765–35.
Printed Excerpt in *Doctor of Revolution*, p. 232.
Text From original MS.
Notes

(1) Tom Wedgwood, 'the first photographer' (1771–1805), was the third and most brilliant of Josiah's sons. This is the first of eleven letters from Darwin to Tom, preserved in the Wedgwood Archive at Keele University Library. Darwin had treated Tom since he was a child, and was well aware of his great talents and neuroses. This letter calms Tom's fears about his eyesight, throws around some medical terms for him to bite on, and provides him with the medicine most acceptable to him – a purgative. Darwin also suggests chemical experiments to keep him occupied and supply him with even more potent purgatives, though '5 or 7 stools a day' seems excessive.

(2) Gutta serena, or amaurosis, is partial or total loss of sight without any external change in the eye, due to disease of the optic nerve.

(3) The idea of 'sympathy' between intestinal worms and facial symptoms was common in the eighteenth century.

(4) Harrogate water is rich in magnesium sulphate.

(5) Liver of sulphur (or hepar sulphuris) is a metallic sulphide, made as Darwin specifies. Bay-salt is common salt. Magnesia vitriolata is magnesium sulphate. Salt of tartar is potassium bitartrate.

(6) An issue is a treatment in which the skin was divided and the wound filled with a small ball of ivory, gold or silver which set up a chronic inflammation. It was applied at a site away from the seat of disease and was thought either by nervous or humoural mechanisms to reduce the inflammation at the primary seat.
(7) The note on tremella nostoc is in *Loves of the Plants* I 373.
(8) 'Mr Darwin' is Erasmus junior.
(9) 'Your brother' may be either John Wedgwood (1766–1844) or Josiah Wedgwood II (1769–1843). For an excellent biography of the three brothers, see *Wedgwood Circle*, pp. 105–241.

93G. To Robert W. Darwin, [November? 1793]

* * *

[His son Robert owed him a small sum of money, and instead of being paid, he asked Robert to buy a goose-pie with it, for which it seems Shrewsbury was then famous, and send it at Christmas to an old woman living in Birmingham] for she, as you may remember, was your nurse, which is the greatest obligation, if well performed, that can be received from an inferior.

* * *

There are two kinds of covetousness, one the fear of poverty, the other the desire of gain. The former, I believe, at some time affects all people who live by a profession.

* * *

[When Robert urged him to retire from professional work, he replied]it is a dangerous experiment, and generally ends either in drunkenness or hypochondriacism. Thus I reason, one must do something (so country squires fox-hunt), otherwise one grows weary of life, and becomes a prey to ennui. Therefore one may as well do something advantageous to oneself and friends or to mankind, as employ oneself in cards or other things equally insignificant.

* * *

Original Not traced. In the possession of Charles Darwin, 1879.
Printed C. Darwin, *Life of E. Darwin*, pp. 62, 66, 52.
Text From above source.
Notes
(1) In the first excerpt, the sentence in square brackets is Charles Darwin's introduction to the quotation. Charles says it was written in 1793, and presumably not long before Christmas, whence my dating. As Charles says, the letter shows Erasmus 'was kind and considerate to his servants'. The fact that he knew the address of Robert's nurse is itself revealing.
(2) The second excerpt may not be from the same letter. Charles says it is from a

'letter not dated, but written in 1793'. Apparently both Darwin's sons by his first marriage, Erasmus junior and Robert, were becoming worried at growing rich. Erasmus junior wrote to Robert on 12 November 1792: 'I am not afraid of being rich, as our father used to say at Lichfield he was, for fear of growing covetous; to avoid which misfortune, as you know, he used to dig a certain number of duck puddles every spring, that he might fill them up again in the autumn'. (C. Darwin, *Life of E. Darwin*, p. 66).

(3) In the third excerpt, the sentence in square brackets is my paraphrase of Charles Darwin's introduction. The third excerpt may be from a different letter: it is dated 'about the year 1793' by Charles Darwin. The wording is similar to 92B: Erasmus took his own advice and remained in practice until 1800, though gradually reducing his commitments.

93H. To Thomas Wedgwood, 30 November 1793

Dear Sir,

A factitious Harrowgate water may do as well as the natural, if it be conducted in a similar manner. The ingredients are similar to those of sea water with a greater proportion of Magnesian vitriol (sal catharticus amarus) and some calcareous Hepar Sulphuris. The saline purge, I have frequently had recourse to, has been much stronger than Harrowgate water, as follows. Mix half a pint of warm water with two ounces of bitter purging salt, and two drams of sea-salt. Take one third or half this for a dose mix'd with a teaspoonful of a solution of Hepar sulphuris, and repeat it every two hours, or every hour, so as to induce 4 or 6 stools a day. Begin at eight in the morning, and take your breakfast at nine – and walk in the cool air.

The solution of Hepar sulphuris is made by roasting one ounce of flowers of sulphur with two or three ounces of salt of tartar, till it becomes of a liver-colour, and then dissolving it in half a pint of water and after it has settled, pour of[f] the clear for use.

If you are easily purged, you will please to dilute the quantity above prescribed with double the quantity of water.

A young girl in this town, who had long had inflamed eyes, got well of them by taking a purge of this kind for 2 or 3 weeks; I believe she had ascarides, but am not quite certain, as the evidence did not appear to me quite clear.

If I was certain, that you had ascarides, I should advise you to take two drams of an amalgama of tin and mercury, twice a day, for a fortnight along with the factitious Harrowgate water.

All of us send best wishes and compliments to all of you. Adieu

E Darwin.

Derby,
Nov. 30 –93

Addressed {Thos. W}edgwood Esqr., {at Josiah} Wedgwoods Esqr., Etruria, Newcastle, Staffordshire.

Original Keele University Library, Wedgwood Archive, 26766–35.
Printed Unpublished.
Text From original MS.
Notes
(1) Though it is easy to condemn Darwin for prescribing such potent purgatives, it can be said in his defence that (a) he was trying to kill the worms without (quite) killing the patient, and (b) patients were attuned to suffering frequent purgation and reacted quite calmly.
(2) For 'magnesian vitriol' etc, see notes to letter 93F.

93I. To Thomas Wedgwood, 6 December 1793

Derby
Dec. 6 –93

Dear Sir,

As the saline solution produces so much thirst, you will please either to dilute it, or take gruel or barley water between the doses. I rather think you possibly take too much of it, as 5 or 6 stools will be sufficient, instead of 8 or 10.

Flowers of Sulphur may be taken at the same time. I think you had better take the sulphur at night, suppose a dram. And purge it off in the morning with the saline solution.

Mrs Darwin joins in our congratulations to your Brother and Sister on the increase of their family. Compliments to all at Etruria. Adieu

E Darwin

Addressed {Thos. Wedg}wood Esqr., {at} Josiah Wedgwood's Esqr., Etruria, near Newcastle, Staffordshire. (*Postmark* Derby.)

Original Keele University Library, Wedgwood Archive, 26767–35.
Printed Excerpt in *Doctor of Revolution*, p. 233.
Text From original MS.
Notes
(1) The ghastly saline purge has proved too strong. Even so, Darwin still wants Tom to continue with it.
(2) Tom's brother Josiah Wedgwood II (1769–1843) had married Elizabeth Allen (1764–1846) in December 1792 and their first child Sarah Elizabeth (1793–1880) had just been born.
(3) The heading, 'Derby, Dec. 6 –93', is written upside down on the MS.

93J. To JAMES WATT, 13 December 1793

Derby Dec 13 –93

Dear Sir,

I am truly happy to hear that Miss Jessey continues to get stronger, and keeps free from those alarming affections of her lungs, or side, to which she has so long been in a greater or less degree subject. As the chalybeate in such small quantity, and the laudanum in such small quantity, seems to have been of service to her, I think I should advise the continuance of them even for 3 months longer; I mean, till the normal warmth approaches – unless any unforeseen accidental inflammatory cough or pain should make it dubious, whether the chalybeate should be continued.

I have seen people so essentially weaken'd by being cover'd with flannel next to their skin, which by the stimulus of its points keeps the cutaneous vessels in too great action, and thus wastes the general quantity of animal power, similar to what happens in the scarlet fevers; that I generally recommend the flannel to be put on the outside of the shirt or shift – and to be disused in the warmer months.

The cold air in winter should be used as a cold bath. Miss Jessy should go into it 3 or 4 times a day, but should remain in it but a few minutes at a time.

In other respects employment, which produces change of posture frequently, in the house in the cold season; and exercise frequent but gentle; such as dancing, swinging, shuttle-cock, and particularly going about the house to fetch what she wants to be brought to her, are good winter amusements – to which should be added, that humble and too much neglected yet delightful contest at Taw vulgarly call'd playing at marbles – so much more wholesome for rendering the joints supple, and varying the attitudes of the body, than its proud son and successor, Billiards.

Ye days, which are past! – when I could have pursued the rolling Taw with spirits light as air; and limbs supple as the bending grass-blade; when I could have crept through an Alderman's thumb-ring! – where are ye! – Corpulency of body, hebitude of mind, or in one word old-age I feel your irresistable approach – but care little about it.

Law is a bottomless pit! – I wish you well out of it! Now I come to conclude, I beg you will present my kindest compliments to Mrs and Miss Watt, in which Mrs Darwin begs to join, and to say that she does not despair of giving them a call next summer.

from your affect. friend
E Darwin

Lead-weights brandish'd in the hand, call'd dumb bells, I think a bad exercise; as their weight compresses the intervertebral cartilages, and impedes the growth of young people, like the carrying of other burdens. But ringing a church-bell on the contrary I should esteem a wholesome mode of exercise for young people, as it imitates swinging by the hands and I see no reason why young ladies should not occasionally ring a peal; – a bob major, or a tripple grandsire.

Addressed James Watt Esqr., near Soho, Birmingham.

Original Privately owned.
Printed Unpublished.
Text From original MS.
Notes
(1) This is the first of eight letters to James Watt and his wife about the illness and eventual death of their only daughter Janet (1779–1794), who was known as Jessie, though Darwin's spelling of her name is variable. She died of consumption, and Darwin's letters provide a harrowing and arresting picture of his vain efforts to treat the disease (see also letter 93C).
(2) A chalybeate was a medicine impregnated with iron. This and subsequent letters show that Darwin was trying out a succession of medicines, replacing each as soon as its ineffectiveness was apparent, and keeping up the spirits of the patient and her parents by implying that the new one might do the trick.
(3) The use of air as a cold bath was recommended and practised by Benjamin Franklin. Gentle exercise was one of Darwin's favourite prescriptions. A Taw was a large fancy marble, with which the player shoots, and marbles was often known as Taws. Hebetude is dulness.
(4) For Mrs Watt, see letter 94D.
(5) A bob and a grandsire are names for ways of ringing changes on a peal of bells.

93K. To Thomas Beddoes, [1793?]

* * *

It seems strange that the metallic oxyds, as of mercury and lead, should so much contribute to heal ulcers when externally applied, if oxygene could occasion them in the lungs of consumptive people.

* * *

Original Not traced.
Printed J. E. Stock, *Life of Beddoes*, p. 85.
Text From above source.
Notes
Stock states that this is a comment on Beddoes's *Observations on the nature and cure of Calculus, Sea Scurvy, Consumption, Catarrh and Fever* . . . , published in 1793. Cf. letter 93C.

94A. To James Watt, 1 January 1794

Dear Sir,

I think it is always useful to cool those parts of the body, which are too hot; and to warm those parts at the same time, which are too cold. Hence in eruptive fevers I have often with apparent advantage exposed the bosom and face to cold air, and wrap'd the cold feet in flannel. This may be done by sponging those parts with cold water, which are too hot, or by exposing them to the air, or under a sheet only, taking care to stop before any chilness is induced.

{1.} Ulcers spread from the saline acrimony of the fluids secreted into them, as is seen in herpes, and some flabby scrophulous ulcers. 2. the saline acrimony is owing to deficient absorption in the ulcer, for in all secretions even of mucus, and urine, the saline aqueous acrimonious part is naturally <u>immediately</u> resorbed. 3. absorption in ulcers is most of all produced by salt of lead, as white lead, secondly by dry application of powder'd bark. Absorption from the lungs is increased <u>miraculously</u> by foxglove, next by other emetics, and <u>particularly</u> by swinging circularly, so as to be in part intoxicated and sickish, as in a ship.

Common air and fix'd air mix'd may be of service both on the theory of Dr Beddoes, and on the antiseptic theory.

I should recommend zi of tincture of foxglove twice a day to be increased gradually to two drams.

And to swing horizontally for ½ an hour or longer 4 or 6 times a day till she becomes vertiginous and sickish.

Internally opium much contributes to thicken matter, that is to produce the absorption of its saline aqueous part, and should be given from 5 to 10 drops at certain regular hours twice a day, and occasionally besides if the hysteric symptoms recur. Tincture or decoction of bark may also be serviceable.

Yeast is believed to be antiseptic, and is said to have been used in putrid fevers with advantage.

Pray make our best wishes to Mrs and Miss Watt, and pray tell Dr Beddoes I have directed a book to be sent to him at Bristol as soon as possible. Adieu from dear Sir

Your affect. friend E Darwin

Pray let me hear how you go on, and with what advantage.

Derby Jan 1 –94

Addressed James Watt Esqr., Heathfield, near Birmingham. (*Postmark* Derby.)

Original Privately owned.
Printed Unpublished.
Text From original MS.

Notes

(1) Jessie's health seems to have worsened during the 2½ weeks since the previous letter. Darwin is now speaking of 'ulcers in the lungs' which he regards as being caused by the 'saline acrimonious fluids' found in ulcers like herpes (as in mouth sores). He believed the ulcers would absorb normally if treated with salts of lead. Foxglove (digitalis) had a 'miraculous' effect in some cases of dropsy when there was water in the lungs, as Darwin reported in *Medical Transactions*, **3**, 255–286 (1785). He also believed that making the patient sick, either by emetics or by swinging, was helpful to consumptives, and this is a recurrent theme in the next five letters.

(2) Although Darwin hints at the use of gases, he does not seem sure what to recommend. The 'antiseptic theory' refers to the idea that certain substances such as vinegar could stop the formation of pus. Hence Darwin's comment on yeast later in the letter.

(3) zi means 1 dram, equal to ⅛ of a fluid ounce, or 4 cubic centimetres.

(4) For Mrs Watt, see letter 94D.

(5) The book to be sent to Beddoes was probably *Zoonomia*, Volume I, published in April 1794.

94B. To Thomas Wedgwood, 6 January 1794

Derby Jan. 6 –94

Dear Sir,

As you have found no benifit from the issue, I think you do right in discontinuing the use of it; and the same of the rhubarb and antimony, except you find, that you have not a daily regular stool without them. You will certainly do right by being much in the open air, and the better if you have some motive of action – in regard to your manner of living I recommend flesh-meat twice a day, with garden stuff or fruit, and small beer at your meals, with one glass of wine, or occasionally two. More fermented or spirituous liquor would in time do you injury. I should certainly advise you to eat some supper, and of flesh-meat in part.

That lassitude or want of spirits, you complain of, may in some measure be owing to want of employment, you should find out something to teaze you a little – a wife – or a law-suit – or some such thing. I think Virgil sais that Jupiter found it necessary to send Care amongst mankind, to prevent them from feeling less evils

– curis acuens mortalia corda,

Georgics

You will see, that I do not esteem your complaints dangerous, but that they are, what you well express, owing to nervous debility, and that time and occupation will cure you.

If you have any ascarides, I should wish you to go to Harrowgate this summer.

I am, dear Sir, with true esteem your obed. serv.

E Darwin

I expect Mr and Mrs Jos. Wedgwood next Monday and Miss Brigstoke, I wish you were of the party.

Mr   of Ashenhurst near Leek I hear designs to sell his house and estate of about 1000 a year – and am told the situation is an agreable one, but never saw it.

Addressed Thos. Wedgwood Esqr., No. 22 Devonshire Place, London. (*Postmark* Derby JA 94.)

Original Keele University Library, Wedgwood Archive, 26768–35.
Printed Excerpt in *Doctor of Revolution*, p. 233.
Text From original MS.
Notes
(1) There is no mention of rhubarb and antimony in the previous letter (93I), so probably there is at least one letter missing from the sequence.
(2) Darwin's prescription of 'something to teaze you' was a good one. With an active mind and nothing to do, Tom tended to luxuriate in hypochondria. Darwin remembered his schoolboy Latin remarkably well, and inserted many apt quotations in *The Botanic Garden*. Here he is ready with another quotation, from Virgil, *Georgics* (I 123): 'with cares sharpening human hearts' is a literal translation.
(3) 'Mrs Jos. Wedgwood' (see letter 93I) was the daughter of John Allen of Cresselly, near Tenby; so 'Miss Brigstoke' may be one of the Brigstock family of Carmarthen, related to Rev. Tudor Brigstock (1741?–1808). The nameless man may possibly be Rev. George Salt (born 1735), whose wife Frances (née Stanley) inherited Ashenhurst Manor. See J. Sleigh, *History of Leek*, 2nd edn. (Bemrose, 1883), p. 109.

94C. To Thomas Beddoes, 6 February 1794

Derby. Feb 6 –94

Dear Sir,

I congratulate you on your being on the road to happyness – and I hope you will contribute to repeople the world, which our politicians are endeavouring to desolate!

I do not dislike what you say about Mr Kenni's[?] case, and when I answer'd his father's letter I urged him to try a lower atmosphere. Your last book I have not yet quite finished (I order'd it for our society some time ago) but do not see an account of an apparatus and manner of procuring these airs, which would put the experiments into many other hands.

You accuse me of having recieved two letters with observations on Zoonomia, I recollect but one – and what you wrote in a letter to my son, and did not see till yesterday.

I do think you did injudiciously in lending my papers to a man, who is a perpetual and voluminous writer; and who of all others is likely to answer my opinions before the book is printed. It was on such an account, that I was not easy till I had them back from Ireland. I shall send you the rest as soon as I get them, but must desire you will not let them go out of your own hands to any body.

What you say of some words not suggesting any idea, as the words Virtue, Wisdom, which the Poets personify, I by no means allow; and, if I forget not, that opinion has been refuted by Berkley and Hume, both of whom I think allow them to suggest particular ideas, as the words a Man or a horse. I believe I have spoken on this subject at the end of the third section – "nothing can come into the intellect but through the senses". How do we get an idea or meaning to the word Virtue? by some of the senses, and to say we have no idea to it, is to say we communicate nothing to our hearer, when we speak the word – which I feel is not just.

By complex ideas I mean comparitively with others less complex, as when I see the front of a fine house, the idea or configuration or motions of the nerve may be divided into many windows and doors. When I look upon the blue sky the idea is less complex, the same of abstracted ideas, when that word is used only comparatively.

So white may be term'd a simple idea, white and red a complex one. As the retina consists of parts, a configuration or motion of it may be very compounded or complex. Or we can abstract many parts from a visible figure, as red from our idea of the rain-bow, or orange, and yellow, and leave only the other colours.

My idea of space is a particular not a general idea, and is a part of an outline of some group of bodies. My idea of time is a particular idea of a part of the motions of some clock or orary [orrery], which happen'd first perhaps to interest my attention.

Mr Seward's letter I wrote a very frigid answer to, and spoke well of you – so you need not trouble yourself about it, which I always think the best way when I hear myself calumniated.

adieu from your sincere friend
E Darwin

Addressed Dr Beddoes, Shiffnal, Shropshire.

Original Bodleian Library, Beddoes collection. Shelfmark Dep. C.134.
Printed Unpublished.
Text From original MS.
Notes
(1) Dr Beddoes married Edgeworth's daughter Anna in April 1794. Darwin's first paragraph refers to this event and also gives an indication of his growing estrangement from the 'establishment' politicians, who were waging war against France.
(2) The second paragraph refers to the experiments in treating diseases by administration of gases: see letter 93C. 'Your last book' is probably the *Observations . . . on Calculus* (see letter 93K).
(3) The papers which Beddoes injudiciously lent to the 'perpetual and voluminous

writer' in Ireland were probably proofs of *Zoonomia*, which Beddoes read for Darwin.

(4) Darwin's discussion of his ideas about ideas is a usefully terse summary of theories propounded at length in *Zoonomia*.

(5) 'Berkley' is the philosopher George Berkeley (1685–1753), who maintained that material objects only exist through being perceived. Hume is of course the great sceptical philosopher David Hume (1711–1776). The ideas of Berkeley and Hume are lucidly set out in Bertrand Russell, *History of Western Philosophy* (Allen & Unwin, 1946), pp. 673–700.

(6) Mr Seward is probably William Seward: see letter 95E.

(7) Beddoes's reply to this letter is printed by J. E. Stock, *Life of Beddoes*, Appendix 6, page xlii.

94D. To Ann Watt, 12 March 1794

Derby Mar 12 –94

Dear Madam,

I am favor'd with your very accurate history of the present state of health of Miss Jesse; and am sorry to hear, that you think her strength and her flesh less, than when I had the pleasure of seeing her. All her complaints, as well as the want of being out of order, I think are symptoms of general debility; and that general debility I suppose to be owing to her not being able to digest sufficient food.

Whatever therefore strengthens her stomach will be likely most to serve her, and even to produce the catamenia, and to relieve the pains of her sides and head.

I mean to direct her a small blister for her side, if she has pain; otherwise for her back; which has at least a temporary power of affecting the stomach by consent of parts, and for a time increases both the digestion, and the energy of the whole system.

I need not repeat, that she should use moderate but frequent exercise; and when the weather is warm, this should be in the open air, as much as may be.

Mr Barr will give you a direction along with the medecines, and I shall be glad to hear in a week or two, if you think her in any respect better.

I beg my best wishes and compliments to my fair patient and Mr Watt, and am, Madam,

Your much obliged
and obed. serv.
E Darwin

Original Privately owned.
Printed Unpublished.
Text From original MS.
Notes

(1) Ann MacGregor became James Watt's second wife in 1776, and by all accounts

the marriage was a great success: Watt himself called it 'one of the wisest of my actions'. They had two children: the brilliant Gregory (1777–1804), who died of consumption at 27; and Janet (1779–1794), the subject of this letter. Ann Watt herself lived until 1832.

(2) Darwin is now treating Jessie for 'general debility', probably because it sounded less alarming than 'consumption'. 'The catemenia' are the menstrual periods.

(3) The stomach and skin were supposed to have a particular affinity: so a blister was believed to stimulate the stomach.

(4) John Barr was a Birmingham surgeon-apothecary who waited on the Watt family. He contributed a letter on scrophula to the second edition of the book by Beddoes and Watt, *Considerations on the Medicinal Use, and on the Production of Factitious Airs* (1795), p. 66.

94E. To James Watt, 25 April 1794

Derby Apr 25 –94

Dear Sir,

I am truly sorry to hear so indifferent an account of your daughter's health, and think you do well in trying the effect of a journey by small stages to any of the southern coasts of the island.

My general idea of her disease is that the symptoms term'd hysterical (and even the cough) and the want of menstruation originate from debility and that this general debility is occasion'd by the particular inability of her stomach to digest a sufficient quantity of nourishment, which last may have been occasion'd by long fasting, or by cold; both which I have seen affect other young ladies.

I have a patient at Tamworth at this time, Miss Wilson, in a situation similar to Miss Watt's, but in very much greater degree of debility, which I ascribe originally to the defect of the power of digestion.

The principal thing, which can be done in these cases, (as our science of medecine is not forward enough to use the transfusion of blood) is by some stimulus a <u>little</u> greater than natural, and <u>uniformly</u> taken for months or years, to invigorate the digestive power. In some I have seen a great benefit by beginning with a pill of half a grain of opium taken at breakfast, and at going to bed (or at tea in the afternoon) every day as a habit; and after a few weeks to increase it to a grain-pill twice a day – and after some months to a grain and a half twice a day. Pills are prefer'd to Laudanum.

30 drops of chalybeate wine in a glass of warm water along with 20 drops of tinctura cantharidum may be taken occasionally for a month, and then discontinued alternately.

And extract of bark from 20 to 30 grains twice a day for a fortnight occasionally might be of service.

One stool daily or alternate days should be produced by rhubarb or by aloe.

When her fits come on, or threaten to come on, she should take 30 drops of a mixture of equal parts tincture of castor and of laudanum, put into a two-ounce phial – besides the daily allowance of the opium in pills above recommended – ether may be used but is of less efficacy than the above – the same of assafoetida.

The above is the general method, which is most likely to serve her; your surgeon will please to put the medecines in such forms as she can best take them. But the opiate I should wish to be given in a very small pill and uniformly persisted in.

In respect to diet, flesh-meat and small beer are recommended twice a day, but as her digestion is weak, her palate must be consulted in every thing both of what she eats and drinks. The young spring herbs, as asparagus, cabages, etc are also recommended; and wine and water instead of small beer, if she prefers it.

She should go quite loose in her stays.

And should lie down after her dinner for an hour if it be agreable to her – and go to bed early.

The drops of tincture of castor and laudanum must be repeated in ½ an hour, if the first dose does not remove the hysteric spasms.

The medecines I last directed may be continued till you see if they are of advantage, but I wish her to begin the opiate pill immediately – and the drops of castor and laudanum.

Mrs D begs to be remember'd to Mrs Watt, to whom and Miss Jessey you will please to make my best compliments, and sincere good wishes, and believe me, dear Sir,

Your affect. friend
and obed. serv. E Darwin

Addressed James Watt Esqr., Heathfield, Birmingham. (*Postmark* Derby.)

Original Privately owned.
Printed Unpublished.
Text From original MS.
Notes

(1) Darwin is still holding to the diagnosis of 'debility of digestion'. Blood transfusion was a favourite idea of his, but he apparently never dared to use it: see Seward, *Darwin*, pp. 109–14. His prescription of opium, increasing to 3 grains a day, was also fairly desperate, but perhaps justifiable in such a desperate case.

(2) For cantharides, see letter 83D. Assafoetida was a gum smelling of onions used as an antispasmodic to combat the 'hysterical symptoms'.

(3) Darwin's own assessment of these and other drugs given to Jessie Watt may be found in the section on Materia Medica in *Zoonomia* ii 657–762.

94F. To James Watt, 25 May 1794

Sunday May 25 –94

Dear Sir,

If Miss Jessy does not soon find some relief in respect to the inflammatory symptoms, I mean "the sharpness and the quickness join'd" of her pulse, by the use of the antimonial, I should wish to try two or three emetics of ipecacuhan, suppose ten grains of the powder in a tea-cup {of} warm water, taken in a morning for three successive mornings – or longer if they seem of advantage.

The use of emetics is to promote absorption from the lungs, which is so evident in the exhibition of foxglove. Now no ulcers can heal, except the absorption is at least equal to the deposition in the ulcer.

I once saw Foxglove heal ulcers in the lungs as evidently as one case (or indeed two) could show; and Mr Hunt, a surgeon of Loughboro', assures me, he has cured as he believes 3 or 4 by the same remedy. But I don't believe it succeeds so well unless it produces sickness. The way I now use it is to put 4 ounces of leaves of foxglove nicely dried and coarsely powder'd into 8 ounces of proof spirit, that is into 4 ounces of rectify'd spirit mix'd with 4 ounces of water and set by the fire to make a tincture, which must be pour'd off from the powder in a day or two. Two drams of this is a dose mix'd with an ounce of pepper-mint water – twice a day.

As the foxglove is liable to operate too harshly, I should advise to try the ipecacuhan for 3 or 4 or 5 mornings <u>immediately</u> if she has not recieved benefit by the James' powder <u>already</u>.

If swinging could be perform'd by being placed on a chair, and whirl'd circularly and horizontally, so as to induce sea-sickness once or twice a day, even without vomitting by it, it might like real sea-sickness promote absorption – which is the means of curing self-spreading ulcers.

If you try these two things, the emetic of ipecacuhan, and the circular swinging for three days, you may possibly note some advantage.

If you think afterwards of the foxglove, it should be taken for 2 or 3 days in doses of one dram twice a day, and then in zi/{ii} if no effect is percieved – and then in zii.

Mrs Darwin communicated your letter to Dr Beddoes – I was from home. He goes to Mr Keir's to day, or tomorrow.

The Bark-dust may be tried, or the flores zinci along with the emetics above recommended, as they act externally by stimulating the surface of the ulcer into absorption.

Pray let me hear from you in a few days.

Dr Beddoes has been inform'd of acid applications inducing scrophulous ulcers to heal. How? – by the stimulus of the acid. Hence I suspect that carbonic acid gas and atmospheric air of each equal parts would tend to heal pulmonic ulcers – and that oxegene gas $\frac{1}{4}$ to atmospheric air would stimulate pulmonic ulcers into healing more certainly than azote or hydrogen, according to some foreign experiments. Pray converse with Dr Beddoes on this circumstance, which is contrary to his own theory. The inflammatory state, and ulcerous state may require different treatment.

Present my most respectful compliments and best wishes to the ladies, and believe me.

Your much obliged
and affect. friend
E Darwin

Addressed J. Watt Esqr., Heathfield, Birmingham.

Original Privately owned.
Printed Unpublished.
Text From original MS.
Notes
(1) Darwin had probably visited Jessie at least once since the previous letter: she is now trying antimonials, i.e. medicines containing the metal antimony. Darwin returns to his idea that the healing of 'ulcers in the lungs' is promoted by emetics, and is now resorting to the well-tried ipecacuanha, in addition to swinging.
(2) Mr Hunt, 'a surgeon of Loughborough', was a member of the Derby Philosophical Society in 1787. Probably Thomas or John Hunt: see R. P. Sturges, *Midland History*, **4**, 228 (1978).
(3) James' powder was a famous 18th-century 'fever powder'. Named after the physician Robert James (1705–1776), its active ingredients were antimony oxide and calcium phosphate.
(4) Darwin somewhat altered his treatments for consumption during 1795, as a result of the work of Beddoes and Watt on treatment with gases. In *Zoonomia* (ii 287–98) Darwin still advocates emetics, but places more emphasis on the 'pneumatic' treatments.

94G. To James Watt, 29 May 1794

Derby May 29 –94

Dear Sir

All the medecines, which have had evident effect in producing pulmonary absorption, and a consequent healing of pulmonary ulcers, have produced it, I believe, by inducing sickness, of these I believe Foxglove to be the most efficacious, I suppose sea-sickness. The former may be managed by a tincture, as I before wrote, and the latter by a circular horizontal swing, as by twisting a rope to which a chair is suspended. I suppose blue vitriol, if often repeated, may probably do the same, but I have more experience on Digitalis – and a swing is well worth trying.

What a suffusion of warm water may do, I should doubt, it seems too feeble an engine.

The powder of bark may be tryed easily; if it does not increase the cough, it may be used with advantage – and the fix'd air is worth giving a due trial to.

Mrs Darwin begs to give her sincere good wishes and compliments to the ladies, and yourself, and I am your affect. friend

E Darwin

I hope to hear how you proceed. I prefer the tinct. Digitalis, as the effect is so much longer continued. But I suppose any sickning drugg has effect – as the scilla has been so long in vogue in pulmonary cases of many kinds.

Addressed James Watt Esqr., Heathfield, near Birmingham. (*Postmark* Derby.)

Original Privately owned.
Printed Unpublished.
Text From original MS.
Notes
(1) To judge from this letter, the Watts seem to have rejected Darwin's prescription of swinging, probably thinking it too cruel a treatment for their dying daughter, though the drugs were equally cruel and no more useful.
(2) Blue vitriol is copper sulphate. Fixed air is carbon dioxide. Scilla, or squill, was a well-known drug made from the bulb of a sea-onion (*Scilla maritima*).

94H. To James Watt, 6 June 1794

Derby Jun 6 –94

Dear Sir,

I do not think I can suggest any thing further, than what I have before said viz. that as I believe sickness to be the cause of healing pulmonary ulcers, and that a swing so as to induce it is preferable to all other modes of inducing it – and next to that Foxglove I esteem of more certain effect in pulmonary ulcers, than blue vitriol. But the rotatory swing is most managible, least dangerous, and most efficacious (I believe) of all means hitherto essay'd in pulmonary ulcers – and where absorption from the lungs is indicated.

Pray let me hear from you, how you go on.

With compliments and good wishes, I am

Your affect. friend E Darwin

Original Privately owned.
Printed Unpublished.
Text From original MS.
Notes
(1) Darwin must have known the end was near, and has nothing more to suggest.
(2) Though Watt did not subject Jessie to swinging, he did later make a detailed engineering design for a 'movable rotative couch' for human centrifugation. See *Doctor of Revolution*, p. 257 and Plate 12A.

94I. To James Watt, 11 June 1794

My dear friend,

I have not felt so poignant a grief of some years, as I feel now for the very affecting loss, which Mrs Watt and yourself have experienced! – a young companion of such uncommon beauty, of such amiable character, and of such ingenious and active mind as are rarely equal'd, never perhaps exceeded! – Mrs Darwin has wept two or three times for the sufferings of Mrs Watt and yourself, tho' she only knew your daughter by my description.

There is nothing I can say to console you, which your own strength of mind has not already suggested, and by which you have endeavour'd to alleviate your own grief and that of Mrs Watt.

"When I sail'd along the coasts, and contemplated the desert situations, where Troy, Thebes, Palmira and other great cities had flourish'd, and were now extinct ["], said Sulpicius to Cicero[, "] I sigh'd indeed, but shed fewer tears, than you do for one delicate and frail woman". – I see nothing in this remark, but that we are all mortal; and as we are necessitated to bear our evils, we should endeavour to bear them with fortitude.

Another argument with which Sulpicius endeavours to console Cicero is from the badness of the times, which is well applicable to the present appearance of things – Had she lived he sais in these ominous times, might not unhappyness have been her share? Where could you have found a man worthy of her, and with whom you could safely have intrusted her happyness?

Now I beg you will present mine and Mrs Darwin's best compliments to Mrs Watt, and tell her that if it can any way alleviate or divert her mind, and your own, from the sorrow, which you labour under, that we shall be glad to see you to make what stay may be agreable to you with us at Derby, and will do all we can to make our house agreable to you.

I am, dear Sir,
Your affect. friend
E Darwin

Derby Jun. 11 –94

Original Privately owned.
Printed Unpublished.
Text From original MS.
Notes
(1) Jessie Watt died about 8 June 1794, and Darwin again had to face the depressing truth that his treatments of consumption were useless.
(2) Darwin's 'quotation' from the letter of Sulpicius to Cicero, though faithful to the sense, is a very free paraphrase. Melmoth's translation, which is the likely source (see letter 80C), has the following wording: 'As I was sailing from Aegina towards

Megara . . . on my right I saw Piraeus, and on my left, Corinth. These cities, once so flourishing and magnificent, now presented nothing to my view but a sad spectacle of desolation . . .' (*The Letters of M. T. Cicero to several of his Friends*. 3 vols., 1772, iii 6–13.)

(3) Watt wrote a long letter to Darwin on 30 June (quoted in Muirhead, *James Watt* ii 246–7), thanking Darwin for his sympathy, and saying that the best consolation is to turn the mind to another subject. Watt had decided to give his attention to medicinal airs. See next letter for Darwin's reply.

94J. To James Watt, 3 July 1794

Derby Jul. 3 –94

My dear friend,

It gave me great satisfaction both on your account and on that of the public, that you are employing your mind on the subject of medicinal airs, of which indeed Dr Beddoes had before inform'd me. You will do me a great favor by sending me an apparatus, or a description of one, and an account how easily to obtain the gasses; I have now an asthmatic patient would readily try oxygen gas mix'd with atmospheric air.

In your letter to me, you speak of obtaining inflammable air from Zinc. Do you hold the theory, that the inflammable air (Hydrogen) is extracted from the Zinc, as your expression seems [to] countenance, or do you suppose, that it is from the decomposition of water according to the idea of poor Lavoisier?

By the coach tomorrow I intend to send you a copy of my new book, hoping to amuse you, not to instruct you; I think to direct it to Mr Boulton's Warehouse in Birmingham – if this direction be wrong, pray inquire after it. If any new matter on reading any part of my book should occur to you, or any great errors, I should be glad of your criticism towards another edition, tho' I have sold the copy-right, I would correct it if I could.

Mrs Darwin and myself beg our kindest remembrance to Mrs Watt, whom we hope the lenient hand of time, and that strength of mind, which she possesses, and your attentions to her, will gradually procure consolation, tho' not forgetfulness. Adieu from dear Sir

Your affect. friend
E Darwin

Addressed James Watt Esqr., Heathfield, near Birmingham. (*Postmark* Derby.)

Original Privately owned.
Printed Unpublished.
Text From original MS.
Notes

(1) The illness and death of his daughter had impressed on Watt the inadequacy of existing medical treatments, and he became very interested in Beddoes's ideas on

the medicinal use of gases. In 1794 and 1795 Watt spent much time and energy in devising apparatus for producing the gases, and made many suggestions about their medical use. Without Watt's help, Beddoes could never have put his ideas into practice. At least 50 letters passed between Watt and Beddoes in 1794–5 and they were co-authors of the book *Considerations on the medicinal use and on the production of factitious airs,* which went through several editions.

(2) It seems that Watt was still not fully converted to the 'new chemistry', so Darwin quizzes him again on the composition of water (cf. letter 88D). Antoine Lavoisier (1743–1794), the chief pioneer of the new chemistry, was guillotined on 8 May 1794 on a trumped-up charge, the most eminent victim of the dictum that 'the state has no need of savants'.

(3) 'My new book' is presumably Volume I of *Zoonomia*, published 3 months before, but probably not sent to Watt because of his daughter's illness.

94K. To Thomas Beddoes, [July 1794]

Dear Doctor,

I am not surprized or chagrined, that you do not believe any part of my theory of ideas consisting in the motions of the fibrillae of the extremity of the nerves of sense; for if you believed this part of the doctrine, the circumstance, that the part of the extremities of the nerve of touch stimulated into action must resemble in figure the figure of body applied to it follows of course.

The power of correcting one sense by another has been observed by other writers on metaphysics, and was not therefore proposed as new. Your example of the corner of a table I suppose perplexed you, which may be considered as consisting of more planes than one, and may thus be reduced to as great simplicity as the example of the ivory triangle. If you allow an idea of perception to be a part of the extremity of nerve, of touch, or sight stimulated into action, that part must have figure, and that figure must resemble the figure of the body acting on it: if I lay a triangle of ivory on a sheet of paper, the part of the paper pressed upon must resemble the figure of the triangle pressing on it. I suppose where two people accustomed to reason, differ so toto cœlo, that there must be some mistake in the terms. Perhaps your mind is about the idea of solidity, or rather perhaps, you use the word idea in its common meaning, as an immaterial being, and cannot assume form.

Enough of this. – Mr Watt has sent me an air apparatus which I have not yet unpacked. I shall now wait with impatience for your next book to teach us the best way of making and using airs.

Adieu,

From your affectionate Friend,

E Darwin.

P.S. – I believe your use of the words 'things' and 'ideas' as if in opposition to each other, as if ideas were not things, may have perplexed the subject: two things may be similar, whether we think of them or not, but we compare our ideas of them you say when we think of them; but our ideas of them are not the middle term of the comparison, they simply represent them. We can have a reflex idea of the operations of our minds. Now as I look upon ideas to be things (that is substances) in motion, I can readily have an idea of an idea, and can hence compare my idea of an ivory triangle with my reflex idea of that original idea. But I believe I am got into that sublime figure of rhetoric called the Profound. – Adieu.

Original Not traced.
Printed J. E. Stock, *Life of Beddoes*, Appendix 6, pp. xliii–xliv.
Text From above source.
Notes
(1) Darwin has received but not unpacked Watt's 'air apparatus', which he asked for on 3 July (letter 94J) and unpacked on 16 August (letter 94P). So the date of this letter is probably late July.
(2) Beddoes's reply to letter 94C (given by Stock, Appendix 6, pp. xlii–xliii) shows that he rejects Darwin's theory that ideas of objects resemble those objects; so Darwin produces some new arguments to support his theory. Beddoes's own views on ideas, especially of mathematical figures, may be found in his book *Observations on the Nature of Demonstrative Evidence* (Johnson, 1793).

94L. To Thomas Beddoes, [summer? 1794]

Dear Doctor,

I am much obliged to you for your remark about the exterior atmosphere of inflammable air; I think it probable that what I conjectured about its standing higher at the poles may not be well founded; since the gravity of the concentric fluids are not overcome by their centrifugal force like oil and water whirled round in a bottle; but as far as I recollect it does not affect the theory of aurora borealis; since if there be a superabundancy of inflammable air, more than will mix with the common atmosphere, some will rise through it.

I do not think, that you still see that what you say about the figure of an idea (in my sense of the word idea) consists of words without meaning. I shall once more make an essay, as I think the passage is of great consequence, and one of the principal parts of my whole theory, and am therefore unwilling to have it contradicted without argument. I assert that ideas are parts of the extremities of the nerves of sense (which may be conceived like the points of the knap of velvet, and not longitudinal fibres laid horizontally, as may be shown by the images left in the eye – the ocular spectra – not extending beyond their correspondent objects); these parts have extension and consequent figure. Now if an ivory triangle presses the palm of my hand in the light, I can have both a tangible and a visible idea of the triangle; I can also have a visible idea of the part of my organ of touch, which is stimulated (and compressed), and according to your view of the matter I can compare the visible idea of the triangle, and the visible idea of the part of the organ of touch together. But where is the necessity of a medium of comparison? it is no syllogism; it goes no further than to say that a triangle pressed on my hand is the figure of the part pressed.

Lest I should be thought prejudiced, because you had attacked the most favorite part of my theory, I wrote last week to Mr Keir on this subject, and have inclosed you his letter, which I do for no other purpose but to clear myself from the supposition of incorrigible prejudice in respect to my own work.

Your affectionate Friend,
E Darwin

Original Not traced.
Printed J. E. Stock, *Life of Beddoes*, Appendix 6, pp. xliv–xlv.
Text From above source.
Notes
(1) Since the ivory triangle appears again, this letter seems likely to be within a week or two of 94K.
(2) The first paragraph refers to Darwin's prescient theory that there is an outermost region of hydrogen in the upper atmosphere and that the aurora borealis often forms in this region. (*Economy of Vegetation*, Note I.) Both surmises are true, though not proved until the 1970s; hydrogen is dominant above 700 km when solar activity is low.
(3) The second paragraph (which I have repunctuated) is a useful restatement of Darwin's theory of ideas, but it did not convince Beddoes.

94M. To Thomas Beddoes, [summer? 1794]

* * *

I have read a little work of yours about preserving health with great pleasure. You deserve a civic crown for saving the lives of your fellow citizens.

* * *

To make Dr Brown's work agreeable reading, it must be totally written over again; and to shew the excellencies and the errors of it, would require a volume or two. All you can do will be to white-wash the old building as it stands, and to put a neat portico to it by way of preface commendatory, and cover the irregularities by shrubberies of myrtle and orange flower; but not to attempt to pull down and rebuild any part of it; because it would be easier to make a new one.

* * *

Original Not traced.
Printed Stock, *Beddoes*, pp. 102 and 106.
Text From above source.
Notes
(1) In the first excerpt Darwin is commenting on Beddoes's pamphlet 'On the Art of Self Preservation' (1794), a do-it-yourself-without-doctors medical guide, apparently published in March 1794. Stock says Darwin's letter was a few months after, and therefore perhaps between June and August.
(2) The second excerpt may not be from the same letter, but was probably written during 1794, because that was the year when Beddoes agreed to revise the *Elements of Medicine* (1780) of Dr John Brown (1735–1788). For Brown, see *DNB*.

94N. To Thomas Wedgwood, 10 August 1794

Derby Augt. 10 –94

Dear Sir,

I am glad to hear that the climate of Devon, and the chearfulness of your companions, have so much contributed to the melioration of your health; but am, at the same time, sorry to loose you from these frigid parts of the country. I shall answer the particulars of your letter in the order of it. The feverish complaint you mention in the evenings, and which sometimes keeps you watchful at night, may be occasion'd by some trivial mismanagement in your hours of meals, or the quantity of your potation; or your hour of going to bed. All these should be as regular as the instability of human affairs will admit of. For then the habit coincides with the other circumstances of fatigue, or of pleasurable ideas, for the purpose of procuring sleep, see Zoonomia Sect. XXXVI.2.1 and .2.

The swellings of the tonsils, or the little ulcerations, which occasionally affect them, are I believe frequently occasion'd by the putrid matter from decaying teeth, like the swelling in the axilla from the absorption of the poison of the smallpox. The headach, when it affects one side of the head only, is also, I believe, generally if not always, induced by sympathy with some decaying tooth; which you may see further spoken of (if you please) in Zoonomia Sect. XXXV.2.1, with a curious case of hemicrania from a decay'd tooth. On both these accounts, if the tooth, you mention, is much decay'd, you had better extract it. Otherwise you must take 60 grains of powder or (what is better) of good extract of Bark, twice a day for a fortnight, or 3 weeks, either as a bolus, or powder, or diffused in two ounces of water made a little warmish; to, suppose, about 82 degrees.

The change of your skin indicates a degree of plumpness, which shews you have lived better, viz. upon food and liquor of greater stimulus, and have thence been able to take it in greater quantity – which you should therefore continue to do.

The constant employment you mention is also a good thing, as it prevents one from feeling little evils, which otherwise torment us, as further explain'd in Zoonomia V.1, p. 422. Now that you may not want employment, I should recommend to you a lawsuite in Chancery, or to follow the example of your brothers and take a wife – – or to read my book of Zoonomia!

In respect to medecine I have no great opinion of the cicuta, as I have heard of patients taking a quarter of an ounce of the extract, beginning with small quantities, twice a day for weeks without any effect good or bad, and therefore I am induced to think it can be of no consequence when given by a few grains at a dose. About five grains of rhubarb*, and ¾ of a grain, or a grain of opium taken every night for many months, perhaps during the whole winter, I think a medecine of much greater value in your situation. The Bark mention'd above should be taken at the same time for a fortnight, and repeated occasionally, when you find yourself not so well.

I beg leave here to thank you for your generosity in sending me the inclosed note of 5 guineas.

Now for philosophy. No one is more impatient under the smell of bad feathers than myself, yet I doubt if any material can be obtain'd so good for pillows, and feather-beds, as well dryed, and well oxygenated feathers. If you stuff a feather-bed cover with air instead of feathers, it must be included in a bag of oil'd silk, which is not easily to be procured without smell as bad, or worse, than that of feathers, in respect to salubrity, {and} if a crack or pin-hole should occur, you would gradually {sink} down, like a boy swimming on bladders prick'd by a thorn {and} find yourself in the morning on hard cold boards or cord{age}.

For a pillow, Take hog's bladders, roll them out thin, slit them, stretch them while soften'd in warm water, glew them together, make a boulster or pillow according to art. By means of a pipe, you might blow it up every night, and perform "God save the King" at the same time; thus in travelling it would take no room in your portmanteau and you would gain the character of a great loyalist at the same time.

A feather-bed case fill'd with quicksilver would be very soft and pleasant; in winter it might be warm'd by one of Argand's lamps kept burning under it; and the exsudation would at the same time prevent or cure some diseases, that young bachellors are said to be liable to – you see, I love to produce two effects from one cause.

If you fill your bed with air, and it should sink, you might have a pipe adapted, so that you might at any hour in the night blow yourself up again – perhaps Dr Beddoes might thus exhibit some of his medicinal airs in this way.

The valve to the touch-hole of a gun I think a great invention, especially to canon. It must save powder, as much, compared with the present charge, as the diameter of the touch-hole is to the diameter of the gun, I suppose.

* Darwin first wrote 'opium', then (fortunately!) crossed it out.

It has been said, that if a valve opening externally was put on the muzzle of a gun, that it would prevent the greatest part of the sound. I doubt this, as I rather think the sound is made by the clapping together of the sides of the vacuum made by the sudden evanescence of the heat, and condensation of the water, than by its returning into the barrel of the gun.

You will please to present my best compliments to all with you and believe me, your much obliged friend
and obed. serv. E Darwin
Mrs Darwin begs her compliments to all with you.

Addressed Thos. Wedgwood Esqr., Tallaton, Ottery St Mary, Devonshire. (*Postmark* Derby.)

Original Keele University Library, Wedgwood Archive, 26770–35.
Printed Excerpts in *Doctor of Revolution*, pp. 247–8.
Text From original MS.
Notes
(1) This is the most disastrous of Darwin's letters. After trying many medications on Tom Wedgwood, who was now 23, Darwin here decided to recommend opium – 'a grain of opium taken every night for many months'. Tom never recovered from the prescription, and remained an addict until his death at the age of 34. The companion whose cheerfulness helped Tom was his great friend Coleridge – Tom was staying with the Coleridges at Ottery St Mary, as the address shows. Coleridge was 22 and, like Tom, a martyr to obscure illness, probably psychosomatic. Here was a prescription specially sent by the doctor recognized as the finest physician of his time, whose *Zoonomia* had just been published – a medical masterwork by general consent (and a source of many of Coleridge's ideas). At this time Coleridge had a high opinion of Darwin, and said he had 'a greater range of knowledge than any other man in Europe'. There seems no reason to suppose that Coleridge would have resisted the temptation to try Darwin's prescription. So Darwin was probably responsible for his addiction to opium.
(2) In the second paragraph, Darwin recognizes the parallel between the lymph glands of the mouth (the tonsils) and of the arms (the axillary glands).
(3) The prescription of a little trouble to tease him is a repeat of letter 94B.
(4) The cicuta (hemlock) was a popular medicine in the eighteenth century, especially after the work of Anton Stœrck and John Hunter.
(5) Half way through, the letter changes from traumatic tragedy to pneumatic comedy, from Dr Hyde to Mr Jekyll, as it were. Darwin gently deflates Tom's idea of an air bed and retaliates with the idea of a mercury bed: this was no more than a joke, for (although mercury was used to treat syphilis) such a bed would weigh more than a ton. See also letter 94Q.

94P. To James Watt, 17 August 1794

Derby Augt. 17 –94

Dear Sir,

I should have written to you sooner to thank you for your magnificent apparatus, but knew, that you was gone from home; and as my elaboratory was so out of order, that I was obliged to have a new floor to it, I did not open the apparatus, till yesterday.

Now, as old Nick has much to do in the affairs of this world, I imagine himself or some of his imps have stolen the penult. letter you favor'd me with; or, what is more probable, I have laid [it] in some safe place, where I can not now find it. Nevertheless I remember'd it so well as to understand the whole; and think both the joints, and the manner of dropping the water very ingenious, and truly Wattean; but there are two or three things I shall remark. 1. there is no perforation between the horizontal and vertical part of the piece immediately connected with the iron-caldron, there is no hole between a and b, which as I suppose to have been a neglect, I shall get

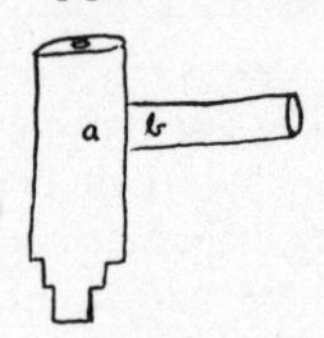

made. 2. should there not have been a means of taking airs out of the water-bellows, without disjointing them from the furnace? Now I think it would be convenient to have another pipe to communicate with the air-bellows for the purpose of taking out the air without breaking the joint at xy.

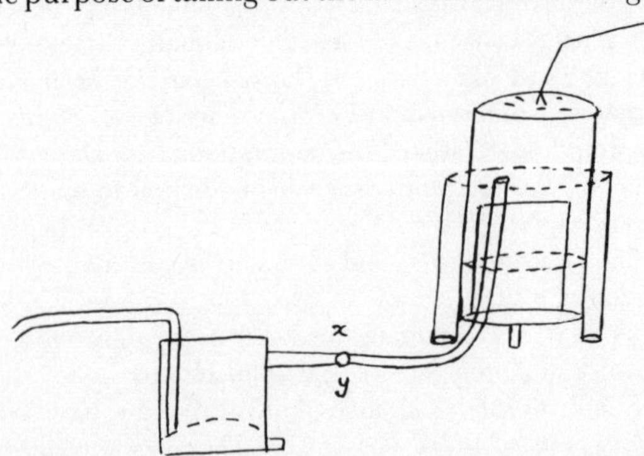

3dly I think a clear pamphlet should be publish'd of directions how to make airs, and in what to recieve them, etc, for I now know how to make the carbonic hydrogen-gas, which you describe as being so sickening, but am at a loss to how to get other airs.

From what material has your Birmingham philosopher obtain'd oxygen, cheaper than from manganese?

I hope you will proceed to supply the world with a new Materia Medica, to be drank in by the lungs. And shall be very impatient to recieve the additional part of your apparatus for oxygene gas. The easy production of azote to mix with atmospheric air would be a great acquisition for consumptive people, if it can be done, so as to blacken the blood a little more, as you observe, instead of healing ulcers by inducing sickness.

Along with the apparatus you have been so kind to send me, there is an iron pot like this, of which I do not know the use.

I have invited all our Derby Philosophers to tea this afternoon, Sunday, to see your apparatus, and I hear a knocking at the door.

Adieu

Mrs Darwin joins in best respects to Mrs Watt with, dear Sir,

Your affect. friend
and obed. serv.
E Darwin

Addressed James Watt Esqr., Heathfield, near Birmingham. (*Postmark* Derby.)

Original Privately owned.
Printed Unpublished.
Text From original MS. (*Diagrams* Redrawn.)
Notes

(1) Watt took up Darwin's suggestion of 'a clear pamphlet [on] how to make airs' and his 'Description of Pneumatic Apparatus' appeared a few months later as Part II of the 210-page book *Considerations on the Medicinal Use, and on the Production of Factitious Airs* by T. Beddoes and J. Watt, from which the diagram below is taken.

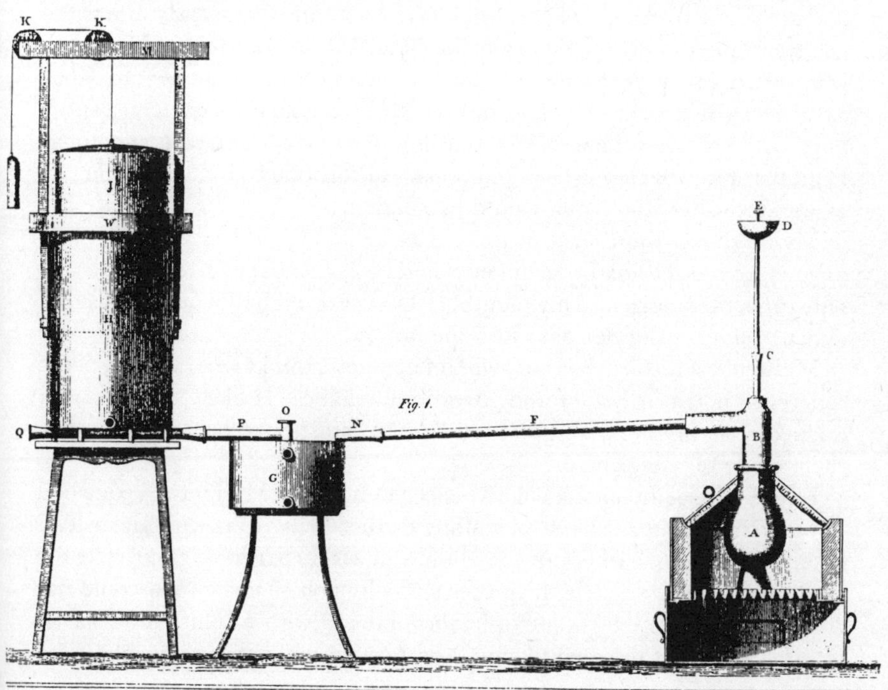

Fig. 1.

In this diagram the alembic A contains a substance to be heated by the furnace beneath it, while DC is a pipe to admit water to A. The cup D is filled with water and E is a screw terminating at C, so that, by rotating E, water can be admitted to A as necessary. G is a refrigeratory to cool and wash the gas. The gas is received and measured in the hydraulic bellows HJ. Watt also added an air-holder to be attached at Q, as suggested by Darwin (see Plate 3 of Watt's 'Description').

(2) The 'carbonic hydrogen-gas' mentioned by Darwin, also called 'carbonated hydrogen' and 'hydro-carbonate', was obtained by heating charcoal red-hot in the alembic and cautiously dropping water on it. The gas produced would presumably have been a mixture of steam, hydrogen, carbon dioxide and carbon monoxide, and Watt admits that it has 'generally a disagreeable smell'. ('Description', p. 34).

(3) Oxygen was made from manganese dioxide, in the form of the mineral pyrolusite, either by reaction with concentrated sulphuric acid (oil of vitriol) or by strong heating in the iron pot Darwin mentions (without addition of water). Watt preferred the second method, and he recommended using the manganese ore found near Exeter, rather than the Mendip manganese, which 'contains much calcareous earth' as impurity ('Description', p. 29).

94Q. To James Watt, 1 September 1794

Derby Sept. 1 –94

My Dear friend

I have been so perpetually from home, that I could not sooner answer the favor of your two last kind letters, nor have I now any thing to say, but to tell you how much I am obliged to you for them. The Hymen of iron I have not yet procured to be perforated but intend it with the first days leizure. Your carbonated hydrogen seems to bid fair to be a valuable medicine, and I hope with you, that it may act by purpling the too scarlet blood, and not as Digitalis does; which produces pulmonary absorption by inducing sickness, as the author of Zoonomia would persuade us.

About giving airs I suspect there may be fallacy, unless the patient be nicely watch'd. I think I remember, when I was a smoaker of tobacco, that I only drew the smoke into my mouth, as I never could puff it out at my nose, and therefore it did not pass into the lungs.

Mr Strutt, who bleaches here with manganese, thinks what you say of the difference between Exeter and Mendip manganese is likely to have been mistaken; as he thinks so great a difference must have been percieved in their mode of using them.

I have not got an oil-silk bag to recieve these airs – Mr Thos. Wedgwood wrote to me about a scheme of stuffing feather-bed-covers with air instead of hair – he thinks feathers always stink, and wishes to rest on clouds like the Gods and Goddesses, which you see sprawling on cielings. Now could not (seriously) a pillow be made of leather imbued with a solution of Elastic gum? – so as not to smell like oil, and to hold air, either atmospheric or carbonate? Thus one might carry one's bed about, and blow it up at night – or blow yourself up as you lye on it, like the philosoph bellows.

Mrs Darwin joins with me in our most respectful compliments to Mrs Watt,

Your affect. friend E Darwin

I shall wait for Dr Beddoes' next book, I think, before I carbonate or hydrogenate, or azotate, or oxygenate any one.

When matter is form'd, as suppose between the muscles of any limb, no hectic fever is produced, tho' the abscess may contain a pint. Now if this absess be open'd by a knife so as to admit the air to the surface of the wound, the hectic fever comes on in a few hours! – this I suppose to be owing to the pus attracting oxygene, as the blood does; and that this oxygene (if the matter be ever absorbed, which is not certain) may produce fever by its stimulus, hence carbonate hydrogen might be of gre{at} service (perhaps) applied to some wounds, and Hydrogen to tendencies to gangreen, which might be done by tying a bladder round the ulcer or over it, and blowing the air under.

Addressed James Watt Esqr., Heathfield, near Birmingham. (*Postmark* Derby.)

Original Privately owned.
Printed Unpublished.
Text From original MS.
Notes
(1) The interchange of ideas and apparatus for pneumatic medicine is flourishing: Watt had evidently written twice in the previous two weeks.
(2) The possibility that the patients might not inhale deeply enough was an important problem in pneumatic medicine. Darwin was presumably a smoker when he was at University: this is the only reference I have seen to his smoking.
(3) Mr Strutt is probably Jedediah Strutt (1726–1797), the pioneer textile manufacturer and friend of Arkwright (Strutt's mill was at Belper, 7 miles north of Derby); or possibly his eldest son William (1756–1830), who was Darwin's closest friend in Derby (see letter 01E). See R. S. Fitton & A. P. Wadsworth, *The Strutts and the Arkwrights, 1758–1830* (Manchester University Press, 1958).
(4) For Tom Wedgwood's pneumatic bed and pillow, see letter 94N.
(5) In the postscript Darwin is becoming over-enthusiastic about pneumatic medicine and is in danger of treating gases as the panacea for all ills.

94R. To Richard Polwhele. 21 September 1794

Sept. 21 –94

My Dear Sir,

The Stanza of Spencer succeeded in Shenstone's School Mistress, but I am not certain whether it is so well adapted to long works; as, like

Sonnets, the rhyme becomes of too much importance, as the ear is obliged to attend to it. It has, however, countervailing advantages; I mean, that a weak line is excused, as it seems necessary for the filling up of the stanza.

The Poem you have favoured me with is beautifully natural, and elegantly descriptive; and I think does better in your kind of stanza, than it would in common heroic measure; yet I doubt the effect of the stanza in a very long work. The word leaden pound for pressing cyder, I should object to; as lead is believed to render cyder not only insalubrious, but even poisonous.

Yours etc [?]
E D

Original Not traced.
Printed R. Polwhele, *Traditions and Recollections*. 2 vols. Nichols, London, 1826. Vol. I, p. 396.
Text From above source.
Notes
(1) For Polwhele, see letter 92D.
(2) The Spenserian stanza was out of fashion in the eighteenth century, but William Shenstone (1714–1763) used it in his poem 'The School Mistress' (1742), which runs to 35 stanzas. Darwin was wedded to the couplet, and his lack of enthusiasm for the Spenserian stanza is not surprising. Since *The Faerie Queene* is extremely long, Darwin's objection to a long poem in the Spenserian stanza presumably arose because he thought Polwhele could not bring it off.
(3) 'The poem you have favoured me with' may have been a manuscript copy of Polwhele's *Influence of Local Attachment*, published in 1796.
(4) The discovery that leaden cider-presses caused lead poisoning – 'the Devonshire colic' – was made by Sir George Baker (see letter 87B).
(5) 'Yours etc' is a form not found in any MS letter of Darwin, and I doubt its authenticity.

94S. To Matthew Boulton, 8 October 1794

Derby Oct. 8 –94

Dear Boulton

I remember you mention'd, that you procured aerated alcaline water either in London or Birmingham, which was better than you could easily make at home. Now I have a patient, who has I think a small stone in his bladder, left after having frequently made bloody urine, and I am giving him aerated water, with half an ounce of sal soda to the pint – but could wish to have it better impregnated with the carbonic acid gas.

Pray send me a line of direction, where it can be had – and any other observations, you may have made on the subject. I hope you are quite well of your own gravelly complaint[;] for my part, I grow old – that is a disease your juvenile spirits will never submit to. Adieu

with compliments to your family

from yours affectly

E Darwin

Addressed Matt. Boulton Esqr., Soho, near Birmingham. (*Postmark* Derby.)

Original Birmingham Reference Library, Matthew Boulton Papers, Darwin 42.
Printed Unpublished.
Text From original MS.
Notes

(1) Boulton replied on 12 October, giving the supplier of 'the aerated alcaline water' as Mr J. Schweppe, 141 Drury Lane. He supplies three types, Boulton says, advising Darwin to prescribe water with 600 grains of dry Salt of Tartar to the gallon, or 960 grains of Sal soda to the gallon: 'I am persuaded it has been of great use to me'. He then discusses his 'gravelly complaint', probably kidney stones. He cannot allow himself to grow old, he says, because he has too much to do: 'I am kept up like a top, by constant whipping'. Boulton concludes with characteristic courtesy, 'by assuring you that there is no good, pleasant, or desireable, thing in this World but what I wish you Mrs Darwin and all your Olive Branches the full possession of. Remaining always with sincere regard, Dear Doctor, Yours affect.ly, Matthew Boulton' (Birmingham Reference Library, Matthew Boulton Papers, Darwin 43).

(2) Jacob Schweppe (1739?–1821), born in Germany, emigrated to Geneva and in 1780 began to develop new methods for manufacturing mineral waters highly charged with carbon dioxide. He came to England in 1792 and began manufacturing in London. Boulton's letter is the earliest reference so far found to the factory at 141 Drury Lane. Schweppe retired to Geneva in 1798, but his name lives on. I am indebted to Mr D. A. Simmons of Schweppes Ltd. for this information.

(3) This letter shows Darwin again taking a good deal of trouble to secure the best treatment for a patient. In *Zoonomia* (ii 45) he recommends aerated alcaline water as a treatment for bladder stone and says it 'is sold under the name of factitious Seltzer water, by J. Schweppe, at No. 8 King's street, Holborn, London'.

94T. To The Duke of Devonshire, 10 October 1794

Derby Oct. 10 –94.

My Lord Duke,

After fully considering your Grace's situation, I can have no doubt, but that the small oozing of blood is from the lungs, either from the pulmonary or bronchial vessels. This blood, if it lies for half an hour unexposed to the air in the cells of the bronchia, becomes dark colour'd; like the blood on the underside of the cake of blood drawn into a bason. Otherwise it is generally florid, when spit from the lungs.

The reason, why these oozings of blood most frequently occur in sleep, is, because the irritability of the lungs is not always then sufficient to carry on the circulation without the assistance of volition, which power is suspended during sleep, as is explain'd in the volume of Zoonomia, which I have already published, in page 294, about the middle.

These oozings of blood are dangerous in young men, between 17 and 27, but frequently occur in gouty habits, or in those who have slight bilious or hepatic complaints, as is the case of your Grace, without danger.

The immediate cause is too profound sleep – but the remote cause is to be saught for in the hepatic complaint, as those patients are liable to other haemorrhages, as from the nose, or from piles.

A cathartic with a few grains of calomel should be taken at night once a fortnight; which is the most efficacious means of removing those hepatic obstructions, which are the remote cause of haemorrhage; and by its being repeated at regular times a new habit will be acquired, which will probably break the bad habit, which your constitution has already acquired.

Besides this cathartic, a draught of the bark with acid of vitriol may be taken twice a day for a fortnight, and repeated occasionally.

To prevent too sound sleep, your Grace should avoid all violent exercise; and should be cautious in the use of much supper, or much wine at night. And as great uniformity, as may be, both in your hours of meals; and especially in the quantity of beer or wine, which you daily drink, should be observed. As nothing is more injurious, than to live high one day and low another[;] that is, to use great stimulus of wine or beer one day, and little another. Another circumstance to prevent too profound sleep consists in lying on a bed rather harder than usual, as on a mattress over the feather-bed.

I do not advise your Grace to live too low, or to abstain from a proper quantity of vinous potation, which must be regulated by every ones previously acquired habits. I only strenuously advise you to take the same quantity of wine or beer every day. Which will contribute long to preserve your health.

The part your Grace saw of my new volume is inclosed, and a prescription, and if any further observations are required, I hope you will honor me with a letter.

I am, with the greatest respect, my Lord Duke,

Your much obliged and obed. serv.

E Darwin

The noise in the ears and starting on going to sleep may depend on the wax or mucus in the internal ear being too viscid, and perhaps obstructing the Eustachian tube on one ear – this is more probable as your Grace complains of noise in swallowing your saliva, and the Eustachian tubes open from the ear into the upper part of the fauces.

A teaspoonful of warm water or a few drops of Ether into each ear on going to bed might prevent this symptom.

Addressed [No address – probably sent by hand.]

Original Chatsworth House, Devonshire Collection, No. 1257.
Printed Unpublished.
Text From original MS.
Notes

(1) This letter is a response to the following letter from the Duke to Darwin, dated 7 October 1794:

'Within the last five years I have sometimes had a shortness of breath without any pain, except the disagreeable sensation of not being able to draw my breath as high as I wish'd. It generally lasted about three weeks, and was continual, except when lying down, and then I was perfectly free from it. I did not perceive that I was releiv'd much by medicines, but it left me by degrees, without my knowing why. I have sometimes had a shortness of breath of a different kind, which was occasion'd by a violent pain at, or a little above the pit of my stomach, which pain generally came suddenly, and was allways entirely remov'd by physick, as soon as the physick began to operate.

About the middle of last February, I one day rode twenty miles pretty fast, and was very well that night and the next day, but the night following I was wak'd by a taste of blood in my mouth, of which I spit about as much as would fill a small tea spoon.

Since that time I have been subject to the same thing, whenever I take a little more exercise than usual, and it has allways happened either during the night or upon first waking in the morning. I have generally a pain which goes through me from my chest to my back, but never any cough.'

(Chatsworth House, Devonshire Collection, No. 1256. From original MS. No address or signature.)

(2) Presumably Darwin had attended the Duke on several other occasions since his previous letter in 1783 (83D), when he had tactfully recommended a reduction in alcohol consumption, to avert hepatic disease. Now he regards the liver dysfunction as chronic, and is concerned to reduce its effects. The extra paragraph on noise in the ears is probably in response to a further note sent by the Duke.

(3) Glossary: the bronchia are the small air passages in the lung; blood was drawn into a bason [basin] when 'bleeding' a patient; Darwin's reference to *Zoonomia* should be to Volume I, page 204, not page 294; the Peruvian bark, cinchona, was a popular eighteenth-century medicine; acid of vitriol is sulphuric acid; 'my new volume' is Volume II of *Zoonomia*, published in 1796; 'the fauces' refers to the cavity connecting the mouth with the pharynx.

94U. To JAMES WATT, 17 November 1794

Dear Sir

I am much obliged to you for both your letters and have this day sent your advertisement and air-paragraph to the newspaper; I sent a shorter one of my own abstracted from Dr Beddoes's the week before, but I think no subscriptions can be got but by personal application. Already I have the following names

Dr Milnes of Chesterfield, Derbyshire	10–10–0
Dr Darwin of Derby	5–5–0
Dr French of Derby	2–2–0
Wm. Strutt Esqr. senr Derby	2–2–0
" " " jnr Derby	2–2–0
Joseph Strutt Esq. Derby	2–2–0
Dr Crumpton Derby	2–2–0
Sam. Fox Esqr.	1–1–0
Mr Liptrot	1–1–0
Mr E. Darwin	1–1–0

I hope to get a few more, but none by advertising I am affraid. In another week I think to insert these names in the Derby paper, and may add your Birmingham list, if you think it will give consequence to it.

Your observations are all ingenious, and curious, I have not time nor activity enough for experiments and am busied in writing my second vol. of Zoonomia, which absorbs all my leisure hours. The Pthisis from the lungs is generally distinguishable from that from suppurated glands of the mesentery by the cough and expectoration, and other symptoms as pain in the bowels. I am [of] opinion, that oxygene united with pus produces a tertium quid, an acid, which causes hectic fever, which fever does not generally exist till the matter is exposed to air. This union of oxygene with other matters, as of the smallpox, great p-x, itch, tinea capitis etc produces other infectious matters, which induce other fevers, or ulcers; but I believe them all to be not fever-producing, till they have been oxygenated. See Zoonomia Vol II not yet publish'd.

Hence cancer is disarm'd by keeping of[f] oxygene, not by any specific virtue in carbonic acid gas. In pulmonary consumption if the air of a room could be lower'd of oxygene, so as to render the expectorated matter mild, and yet the patient not die from want of oxygene, which if he used no exercise, perhaps he might not require much oxygene? Then there might be hopes of these ulcers healing.

Many experiments are wanted, and thence the propriety of a pneumatic institution.

I am call'd away in proper time, not having any thing further to say worthy your attention, except that Mrs Darwin joins with me in our most respectful compliments and best wishes to yourself and family

from your affect. friend

Derby E Darwin

Nov. 17 –94

Addressed James Watt Esqr., Heathfield, near Birmingham. (*Postmark* Derby.)

Original Privately owned.
Printed Unpublished.
Text From original MS.
Notes

(1) Beddoes was now trying to raise money to set up his Pneumatic Institution, where the idea of curing disease by administering gases to patients could be fully tested. He received generous help from Watt, Edgeworth, Tom Wedgwood and Darwin. This letter shows the outcome of Darwin's attempt to raise money in Derby; Watt was doing the same in Birmingham. The full list of subscribers is given in Part III of *The Medicinal Use . . . of Factitious Airs* (1795), pp. 111–2. The Pneumatic Institution was eventually established at Clifton in 1799.

(2) Most of Darwin's subscribers were acquaintances. Dr Richard Milnes (1726–1795) was a JP and a leading doctor in Chesterfield, from the same family as Esther Milnes, who married Thomas Day. See Glover, *Derby* ii 286–7 for the Milnes pedigree.

(3) 'Dr French' is Dr Richard Forester French (1771–1843), who helped Darwin in his experiments on air: see *Phil. Trans.*, **78**, 43–52 (1788) and letter 87I. Forester French was the son of Richard French (see letter 86E). At about this time Dr French became known as Dr Forester, having assumed the name and arms of his maternal grand-uncle William Forester (1715–1768). The French and Forester pedigrees are given by Glover, *Derby*, ii 591.

(4) William Strutt senior (1730–1800) had worked with his elder brother Jedediah (see letter 94Q) in the family textile business since 1762. William Strutt junior (1756–1830), Jedediah's eldest son, was Darwin's closest friend in Derby (see letter 01E). Joseph Strutt (1765–1844) was Jedediah's third son (see also letter 00J). A Strutt family tree is given by R. S. Fitton & A. P. Wadsworth, *The Strutts and the Arkwrights, 1758–1830* (Manchester University Press, 1958), p. 326.

(5) 'Dr Crumpton' is Peter Crompton (1764?–1833), a member of another prominent Derby family. He took his MD at Leyden in 1785, was a member of the Derby Philosophical Society and befriended Coleridge in 1796. In 1798 he left Derby for Liverpool, where he was a strong advocate of Parliamentary reform: see J. A. Picton, *Memorials of Liverpool* (1873) i 385, 418–22. See also next letter.

(6) Samuel Fox (1765–1851) came from another leading Derby family, much intermarried with the Cromptons. Samuel Fox himself, however, married first Jedediah Strutt's daughter Martha (1760–1793) and, second, Ann Darwin (1777–1859), daughter of Erasmus's brother William. Fox also belonged to the Derby Philosophical Society and helped Darwin in some of his experiments on air: see *Phil. Trans.*, **78**, 43–52 (1788). See Glover, *Derby* ii 578–9 and 583 for Crompton and Fox pedigrees.

(7) 'Mr Liptrot' is Isaac Liptrott (born 1771?), son of Rev. Isaac Liptrott (1729–1792), Vicar of Oadby, Leicestershire. Isaac the younger entered Emmanuel College, Cambridge in 1789, and graduated BA in 1793 and MB in 1796; he is probably the Dr Liptrott who married the daughter of John Newton at Bakewell in 1797. See *Alum. Cantab.* iv 177; Nichols, *Leicester* iv 324; and *Gent. Mag.*, **67**, 1128 (1797).

(8) Mr E. Darwin is of course Darwin's son Erasmus.

(9) Darwin is still pursuing his fruitless idea that suppurating matter causes fever only when exposed to oxygen.

(10) The mesentery is the fold of the peritoneum suspending the loops of the intestines, and has lymph nodes or 'glands', which would be much swollen in peritoneal tuberculosis.

(11) Glossary. Tertium quid, third thing. Great pox, syphilis. 'The itch', the modern scabies, is described by Darwin as 'a contagious prurient eruption' often between the fingers or behind the knees (see *Zoonomia* ii 274–5). Tinea capitis is 'a contagious eruption' affecting the roots of the hair, which rouses the lymphatic glands of the neck (see *Zoonomia* ii 279–80).

(12) Darwin regarded cancer as ulcerous, to be treated, like consumption, by cutting down the oxygen supply.

94V. To James Watt, 30 November 1794

Nov. 30 –94
Derby

Dear Sir,

I put into the Derby paper the account of the medical pneumatic institution, and the subscribers copied from the Birmingham paper, but have gain'd no subscribers except those personally applied to – which I indeed suspected. The following is the list of them.

Dr Milnes of Chesterfield	10–10–0
Messrs Strutts ________	10–10–0
Dr Darwin ________	5–5–0
Dr French ________	2–2–0
Dr Crumpton ________	2–2–0
Sam. Fox Esqr ________	1–1–0
Isaac Liptrot Esqr ________	1–1–0
Mr Darwin ________	1–1–0

I also inserted the paragraph you favor'd me with, and another from Dr Beddoes, and therefore think it not worth advertising any more. The money is in the hands of our bankers Crumpton and Newton except two guineas, which are not yet paid.

I was sorry my other engagements would not permit me to pass a few hours with you, when I was lately in Birmingham, but the short days and moon-less nights prevented me.

I have been so much engaged in common business, and in finishing the second vol. of my book, that I have not thought much about airs.

Mrs Darwin begs to be kindly remembered to Mrs Watt and yourself with, dear Sir,

your much obliged friend
and serv.
E Darwin

Addressed James Watt Esq, Heathfield, near Birmingham. (*Postmark* Derby.)

Original Birmingham Reference Library, Boulton and Watt Collection, Box 4D.
Printed Short excerpt in *Doctor of Revolution*, p. 251.
Text From original MS.
Notes
(1) The list of subscribers is nearly the same as in the previous letter, but the amount from the Strutts has increased from 6 to 10 guineas, probably through contributions by Jedediah Strutt (see letter 94Q) and his second son George (1761–1841).
(2) John Crompton, Mayor of Derby in 1792, was a banker and receiver-general for the county; his brother Samuel Crompton (1749?–1810) was also a banker. They were cousins of Dr Peter Crompton. See Glover, *Derby* ii 578–9, and *Universal British Directory* (1791).
(3) The 'Newton' in Darwin's 'our bankers Crumpton and Newton' may be John Leaper (1754–1819), a member of the Derby Philosophical Society who took the surname Newton in 1790. See Glover, *Derby* ii 592.
(4) For the other names, see notes to letter 94U.

94W. To Josiah Wedgwood, 9 December 1794

* * *

Your letter gives me great pleasure in assuring me, what your son Josiah had before mention'd, that you have become free from your complaint – the ceasing of the palpitation of your heart and of the intermission of your pulse is another proof of your increase of strength. In respect to your breath being less free in walking up hill I ascribe to the distant approach of age and not to Asthma. You know how unwilling we all are to grow old. As you are <u>so</u> well, I advise you to leave off the bark and to take no medicine at present.

* * *

Original Not traced. In the possession of Charles Darwin, 1865.
Printed Meteyard, *Wedgwood* ii 610.
Text From above source.
Notes
Wedgwood became ill in the autumn of 1794, with the (for him) familiar symptoms of pain in the right jaw and in what Darwin called his 'no leg', shortness of breath and general debility. He went to Buxton for some time and wrote to Darwin early in December with optimism. Darwin was happy to be able to

respond to the good news with this cheerful letter. But the improvement was illusory. Within a few days Wedgwood's mouth and throat were showing signs of gangrene. Darwin hurried over from Derby as soon as he heard the news, but there was nothing that he or any of the other doctors could do. Wedgwood suffered greatly from pain and fever, and eventually died on 3 January 1795, aged 64. So Darwin lost his best friend, and England lost the man whose achievement in simultaneously advancing art, science and industry has never since been equalled.

94X. To [Elizabeth Evans], 11 December 1794

Dear Madam,

I have sent the books and shall be glad of any observations on them; which you think may be serviceable to mention in my book, etc.

Mrs Darwin begs you will permit her to put another little girl besides Violetta into your chaise on Wednesday to bring from Ashborne along with Miss Evans, and our chaise is to bring all the rest, viz. Miss M. Parker two Miss Bolds and one Miss Madden.

Any thing you think objectionable in the book, pray put the page at the end of the book to refer to it.

Adieu
Your much obliged friend
E Darwin

Dec. 11 –94
Our united good wishes and compl. attend yourself and Mr Evans.

Original Bodleian Library, Oxford, MS Autogr. C.24 fol. 258.
Printed Unpublished.
Text From original MS.
Notes

(1) Mrs Elizabeth Evans (1758–1836), the eldest daughter of Jedediah Strutt, was married in 1785 to William Evans (1754–1796), eldest son of Thomas Evans (see letter 82E), and 'Miss Evans' is their eldest daughter Elizabeth (Bessy) (1786–1880), who was eight years old. Mrs Evans's husband died in March 1796, and in the summer of that year S. T. Coleridge came to stay for five weeks, with the idea of acting as tutor to her children. The scheme fell through, but not before Elizabeth Evans had deeply impressed Coleridge: 'She is without exception the greatest WOMAN, I have been fortunate enough to meet with in my brief pilgrimage through life' (Coleridge, *Letters* (OUP, 1956) i 306). In 1797 she married Walter Evans, half-brother of her first husband. See notes to letters 82E and 94U, and Glover, *Derby* ii 18 and ii 573. For Elizabeth Evans, see Coleridge, *Letters* i 227–34, and Derby Central Library MS 3524.

(2) Presumably Darwin sent Mrs Evans a draft of his book *Plan for the Conduct of Female Education in Boarding Schools*, which was published in 1797, and also copies of some of the books recommended there for further reading (pp. 119–26). Darwin

says that these books 'were recommended to me by ladies whose opinions I had reason to regard'.

(3) Darwin's two daughters Susan and Mary Parker started their school at Ashbourne in 1794 (see *Doctor of Revolution*, pp. 234–7), and the pupils were about to return home for the Christmas holiday. Among them were Darwin's daughter Violetta, now 11, Bessy Evans, the 'two Miss Bolds and one Miss Madden'.

94Y. To Thomas Garnett, 14 December 1794

Derby Dec. 14 –94

Dear Sir,

I am much obliged to you for your communication respecting your table of weather, and, having thought over the subject, am totally at a loss to recommend any particular deductions to be made from them, or any manner of disposing them. I should think it would be best to publish the tables themselves in the manner they were made; so that every one may collate from them, what observations he is in want of to confirm or confute his own theory.

For my own part the observations I should most want would be the succession of winds, from day to day, as which wind preceeded a given wind, and which follow'd it; which could only be learnt but from diurnal tables. Other people may wish for other deductions.

The tables should be printed so as to take up little room as may be, and to fall readily under the eye.

I have not look'd into Dalton or Adams, I suppose my theory of the winds would be too bold for their use. The English Review, as well as the analytical, have criticis'd my Zoonomia with much attention. I shall hope to see your work on herpetic eruptions soon.

The South-East winds prevailing so much at Liverpool, I should suspect must depend on some atmospheric eddy, produced by the situation of the place. I speak this quite from conjecture – as the NW winds are the freezing winds on the eastern coast of north America, and are therefore evidently produced by an atmospheric eddy.

Something might be learnt, if the weather tables in distant parts of Europe were collected and compared. It has been said by somebody, that the barometer at Dublin and Paris and London sinks and rises at the same hour, and about the same proportion or quantity; this is very curious, and would be more so, if the places were still more distant.

I am, Sir with great respect,
your obed. serv.
E Darwin

I don't concieve that the coldness or warmth of our summers any way affect that of our winters; since we had very lately a cold wet summer with a mild succeeding winter.

The knowlege of the winds, their origin or cause, is the principal source (I mean the cause) not the consequence of all the other atmospheric phenomena, in my opinion. All the winds of the N.E. come directly from snowy countries, and as the snow is evaporated by them, great cold is produced; first by the thawing of the snow into water, and then by evaporating the water; which I suppose to be done in one process by the air. Then the S.E. winds are superior currents of N.E. winds driven back. These I esteem to be the source of frost in this country, and how these winds are produced or generated for six weeks together seems to me to be the great desideratum, as I have endeavour'd to shew in the Note in Bot. Garden.

Experiments on freezing and thawing in a perfect vacuum might give light to this subject; as I suspect by the great expansion of ice, that air must be generated in the act of freezing and given up in thawing?

I am so engaged in finishing my second Vol. of Zoonomia that I can not at present attend to this subject. A bit of ice might easily be dissolved in the Torricellian vacuum, to ascertain whether it parted with air in thawing.

I do not give credit [to] a fact mention'd in your book, that the cuticle has more pores than are adapted to the cutaneous secerning and absorbing vessels, as it seems uniformly attached to the cutis.

Addressed Doctor Garnett, Wedderburne House, Harrogate, Yorkshire.

Original American Philosophical Society, Philadelphia, Pa, Misc. MSS.
Printed Excerpts in *Memoirs of the Literary and Philosophical Society of Manchester*, Vol. 4, part 2, pp. 616–7 (1796).
Text From original MS.
Notes
(1) Dr Thomas Garnett (1766–1802) was a physician who practised in Yorkshire, and a talented scientist. In his 124-page paper entitled 'Meteorological Observations' (*Mem. Lit. Phil. Soc. Manchester*, **4** (2), 517–640 (1796)), he praises Darwin's account of the winds in *The Botanic Garden*, from which he quotes at length. He also quotes part of this letter. Garnett was well known for making the first proper analysis of the Harrogate waters, and also as a lecturer. He was professor at the Royal Institution from 1799 to 1801, being succeeded by Humphrey Davy. He then started a medical practice in London, but died of typhus at the age of 36. For further details, see *DNB*.
(2) 'Dalton' is John Dalton (1766–1844), pioneer of the atomic theory, whose *Meteorological Observations and Essays* were published in 1793. 'Adams' is George Adams (1750–1795), a prolific writer on scientific subjects (see *DNB*). Darwin is probably referring to his *Lectures on Natural and Experimental Philosophy* (5 vols., London, 1794), which include 100 pages on meteorology (Vol. IV, pp. 471–570).
(3) *Zoonomia* received some splendid reviews, particularly that in the *Analytical Review*, which ran to 50 pages (**19**, 225–34, 337–50, 450–63; and **20**, 225–37 (July–November 1794)) and described *Zoonomia* as one of the most original productions ever delivered from the press. For discussion of the reviews, see N. Garfinkle, *Journ. Hist. Ideas*, **16**, 376–88 (1955).

(4) Darwin made many important contributions to meteorology. Besides his fundamental discovery of how clouds form, he noted the importance of cold and warm fronts and the sudden wind changes as they pass, and he devised several meteorological instruments: see *Weather*, **28**, 240–50 (1973). Darwin's view that the winds are of crucial significance has much to commend it. Today we tend to think of winds as arising from pressure systems, but the motions of the pressure centres themselves are influenced by winds, particularly those higher in the atmosphere, so the debate is still open.

(5) Darwin's suggestion of collecting weather data at the same time from all over Europe is the germ of the synoptic weather map so familiar today.

(6) The 'mild . . . winter' was the preceding one: February, March and April 1794 were all exceptionally warm; curiously enough, the next month, January 1795, was the coldest month in central England between 1700 and 1973. See G. Manley, *Quart. Journ. Roy. Meteorological Soc.*, **100**, 389–405 (1974).

(7) Darwin's explanation of the chilling of winds by the melting and evaporation of snow is admirably clear and correct.

(8) The experiment to test his speculation that ice might part with air in thawing was taken up by Garnett, who adds a long note at the end of the MS, recording in detail how, on 17 December 1794, he performed the experiment suggested by Darwin of melting ice in the vacuum above the mercury in a barometer. 'When the tube was inclined, the water and mercury entirely filled it, a proof that no air was extricated during the thawing of ice'. He then made the experiment on a larger scale, melting ice within a tall glass jar of water, and again disproving Darwin's conjecture.

94Z. To Thomas Beddoes [late 1794 or early 1795]

* * *

My dyspepsia was not attended with much flatulence nor heartburn, but was very troublesome after eating any strong dish, such as goose, garlic, or cabbage, from a rising of rancid matter from the stomach, perhaps every 5 minutes. This was always <u>immediately</u> checked by a table spoonful of very fine ground charcoal – so much so that the next eructation would be scarcely offensive; and in a little time the stomach was completely set to rights. Several persons in our family have recieved benefit from it in the same way.

Having had no ailment in the stomach for a long time, I cannot say that I have had <u>much experience</u>. Perhaps I may have been relieved a dozen times, and I think never took it without a very sensible effect. I do not believe it has much effect on the bowels; it is aperient, however, rather than otherwise. As to your question of prevention of wind, mine was so little a case of flatulency that I cannot speak very positively of its virtue in that particular. It certainly, however, had this effect to a certain degree. Upon the whole, I have not the smallest doubt of it being a very useful family medecine.

Addressed To Dr Beddoes.

Original Not traced.
Printed T. Beddoes and J. Watt, *Considerations on the Medicinal Use and on the Production of Factitious Airs*. Second edition. Bristol: for J. Johnson, London; 1795. Part I, pages 144–5.
Text From above source.
Notes
(1) The first edition of the Beddoes and Watt pamphlet appeared towards the end of 1794; the second edition, with letters from Darwin and other doctors (supposedly in response to the first edition), has a preface dated 30 March 1795. So January–February 1795 may seem the most likely date. However, since Darwin's remarks seem to have little relevance to the pamphlet, this may be an excerpt from an earlier letter.
(2) Beddoes introduces the letter by explaining that Darwin had devised a 'powder-puffer', a box with bellows which allowed any powder placed within the box to be spurted through a mouthpiece and inhaled by the patient, so that the powder, 'supposed capable of a salutary action' as Beddoes puts it, might act on the lungs (see *Zoonomia* ii 290). One of the substances tried was powdered charcoal, and the mention of charcoal is for Beddoes excuse enough to quote what he calls an 'Observation on the effect of charcoal in correcting rancid eructations'. Darwin's observation was correct, in that powdered charcoal is one of the most efficient absorbers of chemically active gases, and was a vital component in the gas masks carried by everyone in Britain in the second World War.
(3) The printed text ends with 'I am, your's etc', but I omit this because it seems unlikely that Darwin would have written it. Cf. the endings of letters 94C, 94K and 94L.

95A. To William Hayley, 21 January 1795

Jan 21 –95
Derby

Dear Sir,

I have sent your letter to me to Josiah Wedgwood, who resides at Etruria; but I do not know of any intentions about a monument, as I have not seen any of the family since the death of Mr Wedgwood; which I esteem to be a public loss, as well as to me a private one!

The poverty occasion'd, and about to be occasion'd, by this destructive and diabolical war, will depress all the fine arts for a great length of time in this country. – Ah, poor old England!

Is your muse asleep? I have heard nothing lately from either yourself, or Cowper – the press will want food! For my part I am intirely taken up, at my hours vacant from business, with the second vol. of Zoonomia. Afterwards I think to poetize again. Adieu from dear Sir yours most faithfully

E Darwin

Addressed W. Hayley Esqr., at Mr Romney's, Cavendish Square, London.

Original Library of Fitzwilliam Museum, Cambridge, S. G. Perceval Bequest, General Series.
Printed Unpublished.
Text From original MS.
Notes
(1) In the first sentence Darwin is referring to Josiah Wedgwood II (see letter 93I).
(2) Darwin's detestation of war emerges strongly here and in letter 95E, and, as an admirer of the early years of the French Revolution, he was particularly dismayed that the French were now 'the enemy'.

95B. To Thomas Wedgwood, 24 February 1795

Derby Feb. 24 –95

Dear Sir,

My absence from home prevented me from sooner answering your favor. Where there is an opake cornea, and some reason from the shape of the eye not being alter'd, or from the person's being sensible to light, as of a candle, I think it might probably give confined vision by making a small aperture through the cornea by taking a bit out about the size of the smallest crow-quill – and then if so small a scar should be transparent, the patient would see. The operation is nice, and would require a young hand, and a young eye, to perform it. First a steel wire with a sharp point, and a screw must be made thus

with a handle of wax'd thread. This must be suddenly put into the center of the cornea, which will steady the eye; it must then be turn'd round so that the screw may hold the cornea fast; then a fine circular saw, or edge, made of hollow wire, thus

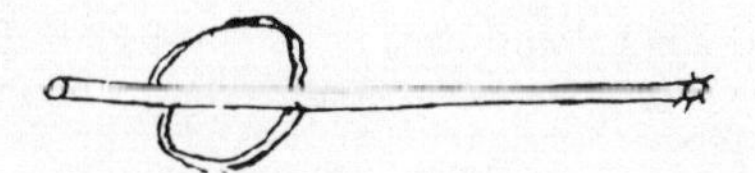

with handles as in the figure must be slided over the other (which other should not have a wax-handle as above described but only a flatten'd one).

Or both these instruments may be made in one piece, like the instrument for trepanning.

When a small circular piece of the cornea is thus taken out the eye must be closed, and in a few days a new scar will be form'd which over so small an aperture I should hope would be transparent.

Your instance of the effect of rhubarb and opium is explain'd in Zoonomia Sect. XII.3.3.

Perhaps the liquid fat you mention amongst your feces was only mucus of the bowels – unless you try'd if it would flame?

A slow pulse is no sign of a short existence, but rather the contrary. On the contrary an habitual quick pulse is a symptom of debility.

Warm bathing is the most agreable stimulus you can use – next to that constant moderate exercise of body and mind – a mixture of animal and vegetable food in such quantity, as your stomach likes – with small beer and one glass of wine as recommended by Saint Paul.

If your digestion is not good half a grain of opium may be taken, or a second glass of wine.

Mrs Darwin begs to be remember'd to you, and all at Etruria with dear Sir

your faithful friend
and serv.
E Darwin

Addressed {Thos. Wedg}wood Esqr, {Etru}ria, {Ne}wcastle, {Staff}ordshire.

Original Keele University Library, Wedgwood Archive 26771–35.
Printed Short excerpts in *Doctor of Revolution*, pp. 253–4.
Text From original MS. (*Diagrams* From original.)
Notes

(1) This letter is an example of Darwin's bold medical strokes – an experimental operation to relieve the blindness caused by an opaque cornea. Darwin seems to be inventing the details as he writes, to judge from the countermanding of his instructions for the waxed handle. The success of the operation would depend on the skill of the eye-surgeon and the transparency of the scar. It was chancy, but even the chance of alleviating blindness was perhaps worth taking. Darwin suggests some improvements in his next letter to Tom (95C).

(2) Later in the letter Darwin again (cf. letter 94E) prescribes opium for poor digestion. Section XII.3.3 of *Zoonomia* discusses the decreasing effects of long-continued doses of opium. Presumably Tom was complaining that it was having less effect – and perhaps he was already increasing his doses?

95C. To Thomas Wedgwood, 9 March 1795

Derby Mar. 9 –95

Dear Sir,

I beg your pardon for not sooner answering your last, but I was considering whether a common cutler, or white smith, or clock-maker could not make the instrument, and would advise you to do as below.

1. Procure some calve's or sheep's eyes, and by cutting them with knives, and sissars, you will come to understand the hardness and toughness of the cornea.
2. The hole can not be made too small.

3. Would not a fine cylindrical file, terminating in a screw – or a fine capillary gimblet – or a cylinder with one cutting tooth –

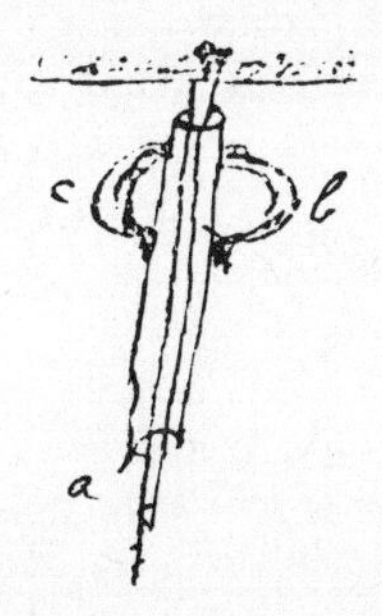

a the cutting tooth, bc the handles of the cylinder, which may be as well made of wax'd thread.

4. Perhaps the cornea may be of such a texture that if a hole is made by distension, as by a sprig-awl in wood, that it would not come together again, and fill up the Hole? This might be learn't by making holes through the cornea of dead calve's eyes.

The piece taken out, or the hole, should be very small, as a small crow-quill, or as a thick bristle or Horse-hair.

Lastly, could a hole be made by a caustic? Suppose a capillary glass tube could be press'd for half a minute on the center of the cornea, with the least possible part of a drop of oil of vitriol in it – the part should [be] wiped and wash'd with warm water as soon as the tube is withdrawn. Could not this be tried on the dead eye, as it is in part a chemical process? –

Or by a capillary syringe to hold up the cornea against the bottom of the glass tube by the pressure of the atmosphere.

ab oil of vitriol

You see it will require a nice operation. Another idea is with a sharp knife, to shave or pair [pare] of[f] the external part of the opake cornea, till it becomes transparent, like scraping ivory or horn quite thin; and try if it would become opake again.

Could not a bit of glass ground into a shape like a shirt-sleave stud be put into the hole made in the center of the cornea, and left there, which would stay, and become a glass eye?

I know you are yourself a very ingenious mechanician, and if you can make these people see by such glass eyes, you will ga{in?} immortal honour, and serve the human animal. I think this last idea bids fair, as the future scar would not signify, whether it be opake or not, and would heal like the holes bored in Ladies' ears. Adieu.

Success attend you, let me hear what you do,

E Darwin

Addressed Thos. Wedgwood Esqr., Etruria, Newcastle, Staffordshire. (*Postmark* Derby.)

Original Keele University Library, Wedgwood Archive 26772–35.
Printed Short excerpts in *Doctor of Revolution*, pp. 253–4.
Text From original MS. (*Diagrams* From original.)
Notes

This letter shows Darwin's continuing inventiveness at the age of 63. He was a physician, not a surgeon, but he was prepared to think out the techniques for this eye-operation from first principles. His idea of putting in a glass stud seems an excellent one, which might well have been effective. But the idea of cutting a hole with concentrated sulphuric acid is rather horrifying. Still, the stakes were high. Anyone who could make the blind see would indeed 'gain immortal honour, and serve the human animal'. As far as I know, Tom did not pursue these eye operations to a practical test: he was fertile in invention but infirm of purpose.

95D. To Matthew Boulton, 13 March 1795

Dear Sir

I take the liberty to recommend the bearer, Doctor Salt, to your acquaintance; who has pass'd through his medical studies at Edinburgh with great eclat, as I have been well inform'd by several of his contemporary students, besides what I can myself judge of his acquirements in his profession. He afterwards went abroard, and has since practised physic for a time at Kidderminster, but thinks it would be more advantageous for him to settle in Birmingham. Any assistance, you would please to afford him, would be gratefully acknowledged by himself, and confer a favor on, dear Sir

Your obed. and faithful serv.

Derby 13th March 1795

E Darwin

Original Birmingham Reference Library, Matthew Boulton Papers, Darwin 44. Only the last eight words, from 'dear Sir' to 'Darwin', are in Darwin's writing.
Printed Unpublished.
Text From original MS.
Notes

'Doctor Salt' is John Butt Salt MD (1768–1804), son of the Lichfield surgeon Thomas Salt and elder brother of Henry Salt (see *DNB*). J. B. Salt became physician to the Birmingham Dispensary, but had to retire because of illness. See *Aris's Birmingham Gazette*, 27 February 1804; Pearson Papers, University College London, 578, Vol. 1, p. 37; A. L. Reade, *Johnsonian Gleanings*, Part IV, p. 133; and letter 96A.

95E. To Richard Lovell Edgeworth, 15 March 1795

Derby
Mar. 15 –95

Dear Sir,

I beg your pardon for not immediately answering your last favor, which was owing to the great influence the evil Demon has at present in all affairs on this earth. That is I lost your letter, and have now in vain look'd over some scores of papers and can not find it – secondly having lost your letter, I daily hoped to find it again without success.

The telegraph you described I dare say would answer the purpose, it would be like a Giant wielding his long arms, and talking with his fingers, and those long arms might be cover'd with lamps in the night. You should place [~~such~~] four or six such gigantic figures in a line, so that they should spell a whole word at once and other such gigantic figures within sight of each other round the coast of Ireland; and thus fortify yourselves instead of Friar Bacon's wall of brass round England, with the brazen head, which spoke, "Time is! Time was! Time is past! . . . ["]

Mr Pit seems to have amused you irish Gentry very nicely, till you granted the supplies – He knows well enough, that he has a great majority in your parlament, who bring their souls to market, like a country woman with eggs in a basket – as Mr Alheed in his comment on Mr Brother's prophecies, so well describes.

I am very glad to hear Mrs Edgeworth is got better of her cough, which in her constitution is always alarming, and for which nothing is yet certainly known, that can benifit her, except a lower'd atmosphere according to Dr Beddoes's plan.

The death of Mr Wedgwood grieves me very much. He is a public as well as private loss – we all grow old, but you! – when I think of dying, it is always without pain or fear – this world was made for the demon, you speak of, who seems daily to gain ground upon the <u>other gentleman</u>, by the assistance of Mr Pitt and our gracious — I dare not mention his name for fear that high treason may be in the sound; and I have a profess'd spy shoulders us on the right, and another on the opposite side of the street, both attornies! and I hear every name supposed to think different from the minister is put in alphabetical order in Mr Reeve's doomsday book, and that if the French should land these recorded gentlemen are to be all imprison'd to prevent them from committing crimes of a deeper dye. Poor Wedgwood told me he heard his name stood high in the list.

America is the only place of safety – and what does a man past 50 (I don't mean you) want? potatoes and milk – nothing else. These may be had in America, untax'd by Kings and Priests.

Mrs Darwin begs her respects to you, and all with you, and we shall be very happy to see you in Derby, with any part of your family. Adieu

from dear Sir

Your sincere and affect. friend

E Darwin

I think you or Lovell write latin verses sometimes – I see Mr Seward (as I believe) has put my head in the European Magazine for the last month, with an account of my life and opinions. I have written the following to put at the end of the next volume of Zoonomia, which is now printing – correct them, or improve, or amplify.

Currus triumphalis Medecinae

Currus it Hygeiae – Medicus movit arma triumphans –
 Undique victa jacent lurida turma mali.
Laurea dum Phoebi serpens sua tempora circum,
 Nec mortale sonans Musa coronat opus;
Post equitet tremulans, sibilatque Senectus in aurem
 Voce cavâ repetens "Sis memor ipse mori".

I kept this letter 3 or 4 days till I heard from Mr Sneyd – and now Mr E. Darwin is from home, I don't know how to send you a legal receipt. – Mr Burns the Quaker once let me a lease, which began with "Friend Darwin, Honest men need no Lawyer. I hereby let thee etc", so I send you my receipt as follows –

Recieved of R. L. Edgeworth Esqr. 104–8–5 by the hands of Mr Sneyd on account

Mar. 17 1795 by me

E Darwin

A legal receipt shall be sent next time, or next week, if this be not satisfactory.

Addressed R. L. Edgeworth Esqr., Edgeworth town, Ireland.

Original Cambridge University Library, Dar. 218.

Printed About half the letter is printed in Edgeworth, *Memoirs* ii 158–9; there are long excerpts (mostly from the other half) in *Doctor of Revolution*, pp. 252–3.

Text From original MS.

Notes

(1) This is one of the most varied and informative of Darwin's letters. It was in reply to Edgeworth's letter of 11 December 1794 (printed in Edgeworth, *Memoirs* ii 157–8), in which Edgeworth describes the telegraph, originally invented by him 26 years earlier and now being offered to the Irish government to help in countering rumours of French invasion. Darwin's comment about the brazen

head is a reference to Robert Greene's play *The Honorable History of Friar Bacon and Friar Bungay* (1594), in which a brazen head speaks the words quoted by Darwin.

(2) 'Mr Brother's prophecies' refers to the notorious book by Richard Brothers, *A Revealed Knowledge of the Prophecies and Times* (1794), purporting to be 'wrote under the direction of the Lord God and published by his sacred command'. Brothers (1757–1824) claimed to be the nephew of God and predicted the destruction of London by earthquake and fire. He was committed to a madhouse in 1795, but he found a staunch supporter in Nathaniel Brassey Halhed (1751–1830), MP for Lymington, who spoke in the House of Commons and wrote many pamphlets in support of Brothers. He is Darwin's 'Mr Alheed'. For details of the controversy, see *Gentleman's Magazine* for 1795, Vol. 65, pages 208, 223–9, 678 and 1097–9. See also J. F. C. Harrison, *The Second Coming* (1979), chapter 4. For a biography of Halhed, see *Gent. Mag.*, **100**, 471–2 (1830), and *DNB*.

(3) Mrs Elizabeth Edgeworth (1753–1797) died of consumption 2½ years later.

(4) The paragraph about spies and not daring to mention the King's name shows a new and unpleasant feature of Darwin's life and of the English social scene. The change was so unwelcome to Darwin that he mentions the idea of emigrating to America – as Priestley had already done, for the same reason, in the previous year. The remark that 'potatoes and milk . . . may be had in America, untax'd by Kings and Priests' shows the depth of Darwin's disillusion with the established order in England. One of the 'professed spies' is probably the attorney Charles Upton (see letter 82E).

(5) 'Mr Reeve' is probably John Reeves FRS (1752?–1829), who was known as 'the prince of spies'. In 1793 Reeves organized an 'Association for preserving liberty and property against Levellers and Republicans'. Darwin may be referring to his pamphlet 'The Malecontents' (1794). See *DNB* for Reeves.

(6) Lovell Edgeworth (1775–1842) was Edgeworth's second son.

(7) 'Mr Seward' is William Seward (1747–1799), the wealthy son of a brewer, who was a well-known anecdotal writer and probably put together the article on Darwin in the *European Magazine* for February 1795 (Vol. 27, pp. 74–7).

(8) Darwin's Latin verses appeared, with some minor changes, at the end of *Zoonomia* (ii 640), and were freely translated by Edgeworth as follows:

The Triumphal Chariot of Medicine

The victor comes! I see his arms from far,
And hear the thunder of Hygeia's car.
Phoebus with laurel binds his brow, and Fame
Sounds from her silver car his deathless name.
Hurrying behind rides Age, with feeble cry,
Eager to tell the sage that he must die.

(Edgeworth, *Memoirs* ii 174).

(9) 'Mr Sneyd' is not Edgeworth's father-in-law Major Edward Sneyd (1711–1795), because he died on 3 January, the same day as Josiah Wedgwood: see *Gent. Mag.*, **65**, 84 (1795). 'Mr Sneyd' is probably either Major Sneyd's nephew John Sneyd of Belmont (1734–1809), or Major Sneyd's son Edward (*c.*1755–1832) of Byrkley Lodge, Staffordshire: see letter 99A.

(10) Darwin's account book for 1771–3 (in the Nottinghamshire Record Office)

shows that he paid £10.10.0 rent to 'Mrs Burnes' on 3 April 1771. So 'Mr Burns the Quaker' is probably Richard Burnes of Aldershaw House, Lichfield, who died in 1767. See Stebbing Shaw, *Staffordshire* i 358.

(11) The letter shows that Darwin was now wealthy enough to have made a large loan to Edgeworth.

95F. To Thomas Beddoes, [April? 1795]

Dear Doctor,

I am much obliged to you for your letter with observations on Zoonomia etc. and am impatient to see your new edition, which I will take care to send for as soon as it is out.

Your correspondent will recollect that I do not enter into the inane dispute about the inanity of the universe, I leave that to be settled by the disciples of Berkeley and Hume; and secondly, for the diffusion of the power of contraction over the body, I appeal to common anatomical experiments; the existence of the body therefore I take for granted; it has nothing to do in my argument, which treats on the 'laws of organic life', which organ and life I therefore take for granted. The other observations in your letter are all good and ingenious but I am too idle to write about them, and so shall only thank you for them.

It is evident that to increase sensorial secretion without at the same time <u>using it</u>, is not easy. I have invented a method of stopping too great exertion of the system in inflammatory fevers, I believe; this great secret, yet untried, you will see in my next volume. Shall I tell it you now? Well here it is. You are to prescribe a wind-mill, lay the patient across the large stone, whirl him about sixty times in a minute, and the brain becomes gently compressed, and the fever ceases, and sleep succeeds; go on, and he dies without pain.

What would be the event if his head was in the centre of motion? Would he become vivacious?

For increasing sensorial power I propose electricity through the brain in fevers, and through the stomach, and warm fomentation of the head, and <u>hot</u> bath.

Now for another secret. If you render the extremity of a nerve torpid (if it be of an organ usually in action), an accumulation of sensorial power occurs, as in cold hands and hungry stomach; both from the want of stimulus. If this torpor of the organ be occasioned by previous excess of stimulus, no accumulation is produced; as after drunkenness or exposure to violent external heat. Now, mind, 1. In some fevers the contagious material is swallowed. 2. The stomach becomes torpid by the great stimulus of the contagious materials. 3. The heart becomes torpid by direct sympathy with the stomach, as shewn by the intermission of the pulse, when the stomach is flatulent from indigestion and in the exhibition of foxglove. 4. The stomach accumulates no sensorial power, because its sensorial power was exhausted by the stimulus of the contagion; but the heart does accumulate sensorial power, because its torpor is owing to the defect of the stimulus or power of association. 5. This accumulated power is exerted on the next link of the associated circle of motions, that is on the capillaries. Hence this fever consists in the torpor of the stomach from defect of sensorial power, the torpor of the heart from defect of the power of association, and the increased activity of the capillaries; the last is shown by the heat and dryness of the skin. All this reasoning wants further consideration.

Adieu,
E Darwin

Original Not traced.
Printed J. E. Stock, *Life of Beddoes*, Appendix 6, pp. xlv–xlvi.
Text From above source.
Notes
(1) 'Your new edition' is probably the second edition of *Factitious Airs* (see letter 94Z). Since the pamphlet was being printed and its preface is dated 30 March 1795, early April seems the most likely date for this letter. Darwin ordered the new edition by 28 April (letter 95H).
(2) The 'observations on Zoonomia' were from a correspondent of Beddoes, who passed them on to Darwin.
(3) The last four paragraphs of the letter are largely a preview of sections of *Zoonomia*, Volume II, which was published in 1796. The therapeutic use of centrifugation was a good idea and may come to fruition on the grand scale if cities in space are ever constructed.

95G. To James Watt, 2 April 1795

Derby Apr. 2 –95

Dear Mr Watt,

I was sorry my other engagements would not permit me to pass a few hours with you last week, when I was at Birmingham – but so the fates ordain'd.

I am now at length about to try oxygene mix'd with atmospheric air in a case of anasarca of the lungs, which has thrice been relieved by foxglove; and am about to construct some silk or linnen bags according to your directions – and am in difficulty how soon to procure Exeter manganese; and trouble you with this long history to ask, if you could send me half a score pounds, till I can get it from Exeter? If you have not any to spare, I think to write to Mr Reynolds of the bank near Colebrook dale, if that be a proper direction.

If you have any japan'd air-holders ready made I should be glad of one, as I think it a convenient machine to measure airs with – or indeed to breathe them from – as by fixing a large bladder with a breathing pipe by adding water into the machine, quart after quart, the quantity breathed might be easily ascertain'd thus:

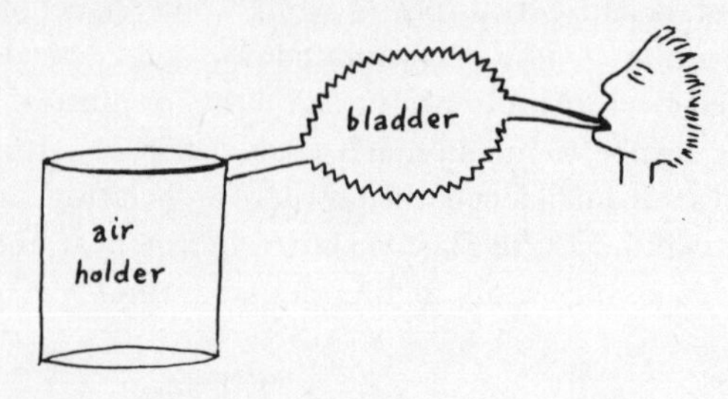

I can get one made here but not japan'd.

It would give me great pleasure to hear that yourself and Mrs Watt are well. I am printing my second Vol. of Zoonomia, which I will not forget to send you, but do not insist on your reading it – adieu

Your affect. friend
E Darwin

Mrs Darwin joins in true respects to Mrs Watt and yourself.

Addressed James Watt Esqr., Heathfield, near Birmingham. (*Postmark* Derby.)

Original Privately owned.
Printed Unpublished.
Text From original MS. (*Diagram* Redrawn.)
Notes
(1) Darwin is still busy experimenting with the medicinal use of gases, and is now trying oxygen to combat fluid in the lungs.
(2) William Reynolds (see letter 87J) lived at Ketley Bank House.

95H. To James Watt, 28 April 1795

Dear Sir,

I am much obliged to you for your letter of instructions, which I have avail'd myself of. And shall be obliged to you further, if you will please to send me, as soon as you can, another ½ hundred of Exeter Manganese, and also another large air-holder and pipe, and another silk-bag with box-wood spiggot and faucet. But if the latter are not ready, please to direct the manganese to be sent, as soon as may be, by the first carrier alone; and the others as soon as is convenient after it.

For I expect a person to come to Derby to breathe oxygene on the 4th or 5th of May and I have a Lady now, who consumes a pound or two of manganese every day. Sometimes she takes it pure, but generally with half common air, and is refresh'd by it, and thinks her headachs much relieved by it, or by a saturated solution of arsenic, which she takes also, from 3 to 5 drops thrice a day.

Arsenic I believe strengthens the stomach, and by sympathy the action of the heart; as I observed it to cause a palpitation of the heart for several days, during the time it was taken. It certainly cures the ague, when the bark fails, and I think relieves Cephalaea, even when decaying teeth are the remote cause. I put any quantity of arsenic in powder into a florence flask, and boil it for an hour gently in less water by half than will dissolve it. I put an ounce of this saturated solution into a two-ounce phial to ascertain the size of the drops. And give from 3 to 5 (and even to ten drops) thrice a day.

I have not yet used your Hydro-Carbonate, but intend to use it in consumption, to promote slight degrees of sickness.

I desired Johnson to send me Dr Beddoes' new edition, but have not yet recieved it. I have recommended your apparatus to the Infirmaries of Nottingham and Shrewsbury.

Mrs Darwin joins in our best compliments to Mrs Watt with, dear Sir,

Your obliged friend
and obed. serv.
E Darwin

Derby Apr. 28 –95

Original Privately owned.
Printed Unpublished.
Text From original MS.
Notes

(1) The order for another ½ hundred – that is, ½ cwt or 56 pounds – of manganese dioxide shows that Darwin was giving the oxygen treatment a lengthy trial.

(2) The second paragraph is confusingly phrased. The lady was 'consuming' the oxygen from 'a pound or two of manganese', not the manganese itself. However, she *was* imbibing the arsenic solution. Darwin's enthusiasm for arsenic may seem rather unfortunate for his patients; but nearly all his medicines were poisons, and he was not to know of arsenic's future fame in this role. From the context, it appears that Cephalaea is cephalalgy, or headache, and Darwin probably mentioned it because Watt was himself subject to headaches (see letter 95K).

95I. To James Watt, 9 July 1795

Dear Sir,

I was much obliged to you for your letter from London, and hope you have got forwards in your law-cause. A law-suit, that pays well to the Lawyers, goes on like a snail creeping up a pole, which slips down again every 2 or 3 inches, as he advances, till he has beslimed the pole all over.

Mr Walhouse has breathed about 7 gallons of pure oxygene a day for I believe 2 or 3 weeks, till He tried the last parcel of manganese, which was sent from your people; the air from which gave him a burning feel in his lungs with something like suffocation, which obliged him to desist, the burning sensation continued many days. On examining the burnt manganese, as it came out of the tube, part of it appeared green. A specimen of this I have sent you, hoping you may detect the kind of impurity, it possesses.

Mr Walhouse has an Hydro-thorax, and has been 6 or 7 times emptied by foxglove. The oxygene seem'd of no avail. I am now thinking of trying Hydro-carbonate with him in small quantities; as his case is desperate without some new aid. He was disappointed in recieving your large furnace etc etc when he only sent for the iron pot etc for the production of oxygene. And one of the air-holders was so imperfectly solder'd, and varnish'd, that water run out through a great length of the joints.

This impurity of your manganese should be inquired into; hoping it may be known by the eye before it is pounded.

One young lady, who was much reduced in strength, recieved great benefit by breathing about 6 gallons of pure oxygene a day for 2 or 3 weeks in this town.

Mrs Darwin begs to be remember'd to Mrs Watt, with, dear Sir, your affect. friend

E Darwin

Derby
July 9 –95

Addressed James Watt Esqr., Heathfield, near Birmingham.

Original Privately owned.
Printed Unpublished.
Text From original MS.
Notes

(1) In the 1790s Watt was involved in several lawsuits over infringements of his steam engine patent. One, against Edward Bull, was heard in June 1793, and the (inconclusive) appeal was in February 1795. Another against Jabez Hornblower was not finally settled until 1799. See L. T. C. Rolt, *James Watt* (Batsford, 1962), Chapter 10.

(2) In offering himself as a supplier of apparatus for manufacturing gases for medical use, Watt was finding that his lot was not a happy one. The customers were usually dying and in desperate urgency; any defects in the equipment would have a bad psychological effect and might accelerate the patient's decline; and, even if the apparatus were perfect, the patients would probably still die. Mr Walhouse seemed to derive no benefit from the oxygen, even before it was contaminated – presumably by sulphur dioxide, from sulphates in the ore. For Darwin too, pneumatic medicine had serious disadvantages. He first had to write to Watt to ask for the apparatus to be sent to the patient, and then, as here, might have to complain to Watt about impurities and imperfections.

(3) Mr Walhouse was Moreton Walhouse (1730–1795), who lived at Hatherton, one mile west of Cannock in Staffordshire. His first wife was Catherine Wilson, heiress of the Okeovers, and their son Edward (1752–1793) took the name and arms of Okeover. His second wife was Frances, sister of Sir Edward Littleton (see letter 84A). As Darwin says, Mr Walhouse's case was desperate: he died on 25 July. See Glover, *Derby* ii 64; *Alum. Cantab.*, Part I, iv 315; and *Derby Mercury* 29 July 1795.

95J. To Maria E. Jacson, 24 August 1795

Derby, Aug. 24 –95

Dear Madam,

According to your desire, Sir Brooke Boothby and myself have been agreably busied for many days in reading and considering your Botanical Dialogues for Children; and much admire your address in so accurately explaining a difficult science in an easy and familiar manner, adapted to the capacities of those, for whom you professedly write; and at the same time making it a compleat elementary system for the instruction of those of more advanced life, who wish to enter upon this entertaining, though intricate study. We think therefore, that not only the youth of both sexes, but the adults also, will be much indebted to your ingenious labours, which we hope you will soon give to the public.

We beg to subscribe ourselves, with true regard,

Dear Madam,

Your obedient Servants,

Br. Broothby

E Darwin

Original Not traced.

Printed *Botanical Dialogues between Hortensia and her four Children*, by a Lady. J. Johnson, London, 1797. Page v.

Text From above source.

Notes

(1) In his *Plan for the Conduct of Female Education in Boarding Schools* (Drewry, Derby 1797), p. 41, Darwin recommends the *Botanical Dialogues* and states that the author was 'M. E. Jacson, a lady well skill'd in botany'. This is 'Mrs' Maria Elizabeth Jacson (*c.*1760–*c.*1840), who also wrote *Botanical Lectures* (1804), *Sketches of the Physiology of Vegetable Life* (1811) and *The Florist's Manual* (1816). All these books were anonymous, but she signs herself 'M. E. J., Somersal Hall' in *The Florist's Manual*, p. 41. Somersal Hall, near Uttoxeter, was the ancient seat of the Fitzherberts and on the death of 'Mrs' Frances Fitzherbert in 1806 the Hall was inherited by her nephew, Rev. Roger Jacson (1753–1826), who was rector of Bebington in Cheshire for nearly 50 years. Jacson married, first, Frances Gibson (19 March 1777) and, second, Mary Johnson (27 May 1801): see Scott, *Admissions* Part IV, p. 330. Maria Elizabeth was Jacson's unmarried sister and, although

Jacson sold the Hall to Lord Vernon (from whom it passed to Alleyne Fitzherbert, Lord St Helen's), she continued to live there: D. & S. Lysons, *Topographical Account of Derbyshire* (Cadell [1818?]), p. 256, state that the Hall 'is now in the occupation of Mr Jacson's sisters'; S. Glover, *Directory of the County of Derby* (Mozley, Derby, 1829) records 'the Misses Jackson' as residents of the village of Somersal Herbert; and S. Bagshaw, *History, Gazeteer and Directory of Derbyshire* (Sheffield, 1846), p. 332, notes that 'Mrs Jackson' gave £20 towards rebuilding the village church in 1836.

(2) Sir Brooke Boothby (see letter 81D) had suffered much in the 1790s. His five-year-old daughter Penelope fell ill and died in 1791, and on her death Boothby's wife left him. Hence the inscription on Penelope's splendid monument in Ashbourne Church: 'The unfortunate parents ventured their all on this frail bark and the wreck was total'. In his *Sonnets sacred to the Memory of Penelope* (1796), Boothby records that Darwin rallied him from a suicidal state. Boothby had close family connections with the Fitzherberts: his grandfather (also named Brooke Boothby) married Elizabeth, daughter of John Fitzherbert (1659–1725) of Somersal Hall.

(3) Darwin's commendation is well composed, saying all the right things in the right order. What is more, the praise was well deserved: the *Botanical Dialogues* sugar the pill most skilfully and are very readable today.

95K. To James Watt, 21 November 1795

Derby Nov 21 –95

My dear friend,

I was much obliged to you for the kind attention, you shew'd me, in sending some pounds of manganese, and for the remainder since; and shall be obliged to you to direct your people to send my account all together, as I can not readily find it piecemeal; and to call on me, if any of them comes this way for the money, if I do not soon find an opportunity to send it.

My daughter Emma is better, but has not taken the oxygene so regularly, as she ought to have done, owing in part to my absence from home; but I intend to manage it better; and if it serves her, I will send you a further history – she takes also 2 grains of opium a day to prevent or lessen a nervous cough, which she has returns of daily about the same hour but which is now much lessen'd by the use of the opiate; one grain of which is given on going to bed and another about five in the morning; as her cough had for many weeks return'd about seven in the morning. She also takes bone-ashes, soda phosphoreta, and powder of bark of each ten grains twice a day, so that the oxygen gas is but a part of the process.

The disease I believe to be "softness of bones" and her quick respiration owing to the ribs of one side not being sufficiently rigid. But why do I tell you all this stuff? – I do it, because you have been so good as to interest yourself about her health.

Dr Menish of Chelmsford writes me word, that a hard brick weighs a great deal more after it is burnt, than when in its dry state, before it is burnt. Does glass acquire oxygene? – I should suppose so from the red lead, and manganese, which so readily induce clays and flint to vitrify. Might not a measure of high degrees of heat be thus made, by weighing certain vitrescent mixtures, or clays, after and before they experience the fire?

The new theory of fever in my next volume will take up 80 pages, I dare say; and will I hope amuse you. It shews, why the capillary vessels of the skin act too violently, producing h{eavy?} perspiration, and great absorption, for 3 or {4} weeks together in continued fevers with weak p{ulse} and why at the same time the heart and arteries act all the while feebler a great deal than natural, tho' their pulsations are quicker.

Now I grow old and not so well amused in common society, I think writing books an amusement – I wish you would write books, instead of having those confounded headachs, which you complain of! Adieu. Mrs Darwin begs to join with me in best wishes and most respectful compliments to yourself and Mrs Watt,

from your sincere and affect. friend

E Darwin

I hope our friend Boulton has not much return of his gravelly complaint, somebody lately told me he was worse, which grieved me much.

I have often thought of a mouth-piece for breathing with two valves, but never could please myself. Mudge made one with a pea for a valve, I have forgot how, but I remember it was too far from the mouth, and therefore left some air to be respired over again.

Addressed Mr James Watt Engineer, Heathfield, near Birmingham. (*Postmark* Derby.)

Original Privately owned.
Printed Unpublished.
Text From original MS.
Notes
(1) Emma Darwin (1784–1818) suffered much illness. Her niece Elizabeth Anne Galton said 'she was very beautiful, agreeable and amiable and most kind-hearted'. See *Doctor of Revolution*, p. 298.
(2) Bone ashes are mainly calcium phosphate, and soda phosphoreta is sodium phosphate, both of which Darwin hoped might strengthen Emma's soft bones.
(3) Dr Henry Menish (*c.* 1735–1810) took his MD at Glasgow in 1763 and was a friend of Joseph Black. He lived at Chelmsford from the 1770s onwards, and was a keen collector of minerals and an authority on geology. Darwin quotes excerpts from his account of the chalk-beds in *Economy of Vegetation*, Note XIX (p. 46).
(4) The theory of fever occupies pages 537–625 of *Zoonomia*, Volume II.
(5) Boulton's gravelly complaint is mentioned in letters 89C and 94S.
(6) 'Mudge' is Dr John Mudge FRS (1721–1793), of Plymouth, who designed an 'Inhaler'. See *DNB* and *Univ. Mag.*, **97**, 42 (1795).

95L. To James Watt, 23 December 1795

Derby

Dear Sir,

I am desired by a patient, who is now in Devonshire, to write to you for a machine for making oxygene gas, the use of which I have recommended to his lady. He wishes to have an apparatus for this purpose only. I must therefore beg, you will please to direct the Iron-pot etc and not the tube nor any unnecessary apparatus. The quantity sent to Mr Walhouse by your people was so much more than was desired, that Mr Babington desires me to specify the pieces, that he may not have more than is necessary simply for procuring oxygene gas. These I think should consist of the iron pot and the thing to set it on, 2nd the bent arm with tin tubes and a refrigeratory or washer. 3. hydraulic bellows and 4th one air holder. 5th a silk bag.

He desires the bill to be sent with them. The direction is Thos. Babington Esqr. Hillersdon near Cullompton Devonshire. When he is at home he lives in Leicestershire.

Emma, whom you was so kind as to ask after, took cold and was for a fortnight worse, but is now much better again, but has not lately taken any more oxygene.

I sent the money to pay my bill lately to Birmingham and I suppose it has or will be soon discharged.

My new vol. will be another month or two, before it sees the light, as the bookseller is reprinting the first to sell along with it. As soon as I get any I will send you one.

Mrs Darwin begs to be kindly remember'd to Mrs Watt with

Your faithful and affect. friend

E Darwin

Dec 23 –95

Mr Babington will be in haste for the apparatus.

Addressed James Watt Esqr., Heathfield, near Birmingham. (*Postmark* Derby.)
Original Privately owned.
Printed Unpublished.
Text From original MS.
Notes

(1) Thomas Babington (1758–1837) of Rothley Temple, near Leicester, was noted for 'integrity and goodness of heart'. A graduate of St John's College, Cambridge, he was MP for Leicester 1800–1818. His wife Jean (née Macaulay) (1764–1845) evidently came to no serious harm as a result of the oxygen treatment. For biography of Babington, see Scott, *Admissions*, Part IV, p. 109. For a four-page family tree, see Nichols, *Leicester* iii 955.

(2) 'My new vol.' is Volume II of *Zoonomia*, published in April 1796.

95M. To Miss S. Lea, [1795?]

Dear Madam

I sent by the carrier last Monday a hamper of roots, which I hope arrived before they were spoil'd. The Meadia should be kept set in a pot and moistish, and when in flower should be shaded from the sun. The Hyacynthus muscari's have flower'd. I will send you as many more in the autumn of different sorts, if you wish to have them. The Meadia, and muscari, and stipa or feather-grass, were wrap'd in papers.

It would give me pleasure to hear from you, to know if you could manage the grain-pills of opium, so as to prevent the returns of your paroxysms of asthma. As in your uncommon complaint I could perhaps be of greater service, if you would write me a particular history of yourself for even one day (or two) like a journal. And mentioning particularly at what hour your fits commence, and how long they continue: and whether a pill of a grain of opium taken before the commencement of a fit (or in one) has render'd it milder, and if you are able to take food. I directed the opium in grain-pills on purpose, that you might have the management of them in your own power – as I hoped by that means you would best learn both the quantity, and the times of taking them, which might be most useful to you.

Your own account of your feelings I should judge better from, than from any other person's description, if writing be not too troublesome to you. I beg my most respectful compliments to all your good family, and am, dear Madam, with sincere good wishes

your obed. serv.
E Darwin

Addressed Miss S. Lea, Shrubbery, Kidderminster. (*Postmark* Derby.)

Original British Library Add MS 29300, ff. 52–3.
Printed Unpublished.
Text From original MS.
Notes

(1) The dating of this letter is arbitrary. The Derby postmark shows that it is after 1782, and the sending of plants suggests a date after the building of Darwin's hot-house, which was probably about 1787 (see letter 88B). Darwin seems still to be active as a doctor, so a date later than 1799 is unlikely, and his reliance on treatment by opium suggests the mid-1790s (see letter 94E and *Zoonomia*). But it could be any year between 1788 and 1799.

(2) The Lea family of Kidderminster was much involved in silk manufacturing: there are four different businesses with the name – Joseph Lea; Henry and Richard Lea; Francis Lea and son; and Samuel Lea – in *Bailey's British Directory* of 1784, all recorded as 'Silk and Stuff Manufacturers'. I have no information about Miss S. Lea.

(3) The letter shows Darwin's interest in practical botany, and gardening. He describes Meadia, with a plate, in *Loves of the Plants* I 61–4 and note.

(4) In the eighteenth century 'asthma' usually meant difficulty in breathing. If Miss Lea's asthma was caused by weakness of the heart, the opium would be helpful; if the asthma was allergic, opium would be useless. The letter shows again Darwin's genuine concern for his patients – and his tendency to dispense opium a little too freely.

96A. To James Watt, 18 January 1796

Derby Jan 18 –96

Dear Sir,

I take the liberty of addressing a line to you to solicit your interest in favor of Dr Salt at the ensuing election of physicians for the hospital at Birmingham. I have been acquainted with Dr Salt almost from his infancy, have attended to the progress of his education, and have since had the pleasure of his company repeatedly, and believe him to be well qualified indeed for the practise of his profession.

Mrs D joins in our best compliments to Mrs Watt. Emma is much recover'd but I can hardly say by what means.

from, dear Sir,
Your sincere friend
E Darwin

Original Privately owned.
Printed Unpublished.
Text From original MS.
Notes
See note to letter 95D, which is a rather similar commendation of Dr Salt, to Boulton.

96B. To Thomas Wedgwood, 23 January 1796

Derby Jan. 23 –96

Dear Sir,

I am much obliged to you for the attention you have given my book, and am flatter'd in finding you have detected so few errors. You return'd my last letter, for what purpose I do not know, as I can not be at the trouble of reading it again.

As the sensorial powers of pleasure and pain, and of desire and aversion can be excited without their causing fibrous motions, so the sensorial powers of irritation, and of association should from analogy exist in this inactive state, tho' they have then acquired no names – (as in the constant small sounds and light in which we are immersed, and which are not follow'd by any observable activity) – hence the existence of these circumstances is not totally gratuitous, as it is at least supported by the stricktest analogy.

The circumstance of the light partitions of a window are explain'd in Sect. XL.3.2.

A blind man has tangible ideas (as I believe every one may have, when he thinks of all the sides of a cube or die at the same time); hence the images in dreams I suspect are easier to understand. Hence an affirmitive answer to his recollecting the figure of Thorp Cloud, would not be conclusive, as that might be a tangible idea.

The mouth of a lymphatic, when it is alive, absorbes; and therefore prevents the return of fluids; but when dead, has no valve, that I know of, to prevent their return.

Surprize, as defined in my book, I believe, does not exist without external stimulus. Admiration may be in common language called surprize, but that is an abuse of words – my book is to be tried by my own definitions.

The tangible vertigo is frequently mention'd in the work, both from intoxication, and from rotation, I believe in a note at the end both of the first and second volume. It is not the chemical changes of the aliment of the stomach, which induces sickness in any kind of vertigo; for it is only unusual motions, which cause vertigo either of the sense of vision or of touch. In the second volume there are many more pages on vertigo.

Birds building nests, and bees building cells, and laying up honey, seem to be acquired arts, deliver'd from parent to offspring, like the arts of mankind.

The appearances of prismatic colours by external pressure in the eye are mention'd by Sir I. Newton, and are said to be like a peacock's tail; and this shows, that light itself is not necessary to this appearance of colours. And are mention'd, Sect. III.4.1 and Sect. XL.1.3.

The experiment at III.3.1 was made by a convex mirror, not a lens.

Your ingenious idea of deep thinking consisting of alternate sleeping and waking is nearly the case, except that in sleep the external senses are closed, or render'd unfit to recieve their objects.

Many of your other remarks had better be defer'd, till we happen to meet, and I hope you will sometime give me your company at Derby.

Adieu from dear Sir
your affect. friend and serv.
E Darwin

My new theory of fever I fear you will find very difficult, but I believe the foundation of it to be true, and the knowlege important, as it points out with more certainty the methods of cure.

I hope this second vol. will appear in March or April, and that you will send me any remarks, which may occur to you.

Addressed Thos. Wedgwood Esqr., No. 22 Devonshire place, London. (*Postmark* Derby 25 JA 96.)

Original Keele University Library, Wedgwood Archive 26773–35.
Printed Unpublished.
Text From original MS.
Notes
(1) The book mentioned in the first sentence is *Zoonomia*, Volume 1, published in 1794.
(2) The 'light partitions of a window' is explained by Darwin as follows: 'If you look for some minutes steadily on a window . . . and then move your eyes a little, so that those parts of the retina, on which the dark framework of the window was delineated, may now fall on the glass part of it, many luminous lines, representing the frame-work, will appear to lie across the glass panes . . .' (*Zoonomia*, Vol. I, Section XL.3.2).
(3) Darwin's discussion of the ideas and dreams of blind men made an impression on Wordsworth, who brought the subject into *Tintern Abbey*. Thorp Cloud is the hill at the entrance to Dovedale.
(4) The pressure on the eye causing a sensation of light is mentioned in Newton's *Opticks*, Query 16.
(5) The experiments in Section III.3.1 are to detect the pressure of light, using a torsion balance.
(6) The idea that deep thinkers are half asleep is a nice one, and at least half true.
(7) Darwin's theory of fever derives from that of the Edinburgh school, particularly William Cullen (see letter 96H). Darwin makes a division into inflammatory fever, due to increased activity of the nervous system, and putrid fevers, due to debility.

96C. To Thomas W. Greene, 25 February 1796

Dear Sir,

As I am subpoena'd on a tryal, which is to take place at Stafford concerning Mr Garick's will, I should be glad, if you would send me a copy of my prescriptions for Mr Garick during the last year or two – if this is refused, I shall be obliged to say in court, that it was refused. – I want these copies, that I may speak with more certainty concerning my attendance on him, if required.

I am, Sir, your obed. serv.

Derby E Darwin
Feb. 25 –96

Addressed Mr Green Surgn., Lichfield, Feb 25 –96. (*Postmark* Derby.)

Original William Salt Library, Stafford, S. MS 478.
Printed Unpublished.
Text From original MS.
Notes
(1) The recipient of the letter seems sure to be Thomas Webb Greene (1763–1842),

the son of the well-known Lichfield surgeon-apothecary Richard Greene (1716–1793) (see letter 90C). Thomas Greene took over the business after his father's death. For details of the Greenes, see 'The Greene Family Chronicle', in A. L. Reade's *Johnsonian Gleanings*, Part VIII, pp. 130–158.

(2) 'Mr Garick' is Peter Garrick (1710–1795), David Garrick's elder brother and a well-known Lichfield resident. According to his gravestone in the precinct of Lichfield Cathedral, he died on 13 December 1795. A draft of his will is in the Lichfield Record Office (Deposit D.15), but I do not know why it was disputed.

(3) Darwin was clearly annoyed to have to endure the discomfort of a 40-mile journey to Stafford for such a trivial reason, and this is one of the most abrupt and impolite of his letters.

96D. To Elizabeth, Lady Harrowby, 21 May 1796

Derby May. 21 –96

Madam,

My absence from home for three successive days prevented me from sooner answering your Ladyship's favor. I am always uneasy, when I think over Miss Ryder's case; as very efficacious medecines have been used without much apparent advantage. The debility of the stomach and bowels has always appear'd to me to be the cause of the {ge}neral debility; and this is believed to be the cause of what is term'd hypochondriasis in men, which is always attended with so much fear, <u>or expectation</u>, of dying, as to induce them to think of nothing but their own health. In this respect Miss Ryder's attention to her own minute sensations are uncommonly great, so as to occupy her whole mind; and produce what would be call'd caprice, if it only existed in a slighter degree.

Her defect of menstruation is also a disagreable circumstance, as it seems to shew the decreasing activity of the system; but is I apprehend a consequence of her general debility, or <u>in</u>irritability, rather than a cause of it.

Her being able to perform those bodily exertions, or mental ones, which she sets her <u>Will</u> to perform, is not contradictory to this general idea of her complaint; as it is not any inability to <u>voluntary</u> exertions, that affects her; but a defect of energy in those actions, which are exerted in consequence of <u>irritation</u> without our consciousness; as those, which perform digestion, and circulation, and secretion.

What then can be done? All those means should be used, which can strengthen her general system, and thus occasion her to digest more food, especially of the animal kind; and thus by invigorating her circulation to restore the deficient catamenia. Secondly, whatever can amuse of [or] employ her without great fatigue, and consequent exhaustion of animal power, should be attempted, for the purpose of relieving her mind from the perpetual attention to her health, or in other words from the fear, <u>or expectation at least</u>, of dying. Which fear or expectation appears to me to give rise to many of those ideas or actions, which are generally ascribed to unmeaning caprice. Thirdly those ideas, which diminish the fear of death may be occasionally suggested to her; of which I have mention'd some in the second Vol. of Zoonomia p. 377 and p. 131. But as all diseases of the mind are best cured by forgetfulness, I believe such amusements, or business, as engaged the attention on other objects are the best means to effect this purpose.

Hence if the trouble of an issue, or her hopes {of} its good effect, could contribute to engage her attention or relieve her vain fears, it might so far be of service; and might be safely heal'd up again in 2 or 3 months, but I do not see that it could be of use in other respects.

Some bitter and some chalybeate medecine may be taken for 2 or 3 weeks occasionally, and then remitted for 2 or 3 weeks, that they may not become habitual.

A pill of half a grain of opium at breakfast, and in the evening, I think I have found benifit from in these cases. Which should be continued some months, as an habitual new stimulus. And if a daily evacuation is not had, about six grains of rhubarb should be taken every night for a long time.

I directed some chalybeate and bitter medecines, when I had the pleasure of seeing Miss Ryder at Shepson[?].

I beg my most respectful compliments to Lord Harrowby and to all your good family, and am your Ladyship's much obliged

and obed. serv. E Darwin

Addressed The Rt Hon. Lady Harrowbye, Park Street, Westminster, London.

Original Sandon Hall, Stafford, Harrowby MSS, Vol. VIII, ff. 222–3. (Privately owned.)
Printed Unpublished.
Text From original MS.
Notes

(1) Lady Harrowby (*c.* 1735–1804) was born Elizabeth Terrick, elder daughter of Dr Richard Terrick, Bishop of London: she married Nathaniel Ryder in 1762, and he became the first Baron Harrowby in 1774 (see letter 98D). 'Miss Ryder' is presumably Lady Harrowby's elder daughter the Hon. Elizabeth Ryder (1767–1830). She lived for another 34 years, though her younger sister Anne died young 'of a fever' in 1801. See *DNB* (under 'Ryder' and 'Terrick') and *Gent. Mag.*, **71**, 675 (1801); **74**, 795 (1804); and **100** (2), 476 (1830).

(2) This letter is a good example of Darwin's firm but kindly treatment of psychosomatic illness. All the medicines and medical terms have appeared previously: see notes to letters 93F, 93J and 94D. *Zoonomia*, Volume 2, page 131, deals with hypochondriasis, a disease of irritation, which 'consists in indigestion and consequent flatulency, with anxiety or want of pleasurable sensation'. Page 377 discusses *Lethi timor*, fear of death, a disease of volition: these patients are sometimes consoled, Darwin says, 'by instances I have related to them of people dying without pain'.

96E. To Burwell Starke, 14 September 1796

Derby Sept 14 –96

Dear Sir,

This last attack of your disease, which you describe, was certainly an affection of your stomach and upper intestines, which is generally call'd a Cholera; and was most probably owing to the cause, you ascribe it to, the distil'd tar-water, which you had taken.

As tar-water consists of a weak vegetable acid with a very small quantity of essential oil, if this was distil'd in an alembic with a copper head, and a pewter worm-pipe, the hot vegetable acid rising from the tar-water would dissolve a great quantity both of the copper and lead, and produce a very noxious fluid indeed! which your stomach has fortunately thrown off.

As the decoction of Logwood was once tried and did not agree, you will please not to use it – otherwise it is an astringent without bitterness, and generally agrees, when simply prepared by boiling.

The eruption on your legs and thighs if it be slight, is probably of no consequence; and might have been produced by your violent and continued efforts in vomitting; and the sympathy between the stomach and the skin.

The warm bath will be of service to you at any time; but if your skin be hot by any considerable degree of fever, the bath should be cooler, so as to render it perfectly agreable to your sensation.

I am, Sir with best wishes,
Your much obliged
and obed. serv.
E Darwin

Addressed Burwell Starke Esq, Beverley, Yorkshire. (*Postmark* Derby.)

Original Historical Society of Pennsylvania, Philadelphia.
Printed Unpublished.
Text From original MS.
Notes
(1) A routine medical letter, showing Darwin the conscientious doctor. I have not identified Burwell Starke.
(2) Tar-water was a popular eighteenth-century medicine, strongly recommended by Bishop Berkeley. Darwin was very familiar with lead-poisoning: see letter 94R and *Zoonomia* ii 375.

96F. To Thomas Beddoes, [1796?]

Dear Dr

I went to Margate, about 3 years ago, which is 200 miles from Derby, and was out 6 days and had a present of 100£. I have twice been to London and had about 10 guineas a day given me and what was supposed to be travelling expences. When a physician goes so far from home he looses the custom of families; as they will not wait for his return, as when he goes but one day's journey. And those families apply to another. Your staying so long from home must have been attended with much loss. I think you should have 10 guineas a day, and a sum for what should be concieved to be chaise-expences.

NB. you say you inclosed a one-pound bank note, which did not arrive!

I like your proposal for a new medical collection.

Mr Creasn's[?] letter on Mr Losh is very sensible, he sees the disease in the light I think is the true one, which is partial insanity, the fear of death, and the fear of poverty.

Adieu, from, dear Dr
Yours very sincerely
E Darwin

Addressed Dr Beddoes, Clifton, Bristol. ['4 Albemarle Row' written on in another hand.]

Original Library of Wellcome Institute for the History of Medicine, London.
Printed Unpublished.
Text From original MS.
Notes

(1) Karl Pearson, in his *Life, Letters and Labours of Francis Galton* (Vol I, p. 46), states that Darwin was paid 100 guineas for a visit to Margate in 1793, to treat Mary Anne Galton. So, if we accept Darwin's '3 years ago', this letter belongs to 1796.

(2) The letter confirms other indications that Darwin rarely undertook medical journeys lasting more than one day, and the reasons he gives are convincing. This is the only written record that he twice visited London on medical business.

(3) 'Mr Creasn' is probably Thomas Creaser (1766–1825), a surgeon in Bath and a friend of Beddoes, Jenner and Watt. Creaser was an advocate of pneumatic medicine and was author of *Evidence of the Utility of Vaccine Inoculation* (Bath, 1800) and other medical pamphlets. Mr Losh may perhaps be James Losh (1763–1833), translator of Constant's *Observations on the strength of the present Government of France* (1797).

(4) The proposed new medical collection may be *Contributions to Physical and Medical Knowledge* (Biggs & Cottle, Bristol, 1799), in which there is a letter from Creaser.

96G. To Thomas Brown, 27 October 1796

Sir,

I am favoured with your letter, and shall be glad to read your manuscript, if it be not very voluminous, and if it be transcribed in a hand easily legible. I well know the power of oratory, and that a man may say many things ingeniously on both sides of almost any question. But in philosophical works my desire is to investigate the truth; and I shall be ready to acknowledge any errors I may have fallen into, or to correct any inaccuracies of language which may have occurred in a work which is in part metaphysical, and therefore ought to be accurate.

I am, Sir, your obedient servant,

E Darwin

Derby, Oct. 27 –96

Original Not traced.

Printed D. Welsh, *Life and Writings of Thomas Brown.* Tait, Edinburgh, 1825, p. 46.

Text From above source.

Notes

(1) Thomas Brown (1778–1820) was a brilliant and precocious young man who entered Edinburgh University at the age of 16, and was, at the time this letter was written, an 18-year-old law student. Later he studied medicine, and in 1810 he became Professor of Moral Philosophy at Edinburgh and a very popular lecturer and writer: his *Lectures* reached a nineteenth edition in 1851.

(2) Darwin's letter is written in reply to a long and polite letter from Brown, dated 24 October 1796 and given in full in Welsh's biography (pp. 44–6). Brown begins: 'In acknowledging the delight which I received from the perusal of Zoonomia, I only agree with the public voice. I am, however, surprised that while every one has been delighted, no one as yet has answered'. Brown noticed that the reasoning was sometimes 'more specious than solid, and the leading principles seemed often to involve inconsistencies. I therefore . . . marked down my observations occasionally . . . I found that my remarks had greatly swelled upon my hands . . . and have since been induced to think of committing them to the press'. He then asks if he might send Darwin a copy of his observations for comment.

(3) Darwin's reply is very civil and reasonable, though if he had known what he was about to receive he might have wished it committed to the flames rather than the press. Darwin was unaware that Brown was so young.

(4) Brown wrote again on 27 November, apologising for the delay and telling Darwin that the manuscript was on its way to him. The letter is given by Welsh (pp. 46–8).

(5) When he received the manuscript, Darwin was obviously appalled. To begin with, it must have been immensely bulky, because the published version, *Observations on the Zoonomia of Erasmus Darwin* (1798), runs to 560 large pages. Darwin was very busy in 1796 as a result of the publication of the second (and longer) volume of *Zoonomia*, and had no time to wade through such a manuscript. Worse still, Brown's comments were often cogent and difficult to answer: Darwin's book *was* riddled with metaphysical flaws, and Brown was already a master of argumentative logic. Darwin had met his match in disputation, and could not hope to answer Brown without spending several weeks on the task. Finally, Brown's own ideas were often rather silly, and not too difficult to tear to pieces, given the time which Darwin did not have. Aware that he was being worsted, Darwin replied somewhat rudely, it seems, on 2 December 1796. Welsh politely says (p. 48) he does not feel justified in publishing this letter. Brown replied on 5 December with his usual courtesy (Welsh, pp. 48–9), saying he was surprised at Darwin's asperity but

would not himself descend to such a tone: 'the angry feelings I have never cultivated, as I do not think they add much to the dignity of our nature'. It is probably just as well that Darwin did not know he was being scolded by a lad of eighteen! After this, Darwin made a more reasonable reply. See next letter.

96H. To Thomas Brown, 20 December 1796

Derby, Dec. 20 –96

* * *

I shall mention those of your observations which I can recollect. You doubt that ideas can exist without being attended with pleasure or pain; and therefore assert, that an idea cannot exist without sensation. I suppose you here use the word sensation in Mr Locke's mode of using it; which has raised a difficulty in your mind, since the figures of the letters which compose words, as we read a book, give us neither pleasure nor pain, and yet they introduce into our minds the sense of those words; and therefore exist, though they are not attended with the sensation of pleasure or pain. The great Malbranche saw this analysis of an idea, but no other metaphysician. Zoonomia, sect. vii. 3.2. In sect. iii. 6.4. it is said, "I ask by what organ of sense you first become acquainted with certain ideas, as of wisdom or benevolence?" This subject has been explained by Berkeley and Hume; who have shown, that what were believed by Mr Locke to be general ideas, are all particular ones. Some of these I have given account of in sect. xv. 1.5. Yet you think, from this question (which was to save repeating what others had said on this subject) that I acknowledged the application to be impossible; and that you exult in saying, that if it would not apply to the whole it could not be a true theory! Something you have said about black when you spoke of ocular spectra, I do not accurately remember; but I think you seemed to hold, that black was a colour, and consequently a stimulus; as the unwise favourers of Dr Cullen's theory have said that cold was a stimulus.

* * *

You would write English well if you would lay aside the nonsense of metaphor. Metaphors, in a philosophical argumentative treatise, are a disgrace, as they consist of unimportant analogies, whereas philosophical argumentation consists of important analogies.

* * *

Original Not traced.
Printed D. Welsh, *Life and Writings of Thomas Brown*. Tait, Edinburgh 1825, pp. 50–1.
Text From above source.
Notes
(1) This is from Darwin's second reply to Brown's comments on *Zoonomia*. See notes

to letter 96G. It is obvious that the subject is speculative enough to rule out any probability of agreement between two disputants who decide to take separate stances.

(2) John Locke (1632–1704) in his *Essay concerning Human Understanding* (1690) denied that we have innate knowledge and argued that the senses feed the mind with the material for 'ideas'.

(3) Nicholas Malebranche (1638–1715), the French Cartesian philosopher, in his *De la Recherche de la Vérité* (1674) argued that only three items are needed to describe the world, namely God, selves and ideas. External objects are not required, being implicit in the ideas.

(4) Darwin's three references to *Zoonomia*, all to Volume I (pages 40, 28 and 129 respectively), are all concerned with ideas and sensations.

(5) William Cullen (1710–1790), Professor of Medicine at Glasgow and Edinburgh, was one of the most influential medical theorists of the eighteenth century, and Darwin's own theory owes much to him.

(6) Brown replied to Darwin on 28 December, his letter being quoted at some length by Welsh (pp. 51–6). Brown is mainly concerned to expound his ideas on ideas and their relation to sensations. Needless to say, he disagrees with Darwin. Despite the gulf between them, Darwin was willing to continue the argument: see next letter.

97A. To Thomas Brown, 12 January 1797

* * *

In my second volume, at the end, in a supplemental page, which perhaps in the Irish octavo edition may be left out (as some are (I see) in the first volume), I have observed how difficult it is to <u>think without words</u>, and still more so, when the words in use have two meanings. This I applied to a person who wrote to me that the motions of epileptic fits were not voluntary motions (as I had asserted), because they were involuntary. Now, I think you and I use the word sensation in totally different meanings; I will therefore repeat what I mean, without the word sensation at all, and will use pleasure and pain instead. Thus, animal motions (viz. of fibres, of muscles, or of the ends of nerves in the organs of sense) are [prec]eded by stimulus, or by pleasure or pain, or by desire or aversion, or by association, and are all also occasionally succeeded by one of them. Now the motion of the retina, or an idea from sight, may be preceded by stimulus of light, and may be succeeded either by pleasure or pain, or by a volition, or by association; and other motions may be produced in consequence of these sensorial powers, but none of them are a necessary part of the idea, or motion of the retina, caused by the stimulus of light, producing irritation and consequent fibrous motion.

In motion produced by the sensorial power of association and of volition, the <u>whole</u> of the sensorium (though a part, the central parts of it, must) may not be affected; because the whole nervous system is an extensive thing, and some branches of nerves may communicate their motions to one part of the sensorium, and other fibres may by association or volition be excited into action by other branches of the sensorium, without affecting the whole of it.

Hence pleasure or pain may ascend from a part of the extremity to a part of the centre of the sensorium, and desire or aversion at the same time, in other branches of the nerves, may pass from another part of the centre to another part of the extremity; but, when almost the whole of the sensorial power, which can be spared from the vital motions, is passing one way (as in great anger, which is excess of volition), little or none passes at the same time the contrary way, as angry people in general fighting do not feel the wounds they receive. These I expressed by saying, when volition acted strongly, pleasure or pain was not felt, as the movements were in contrary direction. In your paper you seem to think this assertion quite general, and were, I recollect, jocular about it. I believe you left out the word strongly.

You ask how reflex ideas are produced? Certainly like general ideas. Ideas are motions of the extremity of some nerve of sense; a general idea is a partial renovation of one of these movements. A reflex idea is also a partial renovation of these movements, and are not exactly the same every time we use the same word, as is evident by the inaccuracy of the word memory, which means either recollection or suggestion.

In the dark, people often see colours, indeed mostly, as when one lies an hour awake; but if muscles are not in action, and not extended by antagonist muscles, they require no counter extension.

What you say about a bell, is treated of in the section on repetition; and of number and time in fifteenth section.

If you think it worth your while to send me a very short outline of your theory of the mind, I will give you a candid opinion of it, or of any other questions which you ask for information only. My second volume has brought me many patients from even London, and distant parts of England, and very many consultation letters from both the faculty, and from others, so that I believe my work meets with daily new readers.

* * *

You do not say what your professional studies are, whether Law, Physic, or Divinity, so that I do not yet know how to direct my letter. You say the stimulus is the same, and the sensorial power the same, etc. You forget that the quantity of stimulus and the quantity of sensorial power is perpetually varying, and hence the fibrous motions may be succeeded by pleasure or pain, or by volition, or by association, or by none of them. Adieu.

From your real friend,
E Darwin

[The following lines are written upon the envelope.]

January 12, 1797

I wrote the inclosed at a public house when I was much fatigued, and I fear you will not be able to decypher it. Your syllogism amused me much, where you say "the stimulus is the same, and the sensorial power the same; the effect of the whole must be the same." Now the stimulus constantly varies in quantity, as the odour of a rose by every breeze; the sensorial power is never the same in quantity, as that depends on the previous accumulation or expenditure. Hence the fibrous motions are sometimes so great as to be attended with pain or pleasure, at other times not so great.

You talk of sensation not to be perceived; that is, which is not attended with pleasure or pain, so that you do not use the word in the same meaning as I do, or as Mr Locke does.

I do not recollect any other of your objections, but I thought them all easily answered.

* * *

Original Not traced.

Printed D. Welsh, *Life and Writings of Thomas Brown* (Tait, Edinburgh, 1825) pp. 56–60.

Text From above source.

Notes

(1) This is a reply to Brown's letter of 28 December 1796: see notes to letter 96G. Although he says he was writing when very tired, Darwin's argument is lucid and lively, and he stoutly maintains his views on ideas and sensations.

(2) Brown replied to Darwin on 21 January 1797 (Welsh, pp. 61–6), but Darwin made no further reply. This is not surprising, since there was little chance of reconciling the disputants and Darwin was probably already busy on *Phytologia*.

97B. To Thomas Wedgwood, 7 February 1797

Derby Feb. 7 –97

Dear Sir,

The ascarides are a dreadful enemy; the water drank <u>at</u> Harrogate I have found (I think) has sometimes succeeded. But I have used clysters of aloe (I think) with success in 2 or 3 cases. Thus, a quarter of an ounce of sowtrine aloes in powder is to be boil'd in a pint or rather more of gruel till it dissolves. To be used as a clyster every third night, or twice a week for 6 or 8 weeks.

Many other injections are described in Zoonomia Vol. II, Class 1.1.4.12, amongst which Harrogate water is advised to be thus applied, but I have used aloe oftener.

Your Ink-stand is very ingenious, but I fear it is too complex for the use of ignorant people. Adieu. Our best wishes attend all with you, I am, dear Sir

yours most sincerely
E Darwin

I am just return'd from a long journey, and the post-hour is ready to begin, and will not wait; otherwise I should have said more about your ink-pot.

Addressed Thos. Wedgwood Esqr., Etruria, near Newcastle, Staffordshire. (*Postmark* Derby.)

Original Keele University Library, Wedgwood Archive, 26774–35.
Printed Excerpt in *Doctor of Revolution*, p. 259.
Text From original MS.
Notes
In this letter we return to the ascarides and the Harrogate waters. See 93F and 93H. Darwin has two pages on ascarides in *Zoonomia* (Vol. II, pp. 55–6), where he recommends various other clysters, including mercurial ointments and tobacco smoke. 'Sowtrine aloes' are presumably Socotrine aloes (from the Isle of Socotra).

97C. To Thomas Wedgwood, 27 March 1797

Derby March 27 –97

Dear Sir,

Ascarides are so nearly transparent and so hair-like, that I never believe their existence, unless the Evidence sees them move. Now as your enemas would probably kill them, I am not surprized at their not being seen in those evacuations.

For the same reason, where one is seen evidently to move, I always believe there are many, which being dead were not seen.

Pains, which are not attended by fever or inflammation, are owing to torpid action (debility) of the part, as of the membranes of the temple or which surround the tooth – hence whatever exhausts your animal power, as much study, or other fatigue, or abstinence, or cold, will contribute to weaken you, and induce a torpor of these weak parts.

The torpor of the membranes of a tooth, which affects the temple, I believe to be of the upper jaw of the same side – but this is not quite certain.

I have given an amalgama in the tape-worm with success, but have not yet tried it in ascarides.

As you are going to Harrowgate, I wish you only to try the water both by the mouth, and by clyster – and the last may be used cold, if not disagreable to your sensations.

I beg to be remember'd to all at Etruria, and am, dear Sir,

your sincere friend
and obed. serv.
E D

Addressed Thos. Wedgwood Esqr., Etruria, near Newcastle, Staffordshire. (*Postmark* Derby.)

Original Keele University Library, Wedgwood Archive, 26775–35.
Printed Excerpt in *Doctor of Revolution*, p. 259.
Text From original MS.
Notes

Darwin has now almost given up trying to find rational explanations for Tom Wedgwood's ailments, and this is his last known letter to Tom. The diagnosis of ascarides, to be treated by the waters of Harrogate, is still being floated as a hypothesis, though Darwin seems only half-convinced by it. His explanation of pain-without-a-cause is of masterly diplomacy, and so is the phrase, 'but this is not quite certain'. Presumably he feared that Tom would have all his upper teeth extracted when he next had a pain in the temple, and would then complain that the pain was just as bad. It was a difficult case.

97D. To James Currie, 22 September 1797

Derby Sept. 22 –97

Dear Doctor,

I had the pleasure of your letter concerning Mr Stirling, and had afterwards an interview with him, and now write to you on the subject. His fits of epilepsy only occur, it seems, in sleep; which I suppose to be owing to the increase of sensibility to internal stimuli at that time. How is this to be avoided? perhaps by a previous increase of stimulus, which may in part exhaust a part of the accumulating sensorial power? In 2 or 3 patients, who had these slighter epileptic fits in their sleep, I have found one grain of opium with six of rhubarb given uniformly every night for months, and even years, has succeeded.

In respect to other medecines, the bark may be taken twice a day for 2 or 3 weeks, and some essential oil, as oleum succini in small quantity – but I should expect rather to prevent the effect, than to remove the cause. adieu

from, dear Sir

Your very sincere E Darwin

Addressed Doctor Currie, Physician, Liverpool.

Original Historical Society of Pennsylvania, Philadelphia.
Printed Unpublished.
Text From original MS.
Notes

(1) Dr James Currie FRS (1756–1805) was a well-known Liverpool physician and man of letters. See *Memoir of the Life, Writings and Correspondence of James Currie*, ed. W. W. Currie. 2 vols. Longman, 1831; *DNB*; and R. D. Thornton, *James Currie, the entire stranger and Robert Burns*. Oliver & Boyd, Edinburgh, 1963.

(2) In June 1793 Currie wrote and courageously published the famous 'Letter from

Jasper Wilson to Mr Pitt', which argued against war with France. Darwin held similar views and may have written to Currie. A long case-history of an illness suffered by Currie in 1784 appears in Darwin's *Zoonomia* ii 293–6.

(3) This letter shows that Darwin was as assiduous as ever in his medical duties. He was also still keen on prescribing opium as a long-term treatment, apparently oblivious to its addictive effects. I have not identified Mr Stirling.

97E. To John Sinclair, 8 November 1797

Derby, Nov. 8 –97

Sir,

I have employed the vacant hours which I could command, in writing a theory on vegetation, applied to agriculture and gardening. The work has proceeded but slowly, and it will yet be some months before I shall think I have sufficiently studied it, to commit it to the press. I believe it will make two volumes octavo. I hope you will permit me to prefix the inclosed dedication to you; as without your instigation, I should not have attempted it.

In respect to a society to collect medical facts, I dare say it might be a very useful institution, if managed with that activity and address which you have so laudably exerted in the Agricultural Society. I am, Sir, with great respect, your much obliged and obedient servant,

E Darwin

DEDICATION

To Sir John Sinclair, Baronet, President of the Board of Agriculture, by whose unremitted exertions such important improvements have been accomplished in the cultivation of the earth, that great source of life and felicity! this Work, which was begun by his instigation, and forwarded by his encouragement, is dedicated, with true respect, by his much obliged and obedient servant,

THE AUTHOR

Original Not traced.

Printed Correspondence of the Rt. Hon. Sir John Sinclair, Colburn & Bentley, 1831, Vol. I, pages 411–12.

Text From above source.

Notes

(1) Sir John Sinclair (1754–1835) was an energetic and influential man. He was the owner of extensive estates in Caithness, and a Member of Parliament from 1780 to 1811. He persuaded Pitt to establish the Board of Agriculture in 1793, and was President of the Board for five years, conducting its business with 'activity and address', to use Darwin's phrase. During the 1790s Sinclair also wrote the valuable *Statistical Account of Scotland* in 21 large volumes, compiled by asking all parish clerks to send details of the population, natural history and products of

their parishes. For biography of Sinclair, see J. Sinclair, *Memoirs of the Life and Works of Sir John Sinclair* (Blackwood, Edinburgh, 1837, 2 vols.) and R. Mitchison, *Agricultural Sir John* (Bles, 1962).

(2) The book on vegetation which Darwin describes is of course *Phytologia*, published in 1800, as a single quarto volume of 630 pages. Darwin began work on it in 1796 after completing *Zoonomia*, and finished in 1799. The published Dedication to *Phytologia* is similar to that given in this letter, but less effusive.

(3) In introducing the letter, Sinclair says 'it was highly gratifying to me, to have prevailed on so able a writer as the celebrated Dr Darwin . . . to draw up a work on Practical Agriculture'. Darwin's *Phytologia* is, he says, 'a most valuable performance', though 'of too philosophic a description' to qualify as Practical Agriculture. However, there are useful practical observations in it: 'in particular, he is one of the first authors who recommended bone-dust as a manure'.

98A. To James Currie, 3 January 1798

* * *

I am in doubt whether I should trouble you with this letter or not; but I think your facts so valuable, that I am grieved to observe that you did not give yourself the trouble to exert the voluntary power which you possess, in reasoning about them, and hope you will again exert yourself in detaching the sources of error in medicine as well as in politics.

If any thing should bring you to this part of the country, your company would give me the greatest pleasure; and I flatter myself I could show you a few, who cultivate reason both in physic and in politics, whose conversation would amuse you, and who are, as well as myself, great admirers of Jasper Wilson. Adieu from,

Dear Doctor,
Your sincere friend,
E Darwin

Original Not traced.

Printed In *Memoir of the Life, Writings and Correspondence of James Currie, M.D., F.R.S. of Liverpool*, edited by his son W. W. Currie. 2 vols., Longman, London, 1831. Volume 1, p. 240.

Text From above source.

Notes

(1) Since the previous letter to Currie (97D), Darwin had been rather upset by the publication of Currie's book *Medical Reports on the Effects of Water cold and warm in Fever* . . . , in which Currie recommended treating fever by dashing or pouring cold water over the body or by drinking large quantities of water. Darwin 'did not express the favourable opinion generally entertained of the work' (*Memoir* i 231), and he wrote what was apparently a rather scathing letter to Currie on 10 December 1797. Currie made a restrained reply in an eight-page letter dated 25 December 1797, given in full in the *Memoir* (i 232–9). Currie defends most of his

conclusions, but admits some errors. Darwin replied in this letter of 3 January 1798, of which only the ending is quoted verbatim in the *Memoir*. In the early part of the letter, W. W. Currie tells us, Darwin says he was induced to reply as his last letter had not disobliged Dr Currie, and he assured Currie that he had no regard for his own opinions more than for those of other people; Truth was his only goddess. He still disapproved of cold being called a stimulus, and of the word 're-action'. Nor could he concur with Currie's idea that heat was repelled to the heart by the spasm of the capillaries, and sent back by the re-action of that organ. He also still rejected the term 'sedative' applied to opium, and the theory that the diminution of heat in fever was the *cause* of perspiration becoming sensible, not the *consequence* of that cooling process. Darwin's letter concludes with the paragraphs reproduced here.

(2) This correspondence with Currie shows that Darwin was not inclined to allow erroneous medical theories to gain ground without protest. Currie made the common error of assuming that if a patient recovered after treatment, it was the treatment that caused the recovery: his water-treatment did have the advantage that the patient remained unpoisoned, and (if able to stand the shock of the cold water) might have a better chance of recovering than a patient poisoned by drugs.

(3) Despite his medical disagreements, Darwin still stands side by side with Currie in politics, as shown by his reference to Jasper Wilson. See letter 97D, note 2.

98B. To Robert W. Darwin, 8 January 1798

* * *

I have not spoken to him [Robert's elder brother Erasmus] on your affairs, his neglect of small businesses (as he thinks them, I suppose) is a constitutional disease. I learnt yesterday that he had like to have been arrested for a small candle bill of 3 or 4 pounds in London, which had been due 4 or 5 years, and they had repeatedly written to him! and that a tradesman in this town has repeatedly complained to a friend of his that he owes Mr D. £70, and cannot get him to settle his account. I write all this to show you, that his neglectful behaviour to you, was not owing to any disrespect, or anger, but from what? – <u>from defect of voluntary power</u>. Whence he procrastinates for ever!

* * *

Original Not traced. In the possession of Charles Darwin, 1879.
Printed C. Darwin, *Life of E. Darwin*, p. 75.
Text From above source.
Notes

In the late 1790s Darwin's eldest surviving son, Erasmus junior, who was in his late thirties, was doing well as a solicitor in Derby. But as he himself remarked, 'I am apt to "put off a little"' (letter to Mrs Hayley, 7 March 1795; in the Fitzwilliam Museum Library, Cambridge), and he seems to have developed a neurosis about settling accounts, which his father regarded as rather a trifling problem – a diagnosis that proved to be tragically in error when Erasmus's unsettled accounts drove him to suicide in December 1799.

98C. To Lucy Galton, 22 July 1798

Derby Jul. 22 –98

My dear Madam,

How could you think of sending me Mr Berington's infected letter? I have had a sore lip, ever since I read it! which I suppose must have been owing to the putrid exhalations from the infected towel, he had used to rub his hands with, transfer'd to the bibulous paper. I hope you will desire him in future to sprinkle all the papers he may send you with holy water, which must wash out all infection.

Or our good friend should exercise his apostolic powers, and exorcise the evil spirit, which resides in the ladies shift, or towel, or in the soap-suds. And lastly he may send it with Buonoparte to be laid in the red Sea; from which no spirit can return, as I am inform'd, till all their earthy stains, and metaphysical contagions, are perfectly washed out, and they return white as snow, and pure as oxygen. So when the Devil enter'd the swine, they were obliged to run headlong into the sea to wash themselves into health.

I ought to be serious! – The power of imagination is well known to all, but should be particularly well known to our ingenious catholic friend, as it has been formerly so successfully used in the mysteries of our holy religion. Do not the bones of Saints cure diseases? I believe those of the mind frequently, and sometimes of the body! About 30 years ago a Saint's tomb in Paris cured so many sick folks, that the Government directed it to be wall'd up – I suppose least the Doctors of Medecine should be starved for want of employment.

I seriously believe that our friend Mr Berington is not desirous to decieve, or easy to be decieved; and that therefore his sister became ill accidentally after puttting on the famous linnen; and that the maid became diseased from imagination; and also in a slighter degree our ingenious friend. And I really suspect, that the ceremony of exorcisement might cure them all!

The shirt of Hercules was moisten'd with oil of euphorbia, or of the root of bryony, or with a solution of arsenic, whence he tore himself to pieces and became all over ulcers. And the Devil is said to have given Job the confluent Smalpox. Medaea gave a gown dip'd in oleum petrae, fossile oil, to Creusa, which took fire, when she came near a torch, and burnt herself and the palace. But none of these seem equal to Mr Berington's towel,. which retain'd its satanical power after so many washings, and interment in the ground for many months.

Finally I conclude that the contagious material of this towel was neither animal, vegetable, nor mineral, but metaphysical, and can only be cured by a counter inchantment, and would recommend the mysterious tractations of animal magnetism. Enough of this!

The ingenious fair lady, who accompanied your letter, is afflicted with a weak digestion, and as all our strength arises ultimately from the food we can perfectly digest, her debility is at present very great, but I hope she may become tollerably healthy, tho' perhaps not very robust.

I don't think I ought to treat this case of our ingenious friend quite so lightly; as the infectious matters of diseases, which were then new, as of the smallpox, measles, itch, {and} the plague itself, must have been as surprizing, when they first arose to afflict mankind. But things so contrary to the known general laws of Nature, as polarity of magnetism, and the shock of the coated electric jar, were not credited, till they were witness'd, and their evidence confirm'd by a thousand testimonies.

Practical education by Miss and Mr Edgeworth will amuse you I dare say – I have been much amused with the parts I have already read of it.

Mrs Darwin begs to unite with me in our best respects to Mr Galton and family, as well as to yourself

who am your sincere
and affectionate friend
E Darwin

Addressed Mrs Galton, Crescent, Birmingham.

Original University College London, Galton Papers, 39G.
Printed Unpublished.
Text From original MS.
Notes

(1) Mrs Lucy Galton (1757–1817) had been a frequent patient of Darwin for nearly 20 years (see notes to letter 93E), and from this letter it appears that Darwin indulged in Lunar-type banter with her as well as with her husband.

(2) Mr Berington was Joseph Berington (1746–1827), an unorthodox Catholic priest who wrote 15 books on historical or controversial subjects. He is said to have been the first Catholic priest in eighteenth-century England who openly dressed in black. Also he apparently attended some Lunar meetings in the 1780s, when he was priest at Oscott, near the Galtons' home at Barr. For further details see *DNB* and *Life of Mary Anne SchimmelPenninck* (Longman, 1858).

(3) The MS of Mr Berington's report on the alleged infection is preserved with Darwin's letter. He states that a lady visitor (apparently his sister), having received her linen from the wash, thought it had a strange smell; soon she developed shivering fits, sweats, dry throat and burning stomach, though with no loss of appetite. Re-washed, the linen still had the same effect. Even burying it in the earth for some days did not remove the taint. A maidservant wore it, with the same effect. Later Mr Berington noticed the smell on a towel, rubbed himself with it and began to sweat, etc. He asks what substance is causing the trouble. Medically, Darwin's sceptical reply is reasonable: it *was* probably imagination,

though it might have been an allergy. Biographically, Darwin's reply is interesting for its sustained banter: although he was 66, he could still be roused to overflowing exuberance even by a professional case. His side-swipes at Christianity, which are difficult to combat because they are integrated into the banter, typify the attitude which earned him the dislike of Mrs Galton's religious daughter Mary Anne (later Mrs SchimmelPenninck), who gave a vivid if inaccurate account of a visit by Darwin to her mother (*Essential Writings*, pp. 45–7).

(4) Darwin's instant command of classical mythology is evident in the two obscure yet apt stories he produces. It was Deianira, his wife, who gave Heracles a polluted white garment when she feared that Iole, whom he had just rescued, might supplant her: the usual story is that Deianira soaked it in the blood of the centaur Nessus, who had tried to ravish her and had been shot by Heracles with a poisoned arrow. See *Dr Smith's Classical Dictionary* (1899), p. 400. Medea was an enchantress who helped and then married Jason; when he left her in favour of Creusa, daughter of Creon, king of Corinth, Medea sent Creusa a gown soaked in petroleum, with the consequences Darwin mentions. See *Loves of the Plants* III 135–78.

(5) 'Animal magnetism', or mesmerism, was introduced as a medical technique by F. A. Mesmer (1734–1815), but he was expelled from Austria in 1778 for practising magic. Though his work created great interest and has led on to modern methods of medical hypnosis, the pioneer himself (as usual) received only obloquy, not thanks.

(6) *Practical Education* by Maria and R. L. Edgeworth (3 volumes, Johnson, 1798) enjoyed great success. See D. Clarke, *The Ingenious Mr Edgeworth* (Oldbourne, 1965), pp. 162–7.

98D. To Lord Harrowby, 9 September 1798

Derby Sept. 9 –98

My Lord,

I believe it was Seltzer's water, which I might mention to your Lordship, as being easily made without an apparatus for aerating it which is as follows: dissolve a quarter of an ounce of crystals of Sal Soda in a wine-quart of water. Make about a quarter of a pint of this solution (pour'd from any sediment, which the previous impurity of the water may occasion), a little warmish, and then add to it about ten drops of marine acid (call'd Spirit of salt) and, as soon as it is stir'd together, it is to be drank while full of small air-bubbles – twice a day.

Spa-water requires Dr Nooth's glass-apparatus, for the purpose of impregnating it with what is call'd fix'd air (or carbonic acid). If to every quart of water in the middle glass of Dr Nooth's Glass-apparatus be added about two drams (two teaspoonfuls) of chalybeate wine; and it is then well impregnated with fix'd air in the usual way (by means of vitriolic acid, and chalk in the lowest glass), in a cool place, and then put into four-ounce phials, and kept in a cellar, one to be taken twice a day: it forms a water in virtue similar to the Spa-water. If rust of steel could be put into the middle glass of the apparatus without obstructing the valve, instead of the chalybeate wine, the water, if well aerated, would more *perfectly* resemble the spa. The quantity of the rust of steel need not be determined, as the aerated water would only dissolve a small portion of it.

The Sea-water about this island is of various strength in respect to the quantity of salt it contains: viz. from $\frac{1}{28}$ to $\frac{1}{30}$ part. The medium is about 29 parts of water to one part of sea-salt, to which there should be added about one eightyth part of magnesian vitriolic salt, commonly call'd Epsom salt, or sal catharticus amarus. Now as all salt-rocks are believed to have been deposited from sea-water evaporated by subterranean fires, the salt from salt-springs (which are produced by water flowing through these salt rocks) is the same as sea-salt; if it be evaporated by a mild heat, as by the Sun, it exactly resembles Bay-salt.

The salt-springs at Shirley-Wich are much stronger than sea-water, and should be diluted with common water either for bathing in, or for drinking, till they become of a similar degree of saline impregnation. This would be easiest done by mixing common water with $\frac{1}{29}$ of common salt (or of bay salt) and with $\frac{1}{80}$th of Epsom-salt sold by the name of bitter purging salt, and then observing its weight by an hygrometer [hydrometer], or such an instrument as Spirit-dealers or excisemen use for determining the weight (or lightness) of spirits, and then by diluting the salt-water of Shirley Wich till it comes to the same specific gravity. It would then be in every respect, I believe, similar to sea-water; and might be used as long as it keeps sweet – or longer, as it then becomes only similar to Harrowgate or Kedleston waters.

I beg your Lordship will present my most respectful compliments to Lady Harrowby and to all your family, and am, my Lord,

Your much obliged
and obed. serv.
E Darwin

Mrs Darwin begs to be remember'd to Lady Harrowby and your Lordship, and family.

Original Sandon Hall, Stafford, Harrowby MSS, Vol. IX, ff. 19–20. (Privately owned.)

Printed Unpublished.

Text From original MS.

Notes

(1) Lord Harrowby (1735–1803), born Nathaniel Ryder, was created Baron Harrowby in 1776 after serving as MP for Tiverton from 1756 to 1776. See Namier & Brooke, *Commons* iii 388–9, and letters 90C & 96D.

(2) This letter shows again Darwin's care and attention to detail in his professional work. He might merely have sent a prescription, but this detailed and readable essay would have a much better psychological effect, even if it was afterwards passed on to an apothecary.

(3) Seltzer water is a manufactured aerated water intended to imitate the natural waters at Selters, near Koblenz. Selters water contains 2250 parts per million of sodium chloride, 800 ppm of sodium carbonate, 260 ppm of magnesium carbonate, 240 ppm of calcium carbonate, and smaller quantities of potassium salts. See *Dictionary of Applied Chemistry* (1937), Vol. 1, p. 156.

(4) Sal soda is sodium carbonate. Marine acid is hydrochloric acid. 'Chalybeate' means 'impregnated with iron'.

(5) Joseph Priestley in 1772 was the first to describe how to make aerated waters by dissolving carbon dioxide ('fixed air') under pressure. His method was criticized and improved by Dr John Nooth (*c.* 1740–1828), who substituted a glass valve for the bladder used by Priestley. See *Phil. Trans.*, **65**, 59 (1775). Priestley discussed Nooth's apparatus in his *Experiments and Observations on Different Kinds of Air* (1774–1777), and eventually seems to have approved of Nooth's method. See J. R. Partington, *History of Chemistry*, Vol. 3 (1962), p. 248.

(6) See notes to letter 93F for magnesian vitriolic salt, Bay-salt, etc.

99A. To Anne Greaves, 18 March 1799

Derby, Mar. 18 –99

Dear Madam,

As the medecines had previously so well agreed with Miss Greaves, and then suddenly produced sickness, and that even when taken in half the dose, one should think some accidental mistake had been made in making them up.

At all events, as she is now considerably better, and has taken medecines so long; I think she had better for a time discontinue the use of them. Or certainly not take any more of this parcel, which has so evidently disagree'd with her.

And by being awhile intermitted, they may have a better effect. The rhubarb she will please to continue, so as to induce a daily evacuation. And unless she continues much better, I shall hope to hear again from you, and will direct something further.

I beg my most respectful compliments and good wishes to Miss Greaves, and am, dear Madam,

Your much obliged friend
and obed. serv.
E Darwin

Addressed Mrs Greaves, Lichfield.

Original American Philosophical Society, Philadelphia, Pa., Getz Collection.
Printed Unpublished.
Text From original MS.
Notes
(1) Mrs Anne Greaves (1739–1819) was the elder half-sister of Brooke Boothby (see

letters 81D & 95J) and the widow of Joseph Greaves (1731–1780) of Aston-on-Trent, Derbyshire, who was High Sheriff of Derbyshire in 1765. They had two daughters: Maria (1763–1848), who in 1783 married Edward Sneyd (*c.*1755–1832) of Byrkley Lodge, Staffordshire, brother of Edgeworth's wives Honora and Elizabeth; and Anne (*c.*1765–1827), who remained unmarried and is the Miss Greaves of this letter. Mrs Greaves, and probably her daughter, lived at Lichfield for nearly 40 years, and Anna Seward knew them well: writing in June 1793 from Buxton, Anna refers to her 'blooming friend, our old acquaintance, Miss Greaves', and to 'dear Mrs Greaves, of our little city' (Seward, *Letters* iii 250). See Glover, *Derby* ii 41–3; *Alum. Oxon.* ii 555; *Staffordshire Advertiser*, 16 Oct. 1819, p. 4; *Derby Mercury*, 10 Aug. 1780; *Gent. Mag.*, **102** (1), 380 (1832) & **30** (NS), 103 (1848); and inscriptions on the Greaves tomb in Aston-on-Trent churchyard.

(2) Though this is a routine medical letter, Darwin's genuine concern for his patient is evident, and would make a good impression.

99B. To Robert W. Darwin, 8 August 1799

* * *

. . . all which I much disapprove. Therefore you will please not to mention it, and I hope it will fall through.

* * *

Original Not traced. In the possession of Charles Darwin, 1879.

Printed In first proofs of Charles Darwin's *Life of E. Darwin*, now in Cambridge University Library.

Text From above source.

Notes

In 1799 Erasmus's eldest surviving son, Erasmus junior, was thinking of retiring from his legal business, though he was only 39. He had the idea of building a cottage at Lichfield on the site of his father's botanic garden, which was probably still jointly owned by his father and himself. This is the idea of which Erasmus 'much disapproves' in this letter, and it did fall through: Erasmus junior bought Breadsall Priory instead.

99C. To Robert Waring Darwin (brother), 7 October 1799

Derby Oct. 7 –99

Dear Brother

Erasmus sends by Miss Hall the deed, and the copy of it, which he assures me is done exactly after your instructions. I feel myself much obliged to you indeed on this subject – and must beg, when you have executed it, that you will return it at any convenient opportunity, and I shall then sign it here. And as I have not yet determined, which of the young men to bring up to the church, I must beg of you not to mention it to any of our family or relations, least it should happen to reach their ears.

We have had no fair weather, I think, since you were here, but our garden has not been again over-flow'd. The mud, which subsided, on the leaves of the rhubarb plants, kill'd most of them; but they have put out fresh ones – there is one leaf of the rheum hybridum escaped, and is now about 3 ft long and above two ft wide.

Miss Hall has also brought for you a beautiful specimen of the Labradore fossil, it seems to have its origin, I suspect, from Mother-pearl shells, as common limestone from other shells – and was probably afterwards converted into flint by some unknown process of nature; as othe[r] calcareous substances have been transmuted into siliceous ones; as I suppose the flint-nodules found in chalk-beds, which I suspect to be occasion'd by the iron contain'd in the animal, which inhabited the shell – because I observe the siliceous chrystals, call'd Bristol-stones, are form'd on calcareous ones, which are colour'd with iron – and the Peak diamonds are siliceous crystals found in a thin bed of iron-stone in crevices of limestone, as I am inform'd.

The Labradore fossil is best seen with your back to the light, as it seems to reflect colours more vividly, than it refracts them. I sent it to be polish'd, but it seems not to have been much worn, so that I suspect it to be very hard.

The colours, like those seen on mother-pearl counters for card-playing, originate, I suppose, from the thinness of the lamina of which it is composed – like those seen on soap-bubbles and explain'd in Sir Isaac Newton's optics.

We are going to get our orange-burgamies to day, as they now waste daily by our garden being so public, and the number of our children. I carried safe home the Verbena trifoliata, and the antirrhinum Peloria, which last is a most uncommon flower, but I suppose a hardy one. Our young ladies have filled the vine-house with pots of plants, most of which, I believe, were originally imported from Elston.

Mrs Darwin begs to be kindly remember'd to yourself, and Brother John, when you see him; and to add, that we should be truly happy to see you both to spend "a merry Christmas" – when the country is less lively than in summer – if it could be made agreable to you.

The chaise is at the door, and my fair postess is about to set out, along with her cousin Miss Vaughton. I assure you, I much admire Miss Hall, as a very sensible, ingenious, and amiable young lady. Adieu, from dear Brother your truly affectionate

E Darwin

Addressed Rob[t]. War. Darwin Esq, Elston, near Newark.

Original Nottinghamshire Record Office, DDA 5/35. (Owner: Mrs V. M. Kindersley, Hamstead Grange, Yarmouth, Isle of Wight.)
Printed Unpublished.
Text From original MS.

Notes

(1) Robert Waring Darwin (1724–1816) was Erasmus's eldest brother, who lived as a bachelor at Elston throughout his long life and was author of *Principia Botanica*, a concise and successful guide to the Linnaean system of botanical classification, which was published in 1787, the same year as Erasmus's second translation from Linnaeus, *The Families of Plants*. Another shared interest was poetry: it was Robert who first roused Erasmus's passion for verse, and the volume of Erasmus's poems in the Pearson Papers at University College London is dedicated to Robert. The two brothers were on very friendly terms, but no other of the (presumably) numerous letters between them has survived.

(2) 'Erasmus' in the first line is Darwin's lawyer son, and the deed he was sending was 'the Engrossment of the Grant of Elston Advowson, in which I have inserted all the Corrections which you have made in the Draft' (Erasmus Darwin junior to Robert Waring Darwin, 7 October 1799, Nottinghamshire Record Office DDA 5/36). The Advowson itself (Nott. Rec. Off. DDA 5/37) is an undertaking to offer the rectorship of Elston to one of Darwin's sons, either Edward (1782–1829) or John (1787–1818), on the death of Darwin's brother John (1730–1805), who had been Rector for 34 years. In fact, Darwin's son John became Rector in 1815, but died three years later.

(3) Miss Hall was Erasmus's great-niece, daughter of his nephew Rev. Thomas Hall (1753–1801), Rector of Westborough (see letter 77D). Miss Hall was probably Elizabeth Hall (1779–1870), later Mrs Welby – unless she was already married and Darwin's messenger was her sister Anne (1781–1846). Thomas Hall's sister Elizabeth (born 1754) married Roger Vaughton, and the Miss Vaughton mentioned in the letter was their daughter Mary (born 1779). See H. F. Burke, *Pedigree of the Family of Darwin* (privately printed, 1888) and A. R. Maddison, *Lincolnshire Pedigrees* (London, 1903) Vol. II, p. 445.

(4) The river Derwent, which formed the eastern boundary of the Darwins' garden at Full Street, Derby, had evidently flooded the garden, presumably after the heavy rain of 15–18 August 1799, which caused widespread flooding: see *Gent. Mag.*, **69**, 730 (1799).

(5) The 'Labradore fossil' is what was later called labradorite, a feldspar from Labrador which shows a brilliant variety of colour: see C. S. Hurlbut, *Minerals and Man* (Thames & Hudson, 1969) pp. 75–7. 'Bristol-stones', also called 'Bristol diamonds' are a type of transparent quartz crystal found in the limestone at Clifton: see R. Bradshaw, *Proc. Bristol Naturalists Society*, **31**, 439–50 (1967).

(6) 'Orange burgamies' presumably refers to Citrus Bergamia. For verbena, see J. Hutchinson, *Families of Flowering Plants* (Clarendon Press, 1973) p. 487. 'Postess', meaning 'postwoman', seems to be a word coined by Darwin.

99D. To Samuel More, 13 October 1799

Derby Oct 13 –99

Dear Sir

Mr Swanwick, a very ingenious philosopher of this town, who teaches a school for writing, arithmetic, and some branches of natural philosophy, has lately turn'd his attention to improve the drill plough. Having observed, that those drill-ploughs now used are very complicate machines, and do not deliver the grain with sufficient accuracy, which prevents them from coming into general use, He has tried many experiments and has finally much simplified and emproved the method of delivering the grain from the seed-box. Which is not done by a revolving axis, as in all other drill ploughs from those first invented by Mr Tull to the present time; but is done in a more simple and more accurate manner by a slider, and is at once accomodated to any kind of seeds in common use from horse-beans to turnip-seed. From its great simplicity it is constructed at less expence, less liable to be out of repair, and much more easily restored, if it be accidentally injured.

I wish to encourage him to send a seed-box to the society of its real dimensions; that they may see its operation in delivering beans and wheat etc. Some other improvements of the machines may be seen by a drawing, as they differ not greatly from those in use. It appears to me to be a thing of consequence, as the intricacy, and expence of Mr Cook's machine for drill-ploughing has, I suspect, contributed to prevent the more general use of the drill-husbandry, and has persuaded many to adopt the dibbling, which is a much slower process.

If you think it is probable, that the society will grant Mr Swanwick a reward, or premium, if it appears to them to possess the excellences I have above described, which may compensate for his time and expenses, I should be glad of a line from you, and will endeavour to prevail on him to send you his seed-box of its full size, and a drawing of some other improved parts of the machine, which I esteem worthy the attention of the society. If you ever come again into Derbyshire, I shall hope to have again the pleasure of seeing you, and am dear Sir with true respect

your obed. serv.
E Darwin

Addressed – Moore Esq., Secretary to the Society of Arts, Adelphi, London.

Original Royal Society of Arts Library, Loose Archives C 13/30.
Printed Unpublished.
Text From original MS.
Notes
(1) Samuel More (1725–1799) was Secretary to the Society of Arts from 1770 until 1799. He was a friend of Wedgwood, and he met Darwin at Derby on 1 July 1785 (S. More, MS Journal, in possession of Miss Ann Turner), and possibly on other occasions. He died on 11 October, two days before Darwin wrote. For biography of More, see *European Magazine*, **36**, 363–5 (1799).
(2) Mr Swanwick was Thomas Swanwick (1755/6–1814), who was a schoolmaster at Derby for 32 years (Memorial in Derby Cathedral). See also letter 00H.
(3) The invention of the drill plough by Jethro Tull (1674–1741) was the most important single advance in the 'agricultural revolution' of the eighteenth century.

'Mr Cook' is T. W. Coke of Holkham (1752–1842), the leading agriculturalist of the late eighteenth century, who improved the drill plough and publicized its virtues. Darwin discusses the achievements of Tull and Coke in *Phytologia*, particularly pages 434 to 444. He also describes in detail an improved drill plough of his own design, incorporating Swanwick's seed-box, in an Appendix which runs to 14 pages with ten detailed drawings (*Phytologia*, pp. 597–612 and Plates X–XII).

(4) The manuscript of this letter is written in a strong and legible hand, and confirms other indications that Darwin was still full of vigour in 1799, at the age of 67.

99E. To Robert W. Darwin, 28 November 1799

[Erasmus junior was about to retire]

* * *

to sleep away the remainder of his life.

* * *

I kept no book, but believed my business to be £1000 a year, and deduct £200 for travelling expenses and chaise hire, and £200 for a livery-servant, four horses and a day labourer.

* * *

Original Not traced. In the possession of Charles Darwin, 1879.
Printed C. Darwin, *Life of E. Darwin*, pp. 76, 25–6.
Text From above source.
Notes

(1) The second excerpt is merely dated 'in 1799' by Charles Darwin and may come from a different letter.

(2) Erasmus junior was persevering with his idea of retiring, and had just bought Breadsall Priory, 4 miles north-north-east of Derby. His letter of 7 October 1799 to his uncle Robert (Nottinghamshire Record Office DDA 5/36) gives the impression that he was overworked, by his standards (the only standards that mattered). So retirement might have been the best solution. His father, who had no premonition of the tragedy that was now only a month ahead, disapproved because he feared solitary retirement would make his son morbid and neurotic.

(3) To pay for the war against the French, Pitt introduced income tax in 1799. As the second excerpt shows, Darwin did not know how much to declare. Charles Darwin says he is surprised that the amount was not more than £1000, but it is likely that Darwin had reduced his practice by 1799.

99F. To Thomas Beddoes, 29 December 1799

Derby Dec. 29 –99

Dear Dr

I am truly sorry to hear Mr T. Wedgwood is in so indifferent a state of health. I well remember, that, when he was a boy, He was troubled with ascarides, and was under the care of a London Quack, who purged him incessantly for some weeks; but that on his return to Etruria, he was still afflicted with ascarides – one or two of which I well recollect to have examined by a microscope, and saw their long fine-drawn proboscis or nose. Now I am inclined to suspect, that these worms are seldom so eradicated, as not occasionally to recur – and would advise Mr T. W. to inspect his evacuations daily; and if they should be seen, I should certainly ascribe his indigestion, and consequent hypochondriacism, to ascarides. And if such animals should be observed, I shall wish to hear again from you concerning him.

In the mean time I should recommend some corroborant medicine, which is not very bitter, and so much rhubarb as may induce a daily evacuation.

And daily exercise for an hour on horse-back and to drink no wine alone, but to mix one part of wine with three parts water.

I shall add a formula, which you will please to alter as you think proper, and pray tell him, that if any thing should bring him into this part of the country, that if He will pass 2 or 3 days at my house, I shall be glad of a visit from him, and will study both his body and mind.

adieu
from dear Dr
Your affect. friend
E Darwin

If he has not a daily evacuation by the rhubarb, He will please to take along with the rhubarb pills at night 2 or 3 aloetic pills.

[There follows a prescription for rhubarb pills and aloetic pills, dated 29 December 1799.]

Addressed Dr Beddoes, Clifton, near Bristol. (*Postmark* Rudgeley.)

Original Keele University Library, Wedgwood MSS, Mosley Collection.
Printed Short excerpt in *Doctor of Revolution*, p. 275.
Text From original MS.
Notes
(1) Darwin is still vigorously pursuing Tom Wedgwood's ascarides, real or imaginary. By now Tom was living in south-west England and was in closer touch with Beddoes than with Darwin. Obviously Darwin suspected psychomatic illness – 'I . . . will study both his body and mind'. But none of Tom's doctors succeeded in helping him much, and he died in 1805 at the age of 34. See *Wedgwood Circle*, pp. 105–30.

(2) This letter is tragically ironical, for on the evening of 29 December, Darwin's son Erasmus walked out to drown himself in the river Derwent (see next letter). 'I am truly sorry to hear Mr T. Wedgwood is in so indifferent a state of health', he had written, unaware of the far worse state of his own son. From the postmark it appears that Darwin was called to Rugeley on 29 December, and he may have stayed the night there.

99G. To ROBERT W. DARWIN, 30 December 1799

[Dear Robert]

I write in great anguish of mind to acquaint you with a dreadful event – your poor brother Erasmus fell into the water last night at the bottom of his garden, and was drowned.

* * *

Original Not traced. In the possession of Charles Darwin, 1879.
Printed C. Darwin, *Life of E. Darwin*, p. 72.
Text From above source.
Notes

At the end of 1799 Erasmus junior made a final attempt to settle the mass of outstanding accounts, and, after working on them for two nights without sleep, he refused to take rest and food on the evening of 29 December, saying, 'I cannot, for I promised if I'm alive that the accounts should be sent in tomorrow'. Later that evening he rushed out of the house and threw himself in the river. His father's anguish at his death was sharpened by the feeling that he, a doctor of supposedly great sympathy and sagacity, should have foreseen and prevented the tragedy. Instead he had regarded his son's failure to settle accounts as a trivial matter, as it was – except to the victim. After his son's death Darwin never really recovered his *joie de vivre*.

00A. To JOSEPH JOHNSON, 10 January 1800

Derby Jan 10 –800

Dear Sir

You will please to do as you please about the Aphidivorous larva with the aphis in its mouth. A few strokes would give it, as in the drawing, which I shall stick on the paper below – as only the legs etc may appear, a part of it being supposed to be swallow'd.

My great affliction for the sudden death of my beloved eldest son, who used to call at your house, has prevented me for a while from sending Bennet's work, but shall be done soon.

Adieu
from your sincere friend
E Darwin

Addressed Mr Johnson Bookseller, St Paul's Churchyard, London. (*Postmark* Jan 11 1800.)

Original Library of Fitzwilliam Museum, Cambridge; S. G. Perceval Bequest, General Series.
Printed Unpublished.
Text From original MS.
Notes
(1) Joseph Johnson was Darwin's publisher (see letter 84G), and Darwin is here referring to one of the illustrations for *Phytologia*. His sketch of the aphidivorous larva, copied here full-size, was beautifully drawn, with the body coloured green. The sketch was printed in Plate IX of *Phytologia*.
(2) Darwin rarely visited London, and Erasmus junior had often acted as an emissary for him, probably sometimes taking the manuscripts of his books to Johnson. Darwin's grief at his son's death permeates this and several subsequent letters.
(3) 'Bennet's work' may have been Abraham Bennet's *New Experiments in Electricity* (1789). See letter 87D. Bennet died in 1799.

00B. To Robert W. Darwin, 8 February 1800

* * *

I am obliged as executor daily to study his [Erasmus's] accounts, which is both a laborious and painful business to me.

* * *

Original Not traced. In the possession of Charles Darwin, 1879.
Printed C. Darwin, *Life of E. Darwin*, p. 74.
Text From above source.
Notes
Darwin's anguish at the death of his son was prolonged for months because he had to work on sorting out and settling the accounts, in the bitter knowledge that it was this very task that drove his son to suicide. Even in a Greek tragedy there could scarcely be a more cruel combination of circumstances.

00C. To Robert W. Darwin, [22] February 1800

* * *

Mrs Darwin and I intend to lie in Breadsall church by his side.

* * *

Original Not traced. In the possession of Charles Darwin, 1879.
Printed C. Darwin, *Life of E. Darwin*, p. 74.
Text From above source.
Notes
(1) Charles Darwin says this letter is 'a fortnight after' that of 8 February. I have taken this to be exactly 14 days.
(2) In this letter Darwin tells Robert about a monument to be erected to his dead brother at Breadsall church. The intention of Darwin and his wife to be buried beside his dead son was fulfilled.

00D. To Matthew Boulton, 23 February 1800

Derby Feb. 23, 1800

Dear Boulton,

There was a few days ago a meeting at Barton to consider about inclosing the Forest of Needwood, in which I understand you are interested, and I trouble you with this line to recommend to your notice, and solicit your interest for my friend Mr Edwards of this town, the partener of my late Son, who wishes to be employ'd as one of the solicitors (if the inclosure proceeds) by the advise of some of his friends, who are interested in the inclosure.

Mr Edwards was partener with my late Son, and I have now a younger son Clerk to him, and I know him to be a man of ability and activity, and of distinct integrity and honor in his profession.

I shall hope to have the pleasure of seeing you either here or at Soho some time in the summer, and am, dear Boulton, your old and affectionate friend

E Darwin

I shall be indeed much obliged to you for this favor, and hope you will consider that my recommending a deserving character is an act of friendship, as well as your employing him.

Addressed Mathew Boulton Esqr., Soho, near Birmingham.

Original Birmingham Reference Library, Matthew Boulton Papers, Darwin 45.
Printed Unpublished.
Text From original MS.
Notes
(1) Darwin's friendship with Boulton had now endured for more than 40 years, but the passage of time had brought great changes. Those enthusiastic letters of the 1760s, full of the 'white-hot technological revolution' of that decade, were now

little more than a memory, and even the Lunar Society was virtually defunct. These later letters are often, as here, acts of kindness to others.

(2) 'Mr Edwards' is the Derby attorney Nathaniel Edwards (1768–1814): for the Edwards pedigree, see Glover, *Derby* ii 570–1.

(3) The son acting as clerk to Mr Edwards was presumably Edward Darwin (1782–1829), who was just 18.

00E. To Thomas Byerley, 18 August 1800

Dear Sir,

I have examined the books of my late Son, and believe the account you have sent me to be right, and have herein transmitted to you a note for the money, 148–10–0.

I see there is a small law account due from Mr Thos. Wedgewood for some journeys etc about an estate, which was to be sold; and shall be glad to know, how I may direct to him. The account is not yet made out, would otherwise have troubled you with it.

You will please to write me word of the receipt of the inclosed from, dear Sir

Your obed. serv.

E Darwin

Derby
Augt. 18 –1800

Addressed Mr Thos. Byerley, Etruria, near Newcastle, Staffordshire. (*Postmark* Derby.)

Original Keele University Library, Wedgwood Archive, 227–2.
Printed Excerpt in *Doctor of Revolution*, p. 274.
Notes

(1) For Tom Byerley, see letter 89J.

(2) This letter shows that Darwin was still at work on the unsettled accounts 7½ months after his son's death. He had now lost touch with Tom Wedgwood.

00F. To Gilbert Wakefield, 19 August 1800

Derby, Aug. 19 1800

Dear Sir,

I am much obliged to you for your severe and elegant satire, which you have so good cause to write, who so long have felt the persecution of these flagitious times! When one considers the folly of one great part of mankind, and the villany of another great part of them, the whole race seems to sink into contempt.

Thank you, also, for your criticisms on the "Botanic Garden", which shew in you a very great sensibility to the elegancies of language, which must be shocked with almost all the poetry you meet with. Most of your observations, I believe, are just; yet perhaps the ambiguity from the numbers of the nominative and accusative cases, both of which precede the verb, could not well be altered in the line you mention, as

"Stay, whose false lip seductive simpers part,"

is not sense, as the word part refers to two lips. But might be thus avoided:

"Stay, whose false lips with smiles seductive part."

I cannot defend myself against many of your other ingenious observations, but think the word "spoke" more in use amongst polite people, than "spake". I esteem it to be an abbreviation of the word spoken. So broke for broken, etc. In your valuable satire you have written drave for drive; which, if it ever was in use, is long since superannuated.

I beg leave to repeat, that I am much obliged to you for your observations, and much admire the sensibility of your taste and accuracy of your knowledge of our common language, and am charmed with the energy of your satire.

I hope, if you are ever released from the harpy-claws of power, that I may some time have the pleasure verbally of more of your criticisms at Derby; and am,

With great respect,
Your obed. serv.
E Darwin

Addressed To the Rev. Gilbert Wakefield, Dorchester Gaol.

Original Not traced.
Printed In *Memoirs of the Life of Gilbert Wakefield*. 2 vols. (Johnson, 1804) Vol. 2, pp. 229–32.
Text From above source.
Notes

Gilbert Wakefield (1756–1801), classical scholar and unitarian, had a taste for controversial writing. In 1799 he published a pamphlet about Bishop Watson that was judged seditious, and the 'harpy-claws of power' imprisoned him in Dorchester gaol. He died soon after his release. While in gaol he read Darwin's *Botanic Garden* and was greatly impressed. He wrote Darwin a long letter, with critical comments on the *Botanic Garden* and a copy of his own poem 'Imitation of Juvenal'. This letter is Darwin's reply. Darwin obviously admired Wakefield for his refusal to be gagged by the harsh laws of the day, and was also well aware that he himself would end up in gaol if he expressed his views too openly. See *DNB* for Wakefield and Watson.

00G. To Matthew Boulton, 9 November 1800

At Mr Galton's Crescent
Birmingham
Nov. 9 –1800

Dear Boulton,

My friend Mr John Gisborne of Holly-bush in Needwood Forest has desired me to write to you to solicit from you a refusal of your allotment, if that forest comes to be inclosed (which I hope it will) if you mean to dispose of it; and I am sure, you can not oblige a more respectable and valuable character who by having married Miss M. Pole, and making an extraordinary good husband, is by many names endeared to me.

I shall inclose his letter to me, and am dear Boulton your old and sincere friend

E Darwin

Addressed Mathew Boulton Esqr., Soho, near Birmingham.

Original Birmingham Reference Library, Matthew Boulton Papers, Darwin 46.
Printed Unpublished.
Text From original MS.
Notes
(1) John Gisborne (1770–1851) married Darwin's step-daughter Millicent Pole (1774–1857) in 1792. Gisborne was author of the 'loco-descriptive' poem *The Vales of Wever* (1797), three cantos of rather Darwinian verse about the countryside around Wootton and the Weaver Hills. He also seems to have been an admirable man, as Darwin implies: 'his geniality, humility, and sympathy made him universally popular', according to the *DNB*.
(2) This is the last of the 36 letters in this volume to Boulton from his 'old and sincere friend'.

00H. To Georgiana, Duchess of Devonshire, [November 1800]

The Galvanic pillar may consist of about 30 or 40 half-crown pieces, as many pieces of Zinc of similar dimensions, and as many circular pieces of cloth, which must be wetted in salt and water. Two thick brass wires, about 2 ft long, communicate from each extremity of the pillar to each temple. The temples must be moistened with brine.

Plates of silver about the size of crown-pieces, and smooth on both sides, are rather better. I have used both. The shock is so great as to make a flash in the eyes, and to be felt th[r]ough both the temples, every time one of the wires is lift'd from the pillar, and replaced. So that 100 shocks may be given in a minute. I have one patient here, a lady from near Scarborough, who has used it daily for giddyness with good success.

I should be extreemly happy to show your Grace the application of Galvanism, the effects of which would surprize you, I am sure. In respect to the invalid your Grace mentions, if his fainting fits be attended with any struggling motions of his arms or legs, or convulsions of his eyes, they are certainly epileptic, and the case of great danger.

a b. Galvanic pillar

[The drawing is very faint in the original, and has been touched up.]

I dare say, I could procure the Zinc plates here for about a shilling a piece; if your Grace wishes to construct a pillar.

There is a very ingenious philosopher here, Mr Swanwick, who teaches a school of writing and Geometry, whom I could perhaps send over to your Grace for one day, if you wish it, who could bring a Galvanic pillar consisting of larger pieces, as large as Crown-pieces, and shew your Grace its effects. The pillar belongs to Mr Strutt, another Philosopher of this place. Mr Swanwick's time is much employ'd in teaching different parts of philosophy. But I shall hope to see your Grace at Derby.

In respect to what your Grace observes about burning the ear, it was at first supposed, that the nerves were burnt and destroy'd – but the fact is, that no such thing is done, tho' pretended by empyrics – but it is the sudden smart of the skin, and the terror of the patient, which destroys small tooth-ach – or by the stimulus of so small a pain and fear, the torpid membrane of the tooth recovers its activity, and the pain from defect of action ceases. This can be of no service in so violent a disease as the tic douloureux. The branching of nerves is in many[?] parts, but not in that part of the ear, which is generally bur{nt} for the toothach, which is only useful in small pains like the smarting of a blister.

I am sure your Grace does me great honour, and I should be happy to see Mr Spencer, but it is so inconvenient to me to leave home.

I am, your Grace's
most obed. serv.
E Darwin

Addressed Her Grace The Duchess of Devonshire, Chatsworth, Derbyshire. (*Postmark* Derby.)

Original Chatsworth House, Devonshire Collection, No. 1537.1.
Printed Unpublished.
Text From original MS.
Notes

(1) Georgiana, Duchess of Devonshire (1757–1806), was the daughter of the first Earl Spencer (1734–1783), who was a great-grandson of the Duke of Marlborough. She married the fifth Duke of Devonshire (see letter 83D) when she was 17, and soon became the 'reigning queen of society'. A woman of great talent and charm, she set the fashion in dress, replacing the hoop with graceful styles, and also took a prominent part in Whig politics and in intellectual life. Her zest for politics emerged in her famous canvassing for C. J. Fox at the Westminster election of 1784, when she gave kisses for votes, while her literary talent appeared in her novel *The Sylph* (1779) and in poems like 'Passage of Mount St Gothard'. She was also keenly interested and knowledgeable in science. After two long visits from her in 1793, Beddoes wrote to Darwin that her knowledge of modern chemistry was superior to what he would have supposed 'that any Duchess or any lady in England was possessed of'. (Stock, *Beddoes*, p. 100.) Her collection of geological specimens was much admired, and she was a patron of mineralogists such as Robert Townson (1762–1827: see his *Philosophy of Mineralogy*, 1798) and White Watson (1760–1835: see his *Strata of Derbyshire* (Sheffield, 1811) pp. 31–2). Darwin obviously also had a high opinion of her scientific understanding. For a selection of her own letters, see *Georgiana*, ed. Earl of Bessborough (Murray, 1955); for the early letters of her daughter Harriet, see *Hary-O*, ed. G. Leveson-Gower and I. Palmer (Murray 1940). For a biography, see A. Calder-Marshall, *The Two Duchesses* (Hutchinson, 1978).

(2) This letter raises several subjects mentioned again in the next letter, which is dated 29 November 1800. So this letter seems likely to be only a few days earlier.

(3) The apparatus described (accurately) by Darwin should be called 'the pile of Volta' rather than 'the Galvanic pillar'. Luigi Galvani (1737–1798), discoverer of animal electricity, was not concerned in the invention. It was Alessandro Volta (1745–1827) who early in 1800 announced that a device of this kind could produce a steady stream of what we now call 'current electricity'. It was an epoch making discovery, which opened the way to all the subsequent uses of electricity. Darwin had used electricity to treat disease for more than 20 years (see *Doctor of Revolution*, pp. 131–2), but he had previously been limited to the sudden discharge of accumulations of static electricity, from electrical machines or the Leyden jar. Darwin came near to discovering the Voltaic cell himself when commenting on Volta's observation that if pieces of lead and silver, touching outside the mouth, are applied to different parts of the tongue, an acidulous taste is noticed 'as of a fluid like a stream of electricity passing from one of them to another' (*Zoonomia* i 120 (1794)). Darwin then comments that zinc is better than lead, and continues: 'If one edge of a plate of silver about the size of half a crown-piece be placed upon the tongue, and one edge of a plate of zinc about the same size beneath the tongue, and if their opposite edges are then brought into contact before the point of the tongue, a taste is perceived . . .' (*Zoonomia* i 120).

This is a Voltaic pile, with the tongue replacing the brine-soaked cloth, but Darwin failed to realize that living tissue was not essential, and merely concluded that the experiment shows 'the great sensibility of these organs of sense to the stimulus of the electric fluid' (i 121). Though he did not make the discovery, Darwin was very quick to use it, and these letters, only seven months after, show that he already had a good deal of experience in using the pile medically, and a theory to explain it. Though the pile was used to treat toothache, Darwin was aware that the treatment did not destroy the nerve, and that any benefits were mainly psychological.

(4) For Mr Swanwick, see letter 99D.

(5) Mr Strutt is William Strutt: see letter 01E.

(6) The tic douloureux is a severe neuralgia of the face, with twitching of the facial muscles.

(7) 'Mr Spencer' was probably a cousin of Georgiana: her nearest male relatives were her only brother Earl Spencer (1758–1834) and her nephew Lord Althorp (1782–1845); but neither of these would be called 'Mr Spencer', and her other nephew was too young.

(8) Despite his high regard for the Duchess, Darwin has no hesitation in refusing the invitation to Chatsworth.

00I. To Georgiana, Duchess of Devonshire, 29 November 1800

Derby Nov. 29 –1800

I endeavour'd to prevail on Mr Swanwick to wait upon your Grace to shew you Galvanism, but he could not with propriety leave his various schools. I have therefore applied to Mr Hadley, who attended your Grace before; He is to day rather indisposed, but if your Grace wishes him to come over to galvanize your eye, he will wait on you any day, you will please to appoint. And will bring Mr Strutt's Galvanic pillar with him.

The Galvanic apparatus of Mr Strutt was made by one of his own workmen, and not for sale. I can not therefore easily procure the silver plates and zinc ones to be made here. But believe they may be had in London.

I am happy to hear your Grace's patient is better, and hope it is the hemicrania sympathetica and not the idiopathica – as the former sympathizes with some diseased tooth, and is thence more readily curable.

I am your Grace's
most obed. serv.
E Darwin

The Galvanic shocks appear to consist of a large quantity of electric ether, which moves more slowly, than the shocks from a coated phial; and therefore it gives no pain in passing from one temple to the other, tho' much light is seen in the eyes. As the fluid matter of Heat (in which all things are immersed) appears either in its loose or uncombined state, as in a warm room or in the summer air; – or in its fix'd or combined state, as in the materials of gunpowder, from which it is set at liberty by a spark, and in other combustions. So the common electric machine seems to collect or condense or accumulate the loose or uncombined electric matter; – but the Galvanic apparatus appears to set at liberty the combined electric matter from zinc and some other metals; which appear to become oxydes by their union with the oxygen from the decompos{ing?} water. Which appears to be decomposed in sma{ll} quantity first by the loose electricity passing from some metals to some other ones, as from Zinc to Silver; as a small spark of loose heat sets at liberty so much combined heat from gunpowder. This is the only theory, which I have been able to investigate of the Galvanic phenomena. Which Mr Hadley will explain to your Grace, if you direct him to come over to galvanize your Grace's eye.

Addressed Her Grace The Duchess of Devonshire, Chatsworth, Derbyshire. (*Postmark* Derby.)

Original Chatsworth House, Devonshire Collection, No. 1537.
Printed Unpublished.
Text From original MS.
Notes
(1) This letter is closely linked with the previous one. Mr Swanwick was evidently unimpressed by the 'queen of society' and declined to wait on her.
(2) 'Mr Hadley' is Henry Hadley (1762/3–1830), a Derby surgeon who attended Darwin at his death in 1802 and later married Darwin's daughter Susan Parker. A footnote on the MS reads: 'Hadley rode from Derby, and was seen in Chatsworth Park, putting a clean shirt on, under a tree'.
(3) Hemicrania is headache on one side of the head. 'Sympathetica' means that it is secondary, caused by some other condition, such as a diseased tooth. 'Idiopathica' means that it is a primary disease.
(4) Darwin's theory of current electricity is extraordinarily prescient, and served as a basis for his theory of electrochemistry propounded in *The Temple of Nature*, Additional Note XII.
(5) The Duchess suffered greatly from inflammation of the eye, and underwent a painful eye operation in 1796.

00J. To [James?] Sims, 4 December 1800

Derby, Dec. 4 1800

Dear Doctor,

From the very accurate account, you have again favor'd me with, I flatter myself, that my Friend Mr Jos. Strutt continues to get better; and, I hope, has no alarming Symptom to delay his recovery; tho a fever of this kind is liable to continue two or three weeks, and sometimes a whole lunation.

In a fever with debility the pulse is generally considerably more frequent than his [has] been at any time, whether in an horizontal or vertical situation. But I learn from his Brother Mr W^{m}. Strutt, that the pulse of Mr Joseph Strutt was always in health much slower than his own, even below 60 in a minute: which may account for its less frequency in fever, than is commonly observed.

If the pains of his head continue; perhaps, if his pulse be full or hardish, the leeches may be again applied – and a small blister, behind his ears, or on the back. So much rhubarb should be daily given as may induce one or two evacuations. And such kind of draughts continued, if you approve, and if they seem to agree.

The antimonial powder consisting of but $\frac{1}{6}$ of a grain of emetic tartar could scarcely produce any sensible effect. Perhaps the clammy sweat, they were supposed to induce, might be occasion'd by his room being too warm, or too close, and perhaps He does not take sufficient diluent fluids.

The long sleep of nine or twelve hours, with the acute shooting pains of his head, seem to shew some affection of the meninges cerebri, and may possible occasion the fever to persist a longer time, and the pulse being more frequent in an erect than in a recumbent posture, I think, indicates debility of circulation; tho' his pulse is never so quick as in most fevers with debility, perhaps owing to its natural slowness.

I wish he could be prevail'd upon to repeat the antimonial powder, the quantity of which may be still less as $\frac{1}{8}$ of a grain of the antim. tartariz.

The frequent sighing is generally owing to the inirritability of the lungs; and the lowness of spirits, I suppose, to general inirritability; which is another expression for debility. Hence I should think a little animal food, as chicken, with weak wine and water might be allow'd, if it be agreable to him, as well as weak fresh broth, cream and water and other diluent fluids.

I am dear Sir with great regard
Your friend and serv.
E Darwin

Addressed To Dr Sims of London.

Original Liverpool City Libraries: Hornby Library.
Printed Unpublished.
Text From original MS.
Notes

(1) The recipient of the letter was probably Dr James Sims (1741–1820), a good-humoured man who was a founder of the London Medical Society and its President for many years. The other possibility is that the recipient was Dr John Sims (1749–1831), a lesser-known London physician. See *DNB* for both.

(2) The patient, Joseph Strutt (1765–1844), was the third son of the textile manufacturer Jedediah Strutt. He lived in St Peter's Street, Derby, and knew Darwin socially. Joseph Strutt became seriously ill in London in 1800 and his family wondered whether to ask Darwin to go to see him. Instead Darwin, now semi-retired, was consulted by letter, and he was as usual conscientious and helpful with his advice.

(3) William Strutt was Joseph's eldest brother and a friend of Darwin (see letter 01E).
(4) The treatment Darwin suggests is in accord with his theory of fever (*Zoonomia* ii 537–625). Treatment by leeches was just becoming popular.

00K. To Samuel J. Arnold, 20 December 1800

Derby Dec. 20 1800

Dear Sir,

I am glad to have the pleasure of a line from you to hear that you are doing well. In respect to an inscription under my Head I do not know what to say: – if a compliment to the head, it should be taken from the author's works; and the design should be to increase the sale of the print. There are two lines in Botanic Garden, p. 11 end of Canto II on Mr Howard, if you think such a compliment would add to the sale of the print I have no vanity of this kind.

Onward He steps, – Disease and Death retire,
And murmuring Demons hate him, and admire.

Botan. Garden.

But perhaps the best would be simply

Erasmus Darwin M.D. F.R.S.
author of Zoonomia, Phytologia, and the Botanic Garden.

Governor Pownal I never saw. He may have forgot our correspondence by this time.

Mrs Darwin desires her compliments to you. I am with great regards
Your sincere friend
E Darwin

Addressed S. J. Arnold, Esq., Duke St. No. 22, Westminster, London.

Original Library of Fitzwilliam Museum, Cambridge; S. G. Perceval Bequest, General Series.
Printed Unpublished.
Text From original MS.
Notes
(1) Samuel James Arnold (1774–1852), the son of the musical composer Dr Samuel Arnold (1740–1802), was educated as an artist, and in his twenties he painted portraits and panoramas. He exhibited portraits at the Royal Academy from 1800 to 1808, and his portrait of Darwin, discussed in this letter, was engraved by B. Pym and published in February 1801: see E. Benezit, *Dictionnaire des Peintres* . . . (1966) i 208. Arnold's main career, however, was as a dramatist and theatrical manager. No. 22 Duke Street was his father's address. See *DNB* for both Arnolds.
(2) Darwin seems embarrassed to know what to say, and at first suggests a bombastic

quotation (*Loves of the Plants* II 467–8), which absurdly exaggerates his medical powers. These lines are on page 81, not page 11, of *Loves of the Plants*.

(3) 'Governor Pownal' is Thomas Pownall, FRS, FSA (1722–1805), Governor of Massachusetts Bay (1757–9) and MP from 1767 to 1780. He knew Franklin and gave advice to the Government during the American war; he also wrote many political pamphlets and antiquarian works. See C. A. W. Pownall, *Thomas Pownall* (Stevens & Stiles, 1908), a 500-page biography; *DNB*; and Namier & Brooke, *Commons* iii 316–18.

01A. To Samuel Tertius Galton, 5 January 1801

As Mr Tertius Galton is liable to have inflamed eyes and difficult respiration on using much exercise in warm weather; and as he is subject to bleeding at the nose; He is advised to take 6 or 10 grains of rhubarb at going to bed uniformly for some months; so as to induce one or two daily evacuations.

He is also advised not to drink any thing stronger than common weak small beer; or one part of wine mix'd with three parts of water:

And especially in the warm season to abstain from violent exercise – but nevertheless at all seasons to use riding or walking exercise in moderate degree.

Jan. 5 –1801 E Darwin

Original University College London, Galton papers, 190A.
Printed Unpublished.
Text From original MS.
Notes

(1) This is half-way between a letter and a prescription, and has been given the benefit of the doubt because of its biographical interest.

(2) Samuel Tertius Galton (1783–1844) was the eldest son of Samuel Galton (1753–1822), Darwin's friend and Lunar Society member. Tertius, now 17, married Darwin's daughter Violetta in 1807, and Sir Francis Galton was one of their sons.

(3) The prescription shows Darwin still advising his normal course of treatment for the slightly ill and the hypochondriacs – moderate purgation, severe temperance and gentle exercise.

01B. To Robert W. Darwin, [February 1801]

[Darwin mentions to Robert that he has received a print of himself]

* * *

well done, I believe – proofs 10s 6d – the first impression of which the engraver, Mr Smith, believes will soon be sold, and he will then sell a second at 5s . . . but the great honour of all is to have one's head upon a sign-post, unless, indeed, upon Temple Bar!

* * *

Original Not traced. In the possession of Charles Darwin, 1879.
Printed C. Darwin, *Life of E. Darwin*, p. 68.
Text From above source.
Notes

(1) Darwin has now received the finished version of the print discussed in letter 00K. It was published on 2 February 1801: hence my dating of this letter.

(2) 'Mr Smith' may be the leading British engraver of the day, Anker Smith, RA (1759–1819), who had engraved the frontispiece of *The Botanic Garden*, Fuseli's 'Flora attired by the Elements'; or he may be John Raphael Smith (1752–1812), who engraved a portrait of Darwin in 1797 and was the son of Thomas Smith of Derby. For Anker Smith, see *DNB*; for J. R. Smith see Bryan's *Dictionary of Painters and Engravers* (Bell, 1905) v 94. However, since the portrait was engraved by B. Pym, Darwin may have been mistaken in referring to either Smith. The opinion of the engraver, be he Pym or Smith, proved correct: there were three impressions of the print. For details of the seven engraved portraits of Darwin, see F. O'Donoghue, *Catalogue of Engraved British Portraits* (British Museum, 1908–25) ii 13.

(3) 'A sign-post' refers to the sign-board of an inn. The macabre comparison with the heads of executed criminals on Temple Bar is the nearest approach to a joke in these last letters.

01C. To Robert W. Darwin, 6 May 1801

Derby, May 6 1801

Dear Robt.

I last night recieved from you £14–9–6, which I believe is the true account betwixt us [*signed*] E Darwin.

I am I think perfectly recover'd, but am like other corpulent old men soon fatigued with walking, especially up hill or up stairs.

I write this day to Elston, and have told your Uncle, that you design to visit him here. Mrs D, and Violetta and Emma are in London with the Strutts seeing the wonders of the world, and do not return these 10 days. Your Brother's cabinet for coins you are welcome to, if you please, but as I intend to give Edward specimens of the English Coins, perhaps if you have a smaller cabinet of your own you could accomodate him with it – this shall rest, till we see you here. Adieu

from
your affect. father
E Darwin

Addressed Dr Darwin, Shrewsbury. (*Postmark* Derby.)

Original St John's College Library, Cambridge.
Printed Short excerpt in *Doctor of Revolution*, p. 283.
Text From original MS.
Notes
(1) Darwin, who was now 69, suffered a serious illness, probably pneumonia, about March 1801, and although he reports himself recovered, his writing is distinctly weaker in this letter.
(2) 'Your uncle' is probably Robert Darwin (1724–1816), who lived at Elston Hall. See letter 99C.
(3) Darwin's daughters Violetta and Emma were now aged 18 and 16 respectively, and his son Edward was 19. For the Strutts, see next two letters.

01D. To Barbara Strutt, [Summer 1801?]

I send a branch of what Dr Jackson says is the sugar maple, and that what I have at the Priory is the ash-leaved maple.

I have brought a flower for a pot, which I don't know, and shall be glad to see you here, or will send it to you, as soon as I have planted it.

E Darwin

I have also a double ragged Robin (Lychnis — —) to shew you, and narrow-leaved Kalmia.

Addressed Mrs W. Strutt [sent by hand].

Original Library of the Wellcome Institute for the History of Medicine, London.
Printed Unpublished.
Text From original MS.
Notes
(1) Mrs Barbara Strutt (1761–1804) was the daughter of Thomas Evans (1723–1814), the Darley mill-owner and Derby banker. She married William Strutt (see next letter) in 1793 and died in childbirth in 1804. (See 'William Strutt', typescript volume in Derby Central Library, 3524 MS, where there is a letter of condolence on her death from Elizabeth Darwin to William Strutt.) The Evans and Strutt families were much intermarried: see letter 94X and Glover, *Derby* ii 18.
(2) The rather weak writing suggests a date after March 1801, and the impression is that the letter was written in summer, with plants and trees in leaf and visits to Breadsall Priory being made. Darwin distinguishes between 'the Priory' and 'here', so the letter was written before he moved to the Priory in March 1802. A date in 1800 is less likely, because Darwin was then much occupied in settling the affairs of his late son Erasmus (who bought the Priory just before his death in December 1799).
(3) The Ragged Robin is *Lychnis floscuculi*, and Kalmia is a genus of evergreen American shrubs, named (1776) after Linnaeus's pupil Peter Kalm.
(4) I have not identified Dr Jackson. He might possibly be Robert Jackson (1750–1827): see *DNB*.

01E. To William Strutt, 6 August 1801

Aug 6 1801

Dear Sir,

1. I wish on Sunday morn. to see the grand effects of your electric apparatus.

2. To learn if positive electricity exists in glass or on it.

3. If glass or sealing wax can be electrised by pressure only? – or by heat, like the tourmalin.

4. Canton's experiments done nicely – with and without being electrised – positive and negative.

5. A Galvanic pile with finely powdered fresh charcoal instead of water. Or with mercury instead of water. Or with the plates soldered together. Or by plates of zinc and silver only. To find whether the decomposition of the water be a cause or an effect of the electric stream. I suspect that alternate plates of quicksilver and glass would produce galvanism if one could easily try the experiment.

I have desired Swanwick to repair my electrometer and send it to you. Shall I send Bennet's Doubler? Shall I invite Swanwick to attend us?

Adieu

E Darwin

Original Not traced. There is a typed transcript in a typescript volume entitled 'William Strutt', part of a Memoir of the Strutt family, in the collections at the Derby Central Library (3524 MS).

Printed Journ. Derby Archaeol. Nat. Hist. Soc., **80**, 63 (1960).

Text From typescript volume at Derby Central Library.

Notes

(1) William Strutt (1756–1830), elder son of the manufacturer Jedediah Strutt, was for many years Darwin's closest friend in Derby. Being near neighbours, however, they rarely wrote to each other. Strutt helped to start the Derby Philosophical Society, acted as chairman whenever Darwin was away, and became President after Darwin's death. Strutt was himself a talented inventor and engineer, being best known as a heating engineer and for his designs of fire-resistant factories: see *J. Derby Archaeolog. Nat. Hist. Soc.*, **80**, 49–70 (1960); *Annals of Science*, **24**, 73–87 (1968); and *History of Technology* (OUP, 1958) iv 475–7. He also founded the Derby Infirmary and, as this letter shows, experimented with electricity. For the personal life of Strutt, see the typescript volume in Derby Central Library (3524 MS).

(2) Although Darwin was 69 and only recently recovered from a serious illness, we see him here embarking on an ambitious research programme into the fundamentals of electricity. And he carried it through, the results being described in the 34-page 'Additional Note XII' to *The Temple of Nature*, entitled 'Chemical Theory of electricity and Magnetism'.

(3) John Canton (1718–1772) was one of the leading electricians of the eighteenth century, and the experiment mentioned is discussed in Note XII to *The Temple of Nature*.

(4) Darwin is keen to exploit the possibilities of the new 'Voltaic pile', or 'Galvanic pile' as he calls it: see letter 00H.
(5) Bennet's Doubler was itself based on one of Darwin's doublers: see *Doctor of Revolution*, p. 117 and A. Bennet, *New Experiments in Electricity* (Drewry, Derby, 1789). For Swanwick, see letter 99D.
(6) A scientific evaluation of Darwin's proposals and his theory is beyond the scope of these notes.

02A. To Edward Jenner, 24 February 1802

* * *

Your discovery of preventing the dreadful havoc made among mankind by the small-pox by introducing into the system so mild a disease as the vaccine inoculation produces, may in time eradicate the small-pox from all civilized countries, and this especially: as by the testimony of innumerable instances the vaccine disease is so favourable to young children, that in a little time it may occur that the christening and vaccination of children may always be performed on the same day.

* * *

Original Not traced.
Printed J. Baron, *Life of Edward Jenner*. 2 vols. Colburn, 1838. Vol. 1, p. 541.
Text From above source.
Notes

Dr Edward Jenner (1749–1823) discovered vaccination in the late 1790s, too late to affect Darwin's medical career: Darwin had to administer the much more dangerous smallpox inoculation – which occupied an appreciable fraction of his time, since inoculated patients had to be visited several times. Darwin was now over seventy, but as this letter confirms, he retained his mental alertness and was still producing good new ideas.

02B. To Jane Borough, 1 March 1802

Derby Mar. 1 – 1802

Dear Madam,

I have sent you the families of plants of Linneus, and also the System of Vegetables of Linneus. The latter is out of print, but this copy is accidentally come to me, and is charged the original price 18/ but is not bound – which may be done by any bookseller in Derby.

If you only know the english name of a plant, you will please to look for it in the english index, which will give you the latin name of the genus, or family; if you look for that in the generic latin index of the system of vegetables, it will refer you to all the species of that genus – or individuals of that family.

If you look [for] the latin name in the generic latin index of the Families, it will refer you to a most accurate description of the genus or family.

I am, Madam,
with true respect
your obed. serv.
E Darwin

The Families of Plants is charged 16 shill.

Addressed Mrs Burrow, Castle Fields.

Original Darwin Museum, Down House, Downe, Kent.
Printed Unpublished.
Text From original MS.
Notes
Mrs Jane Borough (born 1776), *née* Smithson, married the barrister Thomas Borough (1759–1838), and they lived at Castle-fields, Derby until 1803. Her husband had changed his surname from Borrow to Borough: his father Thomas Borrow (1709–1786) was Recorder of Derby for many years. Joseph Wright painted portraits of Thomas Borrow and his wife Anne, who died in 1799. (The portraits are in Derby Art Gallery.) Darwin probably knew the elder Borrows, and was unaware of the altered spelling of the name. See Glover, *Derby* ii 559, and *Alum. Oxon.* i 133.

02C. To [Barbara Strutt?], 16 March 1802

Dear Madam,

I will do myself the pleasure of waiting on you about ten tomorrow morning. Am obliged by the offer of your horses, but have my own here.

Mr G[isborne] has pass'd a better day, and we hope that he daily tho' slowly amends.

I am, Madam, with true respects to Mr S[trut]t and the young Lady
Your much obliged and obed.
E Darwin

Mar. 16 – 1802

Original The Houghton Library, Harvard University, Cambridge, Massachusetts.
Printed Unpublished.
Text From original MS.
Notes
(1) The names Gisborne and Strutt are very faint on the manuscript, but just about legible.
(2) For Mrs Barbara Strutt, see letter 01D. For Mr John Gisborne, see letter 00G. For Mr William Strutt, see letter 01E. The 'young Lady' is probably the Strutts' eldest daughter Elizabeth (1793–1848).

02D. To Richard Lovell Edgeworth, 17 April 1802

Priory, near Derby, April 17, 1802

Dear Edgeworth,

I am glad to find that you still amuse yourself with mechanism, in spite of the troubles of Ireland.

The use of turning aside, or downwards, the claw of a table, I don't see; as it must then be reared against a wall, for it will not stand alone. If the use be for carriage, the feet may shut up, like the usual brass feet of a reflecting telescope.

We have all been now removed from Derby about a fortnight, to the Priory, and all of us like our change of situation. We have a pleasant house, a good garden, ponds full of fish, and a pleasing valley somewhat like Shenstone's – deep, umbrageous, and with a talkative stream running down it. Our house is near the top of the valley, well screened by hills from the east, and north, and open to the south, where, at four miles distance, we see Derby tower.

Four or more strong springs rise near the house, and have formed the valley, which, like that of Petrarch, may be called Val chiusa, as it begins, or is shut, at the situation of the house. I hope you like the description, and hope farther, that yourself and any part of your family will sometime do us the pleasure of a visit.

Pray tell the authoress, that the water-nymphs of our valley will be happy to assist her next novel.

My bookseller, Mr Johnson, will not begin to print the Temple of Nature [Origin of Society?], till the price of paper is fixed by Parliament. I suppose the present duty is paid

Original Not traced.
Printed Edgeworth, *Memoirs* ii 263–4.
Text From above source.
Notes

(1) This letter is unfinished, because Darwin died on the morning of 18 April 1802, probably of a heart attack. When Edgeworth received the letter, the following note was written on the opposite side of the paper by another hand: 'Sir, – This family is in the greatest affliction. I am truly grieved to inform you of the death of the invaluable Dr Darwin. Dr Darwin got up apparently in health; about eight o'clock, he rang the library bell. The servant, who went, said, he appeared fainting. He revived again – Mrs Darwin was immediately called. The Doctor spoke often, but soon appeared fainting; and died about nine o'clock.

Our dear Mrs Darwin and family are inconsolable: their affliction is great indeed, there being few such husbands or fathers. He will be most deservedly lamented by all, who had the honor to be known to him.

I remain, Sir, your obedient humble Servant.

S.M.

P.S. This letter was begun this morning by Dr Darwin himself'. (Edgeworth,

Memoirs ii 264–5). This note was probably written by Mrs Susan Mainwaring, who had acted as foster-mother to Elizabeth, Darwin's wife, when she was a child, and was staying with the Darwins when Erasmus died (see *Doctor of Revolution*, p. 284). The letter was probably begun the previous day, and not 'this morning' – unless Darwin misdated it.

(2) The Darwins had moved out from their town-house in Derby to Breadsall Priory on 25 March and, as this letter shows, Darwin already knew the move was a success. Elizabeth continued to live at the Priory until her death in 1832, and it became known as 'Happiness Hall' to her Galton grandchildren, so much did they enjoy their visits. (The Priory is now a golf club.)

(3) William Shenstone (1714–1763) was a poet of only moderate talent (see letter 94R), but he made a strong mark on eighteenth-century taste by his ideas on bringing out the picturesque qualities of gardens and creating pleasant prospects. He spent much effort in improving his own well-known garden at the Leasowes in Worcestershire, to which Darwin refers. See A. R. Humphreys, *William Shenstone* (CUP, 1937).

(4) *Val chiusa* ('closed valley') may be a reference to Petrarch's *Le Rime Sparse*, No. 303, Lines 5–7,

> fior, frondi, erbe, ombre, antri, onde, aure soavi,
> valli chiuse, alti colli e piage apriche,
> porto de l'amorose mie fatiche.

(5) Edgeworth and his family had already planned to visit Darwin in the autumn of 1802. See Edgeworth, *Memoirs* ii 263, 270.

(6) 'The authoress' is Edgeworth's eldest daughter Maria (1767–1849), who had become the most celebrated novelist of the day with her *Castle Rackrent* (1800). See M. Butler, *Maria Edgeworth* (OUP, 1972).

(7) Darwin's last poem, published under the title *The Temple of Nature* in April 1803, was intended by him to be entitled *The Origin of Society*, and its publication under this title was (prematurely) announced in *The Monthly Magazine* for December 1802. So it seems likely that Darwin would have written 'Origin of Society' rather than 'Temple of Nature', and that the wording was changed by the editor of Edgeworth's *Memoirs*, his daughter Maria, who wished to avoid either puzzling readers, or having to explain the change of title. Maria certainly altered other letters: for example, in letter 90B, line 2, she changed 'The paper' to 'The little Tale'; and in line 8, she changed 'Forgetfulness' to 'Time'.

(8) It is appropriate that Darwin's last letter should have been to Edgeworth. They had enjoyed 35 years of unclouded friendship, and since Wedgwood's death in 1795, Edgeworth had been Darwin's closest friend. Although only four letters from Darwin to Edgeworth have survived, there are 26 MS letters from Edgeworth to Darwin in the Edgeworth Papers at the Cambridge University Library. Their dates are: 1766?; 1768?; 1768; 22 Jan 1769; 16 Sep 1769; 7 Jun 1778; 31 Oct 1778; 1786?; 23 Oct 1786; 19 Mar 1790; 14 Apr 1790; 10 Oct 1791; 15 May 1792; 24 Jul 1793; 7 Sep 1794; 11 Dec 1794; 3 Oct 1795; 27 Feb 1796; 11 Apr 1796; 31 May 1796; 18 Dec 1796; 5 Jul 1797; 11 Sep 1797; 21 May 1798; 12 Mar 1799; 31 Mar 1800. Eleven of these are published, in whole or in part, in Edgeworth's *Memoirs*.

APPENDIX: A LIST OF PRESCRIPTIONS

A large number of Darwin's prescriptions have survived in manuscript. I have not made a special search for them, and the chronological list below merely records those I have noted while seeking letters. I have not tried to identify patients, but their dates are given and a reference added when their names are already familiar. The 34 prescriptions in the Darwin Museum at Down House are all in the Prescription Book; the 7 prescriptions at the Fitzwilliam Museum Library are in the S. G. Perceval Bequest, General Series; the 7 prescriptions at Dunedin Public Library are in the Alfred and Isabel Reed Collection.

Date	*Location*	*Patient*
11 July 1759	Down House	Mrs Roleston [of Derby? see *Univ. Brit. Dir.* 1793]
2 May 1763	Down House	Miss Shuttleworth
10 February 1764	Down House	John Burdett
29 November 1770	Down House	Mr Alderman Bingham
27 November 1771	Down House	Col. Sacheverel Pole. See letter 75G
7 March 1772	Down House	Mr Butterfield
30 August 1772	Fitzwilliam Museum Library	Mrs Sarah Wedgwood (1734–1815). See letter 72C
1 April 1774	Down House	Mr Tatham Eswell[?]
3 July 1776	Down House	Mr Henry Smith, of Allestree, near Derby
14 October 1777	Yale University Library	Mrs Ratcliff, of Ingleby, Derbyshire
29 June 1778	Down House	Sir John Every, Bt. (1709–1779), of Egginton, Derbyshire
22 December 1779	Down House	Atkyn Holden [Possibly of the Holden family in *Burke's Landed Gentry* (1972) ii 303–6.]
8 August 1780	Down House	Mrs Elizabeth Pole (1747–1832), who married Darwin in 1781
14 February 1781	Down House	Ann Hollywell, of Radburn

Date	*Location*	*Patient*
– 1781	Keele University Library, Wedgwood Commonplace Book, p. 122	Josiah Wedgwood (1730–1794) [MS copy]
10 November 1782	Down House	Revd. Mr Barnard
19 August 1783	Down House	Miss Maynal
19 November 1783	Down House	Revd. Thomas Manlove [1730–1802, headmaster of Derby School from 1761 to 1774. See Scott, *Admissions,* Part III, p. 567]
9 July 1784	Down House	Brooke Boothby (1744–1824). See letter 81D. Or his father (1710–1789)
11 August 1785	Down House	Thomas Chadwick
18 October 1786	Down House	Archdeacon Robert Clive (1723–1792). See letter 86I
18 September 1787	Down House	Jos. Flixon
11 May 1788	Down House	Mrs Hardy, of Mickleover, Derby
11 January 1789	Down House	Lewis Buckeridge, of Lichfield (1774–1821)
3 July 1789	Down House	Miss Le Noir[?], of Brereton Hall, Congleton, Cheshire
16 February 1790	Down House	Robert Holden Esq., [of Darley Abbey, near Derby, father of Capt. Robert Holden]
5 January 1791	William Salt Library, Stafford S. MS 478	Miss Baggaly. Addressed to Mrs Holland, Streethay, near Lichfield
25 April 1791	Down House	Mrs Latussier [died 31 July 1791, aged 65]
22 October 1791	William Salt Library, S.603.1, p. 174	Mrs {*name cut out*}
25 January 1793	University College London	J. Simpson
24 September 1793	Down House	Miss Mundy [probably of the family of F. N. C. Mundy of Markeaton: see Glover, *Derby* ii 17]
17 March 1794	Keele University Library, Wedgwood Archive 26769–35	Thomas Wedgwood (1771–1805). See letter 93F

Date	*Location*	*Patient*
9 April 1794	Down House	Mrs Elizabeth Darwin (1747–1832)
9 August 1794	Down House	Miss Emma Darwin (1784–1818), Erasmus's daughter
21 April 1795	Down House	Capt. Robert Holden (1769–1844)
14 December 1796	Dunedin Public Library	Sachaverel Pole (1769–1813), Darwin's stepson
18 December 1796	Historical Society of Pennsylvania	Jer. Strutt Esq [perhaps Jedediah (1726–1797)?]
19 January 1797	Dunedin Public Library	Chas [?] Standen [?]
3 February 1797	Dunedin Public Library	Jer. Strutt Esq.
20 March 1797	Dunedin Public Library	William Strutt Esq. jun. (1756–1830). See letter 01E
23 March 1797	Dunedin Public Library	William Strutt
2 April 1797	Dunedin Public Library	Jer. Strutt Esq.
23 April 1797	Dunedin Public Library	Jer. Strutt Esq.
24 December 1797	Down House	Miss Godwin
8 March 1798	[Advertised in John Wilson's catalogue, Nov. 1979]	Miss Cheyny
18 April 1798	Wellcome Institute Library	William Strutt
18 September 1798	Fitzwilliam Museum Library	William Strutt
26 September 1798	Down House	Robert Holden Esq.
1 October 1798	Wellcome Institute Library	D. Davinson Esq.
24 October 1798	Fitzwilliam Museum Library	Thomas Wilson Esq.
26 October 1798	Fitzwilliam Museum Library	Thomas Wilson Esq.
1 December 1798	Fitzwilliam Museum Library	John Brown
4 December 1798	Fitzwilliam Museum Library	John Brown
10 January 1799	Down House	Capt. Holden

Date	*Location*	*Patient*
16 January 1799	Birmingham University Library, MS L.Add 2449	William Strutt
28 May 1799	Fitzwilliam Museum Library	Mrs Barbara Strutt (1761–1804). See letter 01D
29 December 1799	Keele University Library, Wedgwood MSS, Mosley Collection	Thomas Wedgwood
27 February 1800	Down House	Robert Holden
24 August 1800	Down House	Pen. Sherwood
4 January 1802	Down House	Robert Holden Esq.

NAME INDEX

Eighteenth century and earlier

Page numbers in italic type indicate some biographical detail, however slight.
An asterisk indicates that there is a portrait among the illustrations, between pages 192 and 193.

GENERAL INDEX

Subjects; places; and persons born after 1800